Semiconductor Electronics

Semiconductor Electronics

A.K. Sharma
Ph.D.
Reader in Physics
Maharshi Dayanand University
Rohtak (Haryana)

PUBLISHING FOR ONE WORLD

NEW AGE INTERNATIONAL (P) LIMITED, PUBLISHERS
New Delhi • Bangalore • Chennai • Cochin • Guwahati • Hyderabad
Jalandhar • Kolkata • Lucknow • Mumbai • Ranchi
Visit us at www.newagepublishers.com

Published by New Age International (P) Ltd., Publishers
First Edition: 1996
Reprint: 2011

Branches:

- No. 37/10, 8th Cross (Near Hanuman Temple), Azad Nagar, Chamrajpet, **Bangalore** - 560 018.Tel.: (080) 26756823, Telefax: 26756820 E-mail: bangalore@newagepublishers.com
- 26, Damodaran Street, T. Nagar, **Chennai** - 600 017. Tel.: (044) 24353401, Telefax: 24351463, E-mail: chennai@newagepublishers.com
- CC-39/1016, Carrier Station Road, Ernakulam South, **Cochin** - 682 016. Tel.: (0484) 2377004, Telefax: 4051303, E-mail: cochin@newagepublishers.com
- Hemsen Complex, Mohd. Shah Road, Paltan Bazar, Near Starline Hotel, **Guwahati** - 781 008, Tel.: (0361) 2513881, Telefax: 2543669 E-mail: guwahati@newagepublishers.com
- No. 105, 1st Floor, Madhiray Kaveri Tower, 3-2-19, Azam Jahi Road, Nimboliadda, **Hyderabad** - 500 027. Tel.: (040) 24652456, Telefax: 24652457 E-mail:hyderabad@newagepublishers.com
- RDB Chambers (Formerly Lotus Cinema)106A, 1st Floor, S.N. Banerjee Road, **Kolkata** - 700 014, Tel.: (033) 22273773, Telefax: 22275247 E-mail:kolkata@newagepublishers.com
- 16-A, Jopling Road, **Lucknow** - 226 001. Tel.: (0522) 2209578, 4045297, Telefax: 2204098, E-mail: lucknow@newagepublishers.com
- 142C, Victor House, Ground Floor, N.M. Joshi Marg, Lower Parel, **Mumbai** - 400 013, Tel.: (022) 24927869, Telefax: 24915415 E-mail: mumbai@newagepublishers.com
- 22, Golden House, Daryaganj, **New Delhi** - 110 002. Tel.: (011) 23262370, 23262368, Telefax: 43551305, E-mail: sales@newagepublishers.com

ISBN : 978-81-224-0802-7

₹ 200.00

C-11-04-5465

Printed in India at Nisha Enterprises, Delhi.

PUBLISHING FOR ONE WORLD
NEW AGE INTERNATIONAL (P) LIMITED, PUBLISHERS
4835/24, Ansari Road, Daryaganj, New Delhi-110002
Visit us at **www.newagepublishers.com**

Preface

A textbook on semiconductor electronics should include the topics of fundamental nature on semiconductor, semiconductor devices (p-n junction diode, BJT, FET, etc), circuit analysis, basic electronic circuits (amplifiers, oscillators, etc), and integrated circuits along with their fabrication technology. The present book meets all these requirements. The basic aim of writing this book is to produce a comprehensive book of electronics for the students of Physics, Applied Mathematics, and Engineering at the graduate and postgraduate levels.

This book is based upon my fourteen years of teaching experience about the various problems faced by the students and teachers in the hitherto available books. Therefore, I have tried to develop the subject matter of this text in a systematic way maintaining the continuity in the topics so as to fulfil the students' requirements. It has been my effort to introduce the subject matter through analysis and explanations rather than definitions and descriptions only.

All the chapters are so written as to impart basic and essential knowledge of the topics. Stress has been on highlighting the electronics rather than lengthy mathematics. Only the required mathematical calculations are given in a simplified manner. The circuit and other diagrams are self-explanatory and are well described in the text. Cross referencing is made to help the students understand the subject. In order to make the subject more clear and understandable, the solved problems and questions along with unsolved problems are included in each chapter. Tables and standard parameters are also included wherever required.

The physics of semiconductors is described in the first chapter. Chapters 2, 3, 4, 5 and 6 provide an elaborative description about the basic background of various semiconductor devices, which include construction, fabrication, biasing and circuit models. As the amplifiers and oscillators are the main building blocks of all the electronic circuits, seven chapters are devoted to these topics and detailed discussions have been presented. As the logic gates form the basis of digital circuits, Chapter 14 describes all the basic logic circuits required in digital electronics. I hope that the students will certainly find themselves suitably equipped with a firm grip over the fundamentals of the subject.

During the preparation of the manuscript of this book, I have consulted the subject matter from the available books, which are listed in the references and therefore, I feel grateful to the authors and publishers of these books. I am also grateful to my colleagues for valuable discussions and constructive criticism at different stages of the work. I heartily thank the publishers New Age International (P) Limited for publishing the book.

The author welcomes suggestions for improvement in the contents of this text book.

A.K. SHARMA

CONTENTS

1

Semiconductor: An Introduction

Before the advent of semiconductors, all solids were divided into two groups, viz. conductors and insulators. A conductor e.g. aluminium which is most frequently used, has a very low resistivity of about 10^{-6} $\Omega-$cm at room temperature. On the other hand, an insulating material, e.g. silicon dioxide, which is used in integrated circuits, has a very high resistivity of about 10^{16} $\Omega-$cm; being 22 orders of magnitude higher than of aluminium. For electrical applications, resistivity is the sole criterion to classify materials. The resistivity of conductors lies below 10^{-2} $\Omega-$cm, whereas the materials having resistivity greater than 10^{5} $\Omega-$cm are called insulators. A new class of materials was found, which could not be included in either of these groups. These materials having resistivity in the intermediate region are known as semiconductors or half conductors. These gradually acquired much importance because their resistivity can be varied by design under precise control.

It was also found that the effect of temperature on conductivity of conductors and semiconductors is quite the opposite. In conductors, the conductivity decreases with increase in temperature, whereas the conductivity of a semiconductor increases sharply with temperature. Another important property of a semiconductor is its sharp increase in conductivity by the introduction of a negligibly small amount of certain impurities. Irradiation or bombardment of a semiconductor by a beam of light also causes a rapid increase in its conductivity. Because of these types of peculiar properties, the semiconductor material is used in the construction of all solid state electronic devices. In order to understand the operation of semiconductors and semiconductor devices, it is necessary to have some familiarity with the basic concepts of the atomic structure of matter.

1.1. ATOMIC STRUCTURE OF HYDROGEN ATOM

In 1913, the Danish Physicist Niels Bohr introduced an atomic model using the quantum theories of Planck and Einstein and used it to explain successfully the absorption and emission spectra for hydrogen. The picture of an atom accepted by Bohr consists of a nucleus of positive charge containing nearly all the mass of the atom, with electrons revolving around this nucleus as shown in Fig. 1.1. Bohr formulated two postulates for his atomic model.

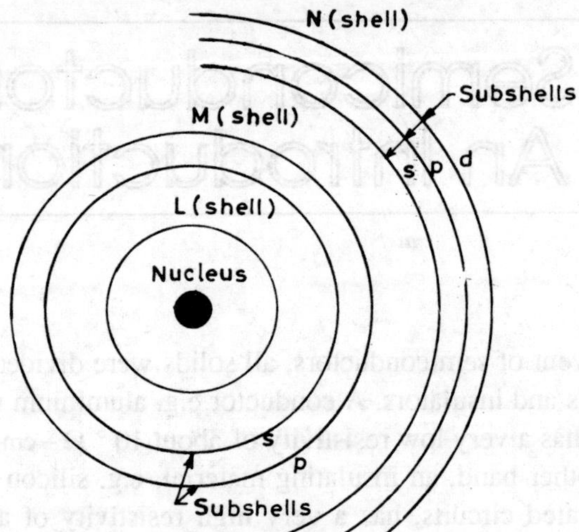

Fig. 1.1. Atomic structure.

Bohr's first postulate explains that an electron can only be in an orbit or shell, for which its angular momentum is a multiple of $h/2\pi$, h being the Planck's constant. Each of these allowed orbits corresponds to a certain energy and is called stationary or nonradiating state as the electron does not emit any energy so long as it remains in this orbit. This postulate can be expressed as

$$mvr = n \cdot \frac{h}{2\pi} \tag{1.1}$$

where $n = 1, 2, 3, \ldots$, is an integer called the principal quantum number and mv is the momentum of the electron in the orbit of radius r.

According to Bohr's second postulate, the absorption or emission of light by an atom occurs only when an electron undergoes a transition from one stationary state to anther. In each such transition, the energy $h\nu$ is absorbed or emitted by the atom and is equal to

$$h\nu = E_2 - E_1 \tag{1.2}$$

where v is the frequency of the radiant energy, and E_1 and E_2 are the energies of the two states involved in the transition. Combining Eqs. 1.1 and 1.2, the energy level in joules of each state is found to be

$$E_n = -\frac{me^4}{8h^2\varepsilon_0^2}\cdot\frac{1}{n^2} \tag{1.3}$$

where e is the charge on the electron and $\frac{1}{4\pi\varepsilon_0} = 1$

Bohr labelled these stationary energy states as L, M, N, and O. The maximum number of electrons that can exist in these states is 2, 8, 18 and 32 respectively. The outermost shell of the atom containing electrons is known as the valence shell and the electrons in this valence shell are called the valence electrons.

Bohr's model was modified by Pauli's exclusion principle, which introduced subshells in a shell. According to the Pauli's exclusion principle, no two electrons in an atom can have exactly the same quantum state. These subshells are labelled as s, p, d and f. The maximum number of electrons that these subshells can have is 2, 6, 10 and 14 respectively.

1.2. ENERGY LEVEL DIAGRAM

The energy level diagram for the hydrogen atom is shown in Fig. 1.2. This diagram is obtained by putting n = 1, 2, 3, ..., in Eq. 1.3 and drawing

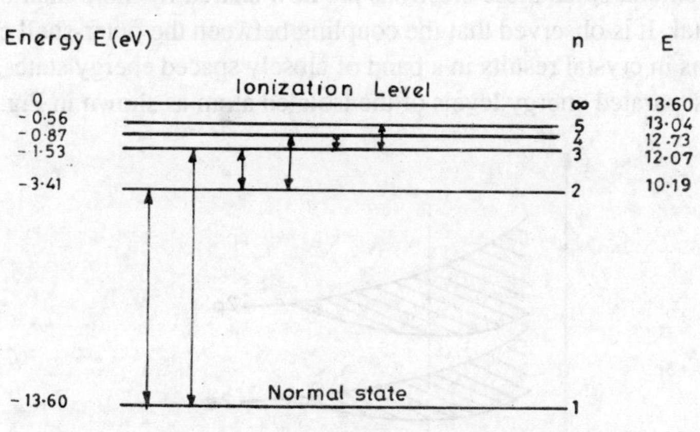

Fig. 1.2. Energy level diagram for hydrogen atom.

horizontal lines arranged vertically in accordance with the energy values, calculated from Eq. 1.3 for n = 1, 2, 3, ... Theoretically, an infinite number of states exist for each atom, but only the first five and the state for $n = \infty$ are depicted. Expressing the energy of stationary states in electron volt using Eq. 1.2, the wavelength λ of the emitted radiation may be given by

$$\lambda = \frac{12,400}{E_2 - E_1} \tag{1.4}$$

The lowest energy state is called the normal or the ground state and the other stationary states of the atom are called excited, radiating, critical, or resonance levels. The energy required to move an electron completely out of the field of the nucleus is called ionization potential and is shown as the highest energy state in the energy level diagram shown in Fig. 1.2. The ionization potential for hydrogen is 13.60 eV.

1.3. BAND THEORY OF CRYSTALS

The x-ray and other studies show that most metals and semiconductors have a crystalline structure, i.e. a regular and repetitive atomic or molecular arrangement in three dimensions. In these crystalline materials, the electronic energy level diagram discussed in Sec. 1.2 is not applicable. This is because the atoms begin to experience the influence of their neighbours, which is, in a way, similar to the influence of an external electric field. As the atoms converge, come closer, the interaction between them becomes stronger. Because of this interaction, the energy levels of the inner-shells' electrons are not affected appreciably, but the levels of the outer-shells' electrons are changed to a marked extent, since these electrons are now shared by more than one atom in the crystal. It is observed that the coupling between the outer-shell electrons of the atoms in crystal results in a band of closely spaced energy states, instead of widely separated energy levels of the isolated atom as shown in Fig. 1.3.

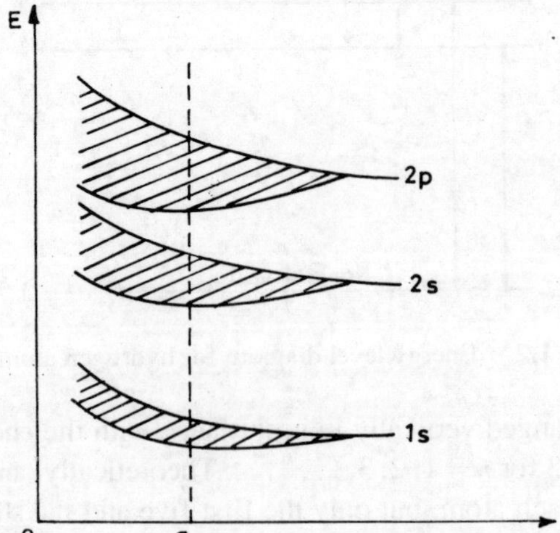

Fig. 1.3. Bands of closely spaced energy states.

In order to understand this band structure, a crystal of N atoms of one of the elements of group IV A of the periodic table e.g. Si or Ge, may be considered. The electronic configurations of these elements are $1s^2 2s^2 2p^6 3s^2 3p^2$ and $1s^2 2s^2 2p^2 3s^2 3p^6 3d^{10} 4s^2 4p^2$ respectively. The outer two subshells for each contain two s electrons and two p electrons. Since the s atomic subshell has two possible states at the same energy, there will be $2N$ possible s states in a crystal of N atoms and the number of s electrons will be $2N$. So, there will be $2N$ electrons completely filling the $2N$ s-subshells. But in case of the p atomic subshell, there are six possible states, therefore, there will be $2N$ electrons filling $2N$ p-subshells (out of possible $6N$ states) at the same energy.

Now, if it is possible to decrease the spacing between atoms without changing the crystal structure of this imaginary crystal (moving from right to left in Fig. 1.4, the neighbouring

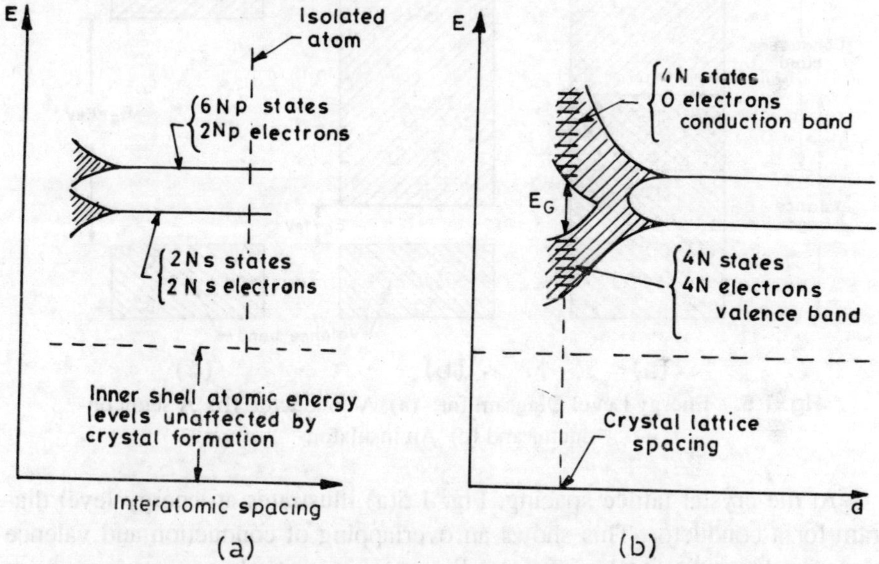

Fig. 1.4. Formation of energy bands in crystalline material.

atoms start interacting and their atomic wavefunctions overlap. As a result, the crystal becomes an electronic system obeying the Pauli's exclusion principle. If the interatomic distance is decreased sufficiently, the $2N$ degenerate s states and $6N$ degenerate p states spread out in energy of the order of several electron volts. These large number of discrete but closely spaced energy states are known as energy bands [Fig. 1.4(a)]. These two energy bands are separated by an energy gap known as a forbidden band. If the interatomic distance is further decreased to the limit of crystal lattice spacing, these bands overlap and $6N$ upper states (p states) merge with $2N$ lower states (s states), giving a total of $8N$ states. As the available electrons are $4N$, half of the $8N$ states are filled. So, each atom contributes 4 electrons to the band and now these electrons no longer belong to the s or p subshell of an isolated atom, but rather to the crystal as a whole. The band occupied by these electrons is called the valence band. Hence, at the crystal lattice spacing, there is a valence band filled with 4 N electrons separated by a forbidden band of energy E_G from an empty band having $4N$ states. This empty upper band is known as the conduction band [Fig. 1.4(b)].

1.4. CONDUCTORS, INSULATORS AND SEMICONDUCTORS

On the basis of the energy band structure, a material may be called a conductor, insulator, or semiconductor.

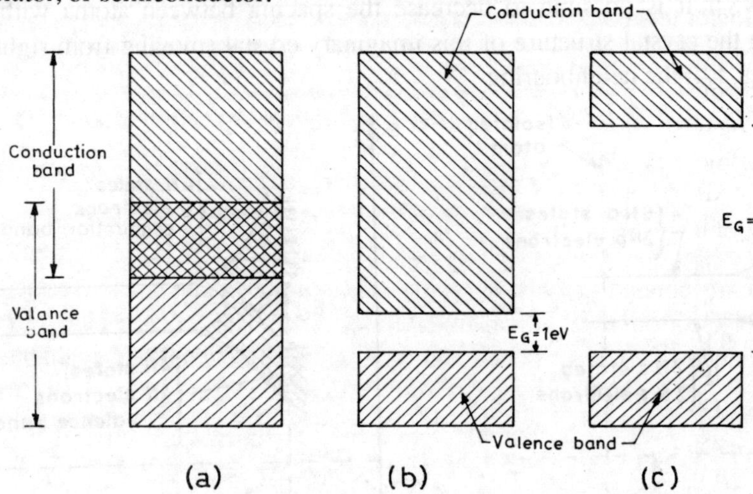

Fig. 1.5. Energy Level Diagram for: (a) A conductor (b) A semiconductor and (c) An insulator.

At the crystal lattice spacing, Fig. 1.5(a) illustrates an energy level diagram for a conductor. This shows an overlapping of conduction and valence bands, implying that it takes very small or no energy for electrons to jump from the valence band to the conduction band and therefore, all the valence electrons are free to take part in the conduction. Hence, the conductivity of a conductor is very high.

Figure 1.5(c) illustrates an energy level diagram for an insulator. In this case, the forbidden energy gap is of several electron volts ($E_G \cong 6$ eV), thus separating the valence and the conduction bands. Since the electron cannot be supplied energy more than 6 eV (say by an electric field), the valence band electrons cannot be raised to the conduction band. Hence conduction is impossible in an insulator.

Figure 1.5(b) illustrates an energy level diagram for a semiconductor. The energy of the forbidden band is approximately 1 eV at room temperature. The most commonly used semiconductors are germanium and silicon, which have this forbidden band of 0.72 eV and 1.1 eV respectively. The energies of this order of magnitude normally cannot be acquired from an applied field. Hence, these semiconductors are insulators at low temperatures. As the temperature is increased, some of the valence electrons acquire thermal energy more than the band. gap energy and reach the conduction band The insulator has now become slightly conducting i.e. a semiconductor.

Actually, the division between insulators and semiconductors is quite arbitrary, because many generally known dielectrics are now used as semiconductors. For example, silicon carbide ($E_G \cong 3$ eV) is used now in semiconductor devices. Even diamond is being investigated for a possible application in the semiconductor technology.

1.5. CRYSTAL STRUCTURE OF SEMICONDUCTOR (Concept of Hole)

The most commonly used semiconductors in electronic devices are germanium (Ge) and silicon (Si). As discussed in Sec. 1.3, in each case, there are four electrons in the outermost shell. In a pure Ge or Si crystal, these four valence electrons are bonded to four adjoining atoms by covalent bonds (formed by sharing of electrons), as shown in Fig. 1.6. At 0°K all valence electrons are firmly bonded in covalent bonds and there are no free electrons available for electrical conduction. From the point of view of band theory, at 0°K a

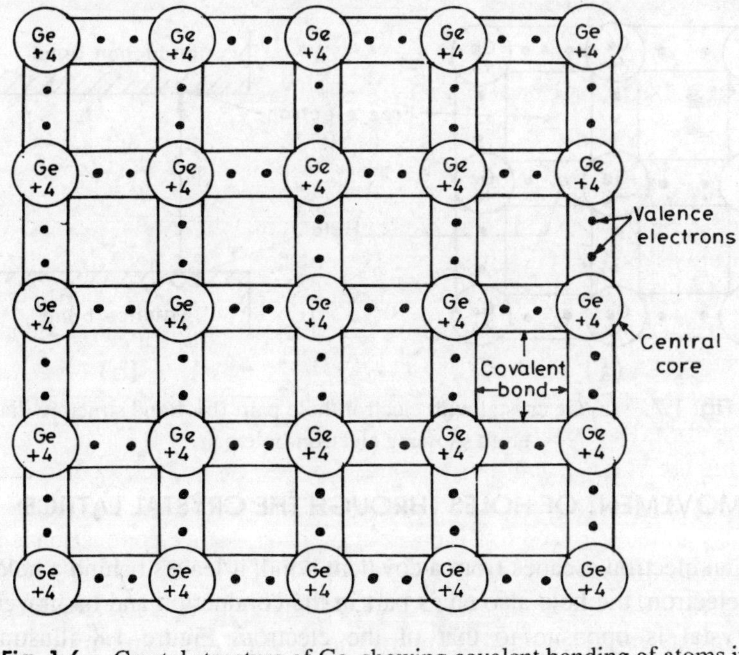

Fig. 1.6. Crystal structure of Ge, showing covalent bonding of atoms in two dimensions.

semiconductor has a completely filled valence band and an empty conduction band, thus behaving as an insulator.

At room temperature, the crystal littice is thermally excited and the valence electrons get additional energy. When electrons receive sufficient energy, more than the energy of the covalent bonds, they rupture the covalent bonds and wander in the lattice space as free negative charge carriers. The energy required to rupture a covalent bond is about 0.72 eV for Ge and 1.1 eV for Si at room temperature. If the temperature of the crystal is increased, more covalent bonds are broken and more free electrons are available in the crystal, showing an increase in the conductivity or a decrease in the resistance. This shows that a semiconductor has a negative temperature coefficient of resistance. When an electron becomes free by rupturing a covalent bond, the absence of electron in its parent position in the bond is represented by a small circle and is known as hole. [Fig. 1.7(a)] Since a hole consists of a missing negative charge *e*, it is accepted as equivalent to a positive charge *e*. Hence, there is always an attractive force between a hole and an electron and at any given temperature, the rate of recombination of holes and electrons is always equal to the rate of production of new holes and electrons. Therefore, the total number of free electrons and holes in a crystal remains constant. On the energy level diagram, the creation of electron-hole pair corresponds with the electron transition from the valence band to the conduction band as shown in Fig. 1.7(b).

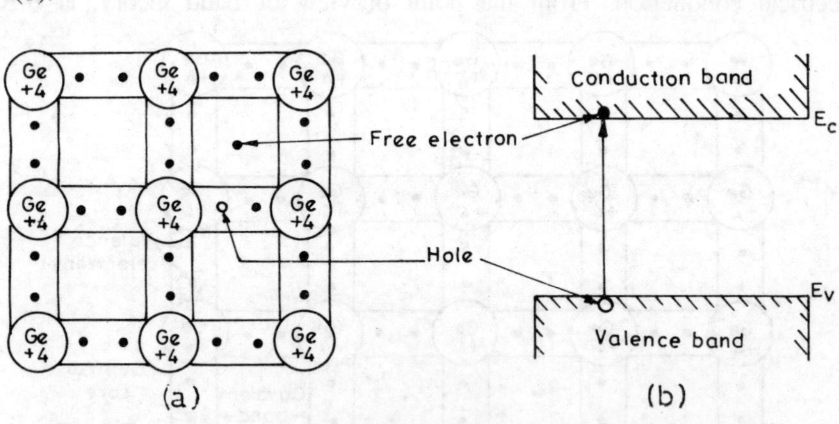

Fig. 1.7. (a) Ge crystal with electron-hole pair (b) Band structure diagram showing electron-hole pair.

1.6. MOVEMENT OF HOLES THROUGH THE CRYSTAL LATTICE

When an electron escapes from a covalent bond, it leaves behind a hole. Similar to electron, the hole also takes part in the conduction and its movement in the crystal is opposite to that of the electron. Figure 1.8 illustrates the

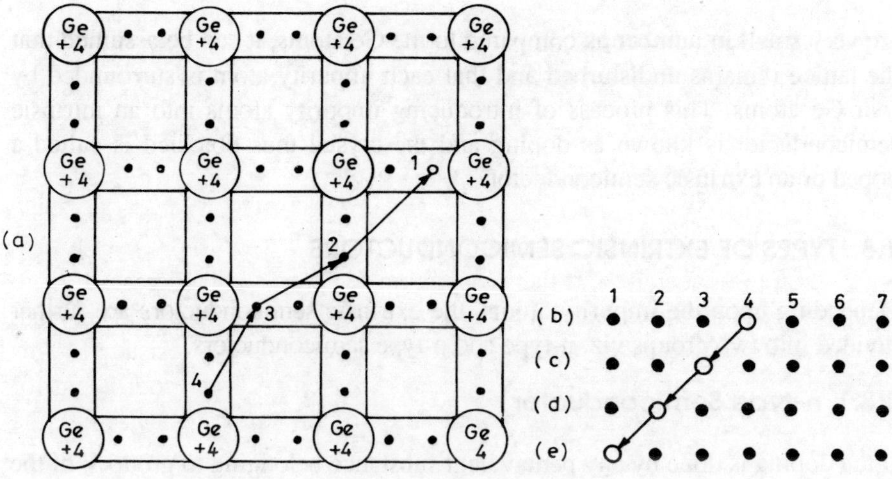

Fig. 1.8. Movement of hole in a crystal.

movement of hole through the crystal lattice. In Fig.1.8(a), the hole has moved from atom 4 to atom 1. The detailed mechanism of this movement is shown in Figs. 1.8 (b, c, d and e). Suppose by the rupturing of a covalent bond, a hole is created in the atom 4, as shown in Fig. 1.8(b). Now, imagine that an electron from atom 3 jumps into the hole of atom 4, so that the configuration of Fig. 1.8(c) results. Now, the electron from atom 2 jumps into the hole of atom 3 and we get the configuration of Fig. 1.8(d). Similarly, Fig. 1.8(e) is obtained when an electron from atom 1 jumps into the hole of atom 2. If we compare these four configurations, it looks as if the hole has moved from atom 4 to atom 1. Actually, the movement of a hole means the electron movement in the opposite direction. Hence, the holes can be considered as physical entities and free positive charge carriers, whose movements also contribute the electric current. It is important to note that the holes travel more slowly than do electrons. This is because the movement of a hole in a particular direction actually consists of a series of discontinuous electron movements in the opposite direction.

1.7. INTRINSIC AND EXTRINSIC SEMICONDUCTORS

Semiconductors are classified as intrinsic and extrinsic or doped. The intrinsic semiconductors are those that are very pure. In these, the properties of the crystal are determined only by the properties of atoms of the semiconductor material itself. The details of intrinsic semiconductors are discussed in Sec. 1.5 and 1.6.

If an extremely small (approximately 1 in 10^8 atoms), carefully controlled amount of an impurity element (generally trivalent or pentavalent atoms) is introduced into the intrinsic semiconductor, say Ge, each of the impurity atoms take the place of one of the Ge atoms in the lattice. Since the impurity atoms are very small in number as compared to the Ge atoms, it can be assumed that the lattice remains undisturbed and that each impurity atom is surrounded by four Ge atoms. This process of introducing impurity atoms into an intrinsic semiconductor is known as doping and the crystal thus obtained is called a doped or an extrinsic semiconductor.

1.8. TYPES OF EXTRINSIC SEMICONDUCTORS

Depending upon the impurity atoms, the extrinsic semiconductors are further divided into two groups viz. n-type and p-type semiconductors.

1.8.1 n-type Semiconductor

If the doping is done by any pentavalent substance belonging to group V of the periodic table of the elements (arsenic, antimony, or phosphorus), four valence electrons of the impurity atoms form covalent bonds with four electrons of the neighbouring Ge atoms and the fifth electron of the impurity atom becomes free to wander in the lattice. This is shown in Fig. 1.9 (a). Since each impurity atom has donated one free electron to the crystal, this pentavalent impurity is known as donor. The crystal is said to be n-type semiconductor as the number of electrons (negative charge carriers) in the crystal is much greater than the number of holes produced by thermal agitation.

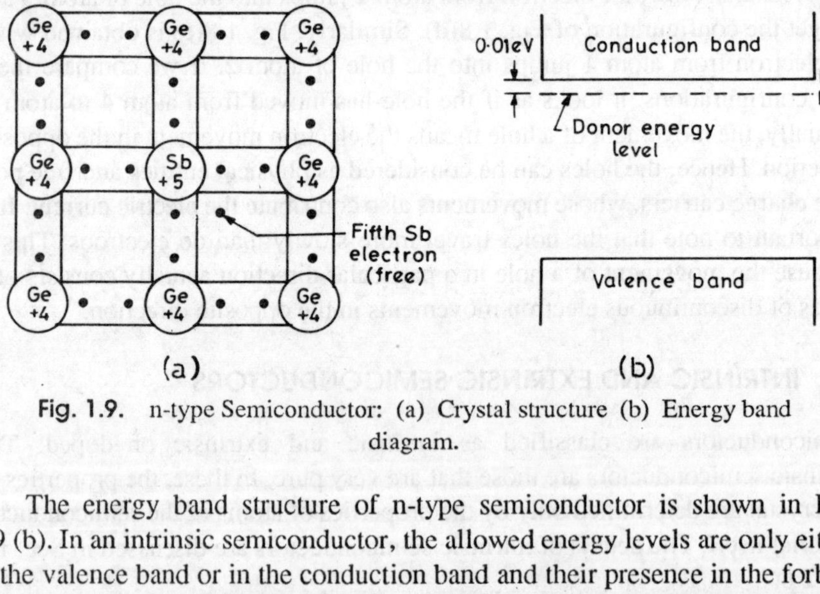

Fig. 1.9. n-type Semiconductor: (a) Crystal structure (b) Energy band diagram.

The energy band structure of n-type semiconductor is shown in Fig. 1.9 (b). In an intrinsic semiconductor, the allowed energy levels are only either in the valence band or in the conduction band and their presence in the forbidden band is ruled out. But, because of impurity atoms, additional allowed energy levels appear in the energy spectrum in the forbidden band and at very small energy below the conduction band as shown in Fig. 1.9 (b). In Ge, the energy of these new discrete allowed energy levels is only 0.01 eV (0.05 eV in

Si) below the conduction band. This is the reason why at room temperature almost all the fifth electrons of the donor are raised into the conduction band. In n-type semiconductor, the recombination rate of electrons and holes also increases because of the presence of an increased number of electrons donated by the donor and as a result, the number of holes decreases.

1.8.2. P-type Semiconductors

If the doping is done by any trivalent substance belonging to group III of the periodic table of the elements (indium, aluminium, or gallium), three valence electrons of the impurity atom form covalent bonds with the three electrons of neighbouring Ge atoms. The fourth covalent bond has a missing electron, thus a hole is introduced for each impurity atom [Fig. 1.10 (a)]. This hole is mobile as discussed in Sec.1.6. Since each impurity atom has a tendency to accept one electron, this trivalent impurity is called an acceptor. The crystal is said to be p-type semiconductor as the holes (positive Charge Carriers) are in the majority.

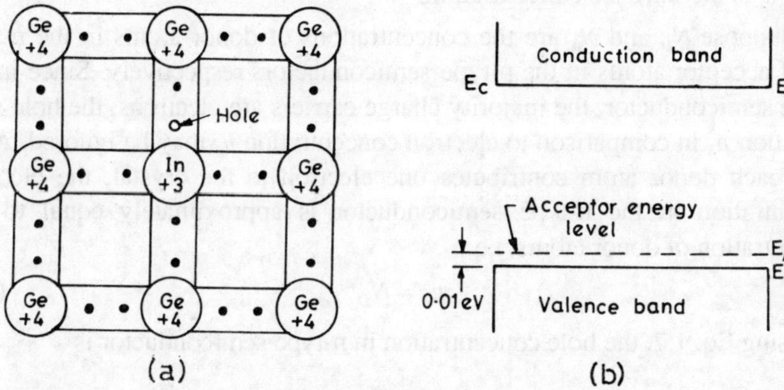

Fig. 1.10. p-type semiconductor: (a) Crystal structure, (b) Energy band diagram.

The energy band structure of p-type semiconductor is shown in Fig. 1.10 (b). The p-type impurity introduces an allowable discrete energy level just above the valence band (0.01 eV) as shown in Fig. 1.10 (b). The presence of this energy level helps the electrons to leave the valence band and occupy this acceptor level because the energy required is very small. The holes generated in the valence band by these electrons constitute the largest number of carriers in the p-type semiconductor.

The n-type or the p-type semiconductor is electrically neutral because each impurity atom is itself neutral. In n-type semiconductor, the electrons are the majority charge carriers and the holes are the minority charge carriers. In p-type semiconductor, the electrons are the minority charge carriers and holes are the majority charge carriers.

1.9. CONCENTRATION OF CHARGE CARRIERS IN n-TYPE AND p-TYPE SEMICONDUCTORS

As discussed in Sec. 1.5, at any temperature the rate of recombination of holes and electrons r(t) is always equal to the rate of production of new holes and electrons p(t), i.e.

$$r(\mathrm{t}) = p(\mathrm{t}) \tag{1.5}$$

Further, the rate of recombination depends on the number of encounters of holes and electrons, therefore, $r(t)$ is proportional to p and n, the densities of holes and electrons respectively, so

$$r(t) = p(t) = knp \tag{1.6}$$

where k is a constant of proportionality. As in pair production $n = p = n_i$ (say),

$$np = n_i^2 \tag{1.7}$$

where n_i is the intrinsic concentration.

Suppose N_D and N_A are the concentrations of donor atoms in the n-type and of acceptor atoms in the p-type semiconductors respectively. Since in the n-type semiconductor, the majority charge carriers are electrons, the hole concentration p_n in comparison to electron concentration n_n may be ignored. Also, since each donor atom contributes one electron to the crystal, the electron concentration in the n-type semiconductor is approximately equal to the concentration of donor atoms, i.e.,

$$n_n = N_D \tag{1.8}$$

and using Eq. 1.7, the hole concentration in n-type semiconductor is

$$p_n = \frac{n_i^2}{N_D} \tag{1.9}$$

Similarly, in the p-type semiconductor, the hole and the electron concentrations are

$$p_p = N_A \tag{1.10}$$

and

$$n_p = \frac{n_i^2}{N_A} \tag{1.11}$$

1.10. TEMPERATURE DEPENDENCE OF CHARGE CARRIER CONCENTRATION

The purity and the temperature are the main factors on which the concentration

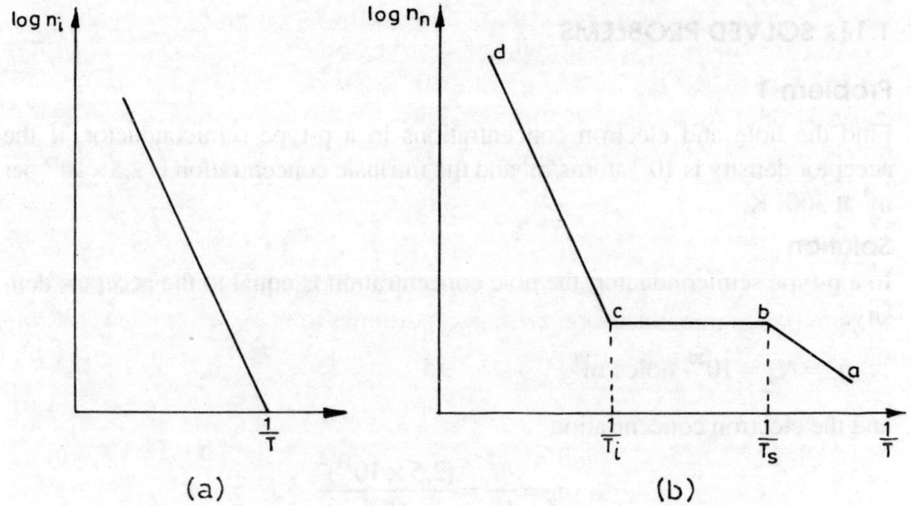

Fig. 1.11. Temperature dependence of charge carrier concentration (a) Intrinsic, and (b) n-type Semiconductor.

of charge carriers in a semiconductor depends. A semiconductor becomes conducting only when the temperature is increased considerably higher than absolute zero. Figure 1.11 shows the variation of the concentration of charge carriers in the intrinsic and the extrinsic (say, n-type) semiconductors. The charge carrier concentration rises with temperature because electrons from deeper and deeper states of the valence band move to the conduction band [Fig. 1.11 (a)]. In Figure 1.11(b), there is not a continuous rise in carrier concentration with temperature. As the temperature is increased above absolute zero, electrons from the donor level being very close to conduction band reach the conduction band much earlier than in the case of intrinsic semiconductor. As the temperature is further increased, more electrons from the donor level reach the conduction band and at a temperature known as depletion temperature, all the donor electrons reach the conduction band. When the temperature is increased beyond T_s, the number of electrons in the conduction band remains constant up to the temperature when the thermal energy becomes sufficient to raise the electrons from the valence band to the conduction band [region b c in Fig. 1.11(b)]. Further rise in temperature increases the number of electrons in the conduction band (region c d). At a temperature T_i (known as temperature of transition of intrinsic conductivity), electrons from the valence band become equal to the electrons from the donor level. Beyond this temperature, the extrinsic semiconductor behaves as an intrinsic semiconductor. This is because in the conduction band the number of electrons raised from the donor level can be ignored in comparison to the electrons raised from the valence band. In most semiconductors, T_i exceeds room temperature considerably.

1.11. SOLVED PROBLEMS

Problem 1

Find the hole and electron concentrations in a p-type semiconductor, if the acceptor density is 10^{20} atoms/m^3 and the intrinsic concentration is 2.5×10^{19} per m^3 at 300° K.

Solution

In a p-type semiconductor, the hole concentration is equal to the acceptor density.

$$\therefore\ p_p = N_A = 10^{20} \text{ holes/m}^3$$

and the electron concentration

$$n_p = \frac{n_i^2}{N_A} = \frac{(2.5 \times 10^{19})^2}{10^{20}}$$

$$= 6.25 \times 10^{18} \text{ electrons/m}^3.$$

Problem 2

If the wavelength of the electromagnetic radiation is 1,000 nm, find the band-gap energy.

Solution

Using Eq. 1.4, the band-gap energy in electron volt is given by,

$$E_2 - E_1 = \frac{12400}{1000 \times 10}$$

$$= 1.24 \text{ eV}.$$

Questions and Problems

1. What is the effect of doping on the resistance of a semiconductor?
2. What is the effect of large pressure on a semiconductor?
3. How is semiconductor defined? What are n-type and p-type semiconductors?
4. How does a hole move in a crystal lattice?
5. Explain the effect of temperature on the charge carrier concentrations in a semiconductor.
6. How does a semiconductor behave at 0°K?
7. How do the impurities change the character of an intrinsic semiconductor?
8. What are the majority charge carriers in an intrinsic semiconductor?
9. Why does a semiconductor have a negative temperature coefficient?

10. Find the hole and electron densities in an n-type semiconductor if $N_D = 10^{22}/m^3$ at 300°K.
11. If the band-gap energy is 1.1 eV, find the wavelength of the electromagnetic radiation emitted by this material.

Semiconductor Junction Diode

As the name indicates, the semiconductor diode is constructed from a semiconductor material i.e. Ge or Si. The semiconductor diode is an electronic device which has a high resistance to the flow of current in one direction and a low resistance in the opposite direction. The diode is one of the fundamental building blocks of the wide variety of electronic circuits used today. The diodes are used as rectifiers, doublers, limiters, clippers, clampers, modulators, demodulators, waveshaping circuits, logic gates, frequency convertors, etc. The semiconductor diodes available in many sizes and shapes, consist of p-n junctions formed in either Ge or Si crystals.

2.1. THE p-n JUNCTION DIODE

In order to obtain a p-n junction diode with controlled and permanent properties, a Si or Ge crystal is doped with donor atoms at one end and acceptor atoms at the other. This semiconductor crystal has both p-type and n-type regions, which make a junction between them as shown in Fig. 2.1. The donor ions are shown as ⊕ because the donor impurity atom becomes positive after donating one electron. The acceptor ions are shown as ⊖ because the acceptor atom becomes a negative ion after accepting one electron. Free electrons and holes are represented as – and + symbols. The formation of the p-n junction is described below.

Initially, holes are present in the majority to the left of the junction, and electrons in the majority to the right of the junction. These holes and electrons are free to move randomly in the lattice. Since there are more electrons to the right of the junction and more holes to the left of the junction, there is a density gradient across the junction. Therefore, more holes diffuse to the right of the junction and more electrons diffuse to the left of the junction. As a result of the electrons leaving the n-material, donor ions are produced on the n-side of the

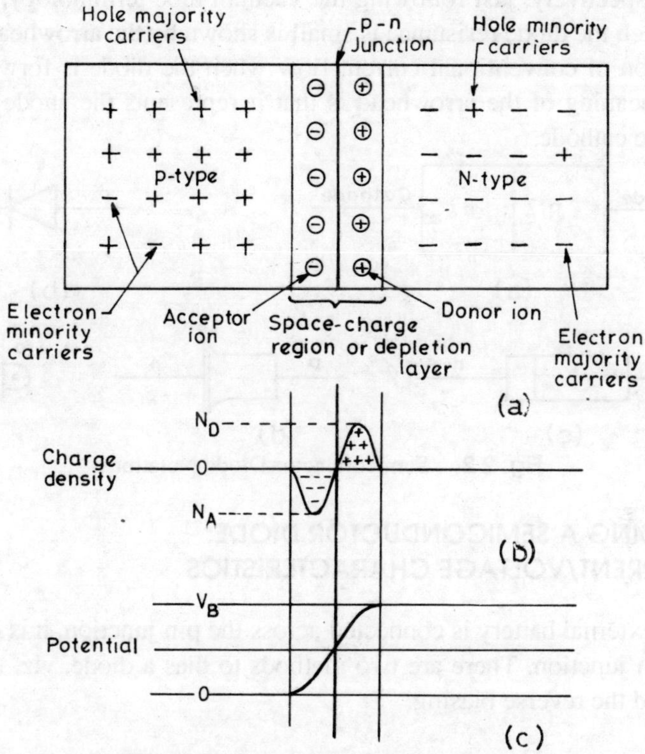

Fig. 2.1. Formation of a p-n junction.

junction. When these electrons fill holes on the p-side of the junction, the acceptor ions are produced. These stationary positive and negative charges along the n- and p-sides of the junction respectively are shown in Fig. 2.1(a) and (b). The space occupied between these ions is called the space-charge region. The movement of holes and electrons across the junction constitutes a current known as diffusion current. Thus, the region just to the left of the junction becomes negatively charged and the region just to the right of the junction becomes positively charged, and a potential difference is established across the p-n junction (Fig. 2.1(c).

Any charge carriers entering this space-charge region, are rapidly swept out of it, and hence this region is depleted of charge carriers and is known as the depletion region. It is a region of high electrical resistance and is approximately 0.001 mm in width.

The region near the junction of p and n materials is thus referred to by many names: the space-charge region, potential barrier, and depletion region; and this device is known as the p-n junction or a semiconductor diode. The various notations generally used for a semiconductor diode are shown in Fig. 2.2. The p-type and n-type portions of the diode are called anode and cathode respectively, just following the vacuum tube terminology. The direction in which the diode resistance is small is shown by the arrowhead. It is also the direction of conventional current flow when the diode is forward-biased. Another meaning of the arrowhead is that it represents the anode and points towards the cathode.

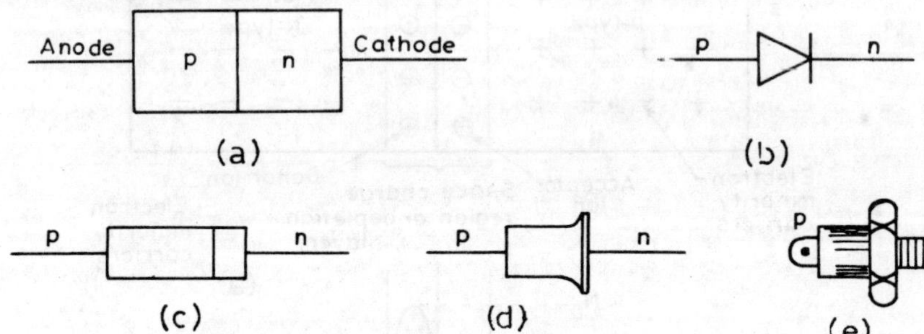

Fig. 2.2. Semiconductor Diode Notations.

2.2. BIASING A SEMICONDUCTOR DIODE: CURRENT/VOLTAGE CHARACTERISTICS

When an external battery is connected across the p-n junction, it is called biasing the p-n junction. There are two methods to bias a diode, viz. the forward biasing and the reverse biasing.

2.2.1. Forward Biasing of a Diode

When the positive terminal of a battery is connected to the p-side and the negative terminal to the n-side of a p-n junction, the diode is said to be forward biased, as shown in Fig. 2.3(a). In this condition, the holes in the p-region and electrons in the n-region of the diode experience a repelling force from the battery and therefore, drift towards the junction. This drift of holes and

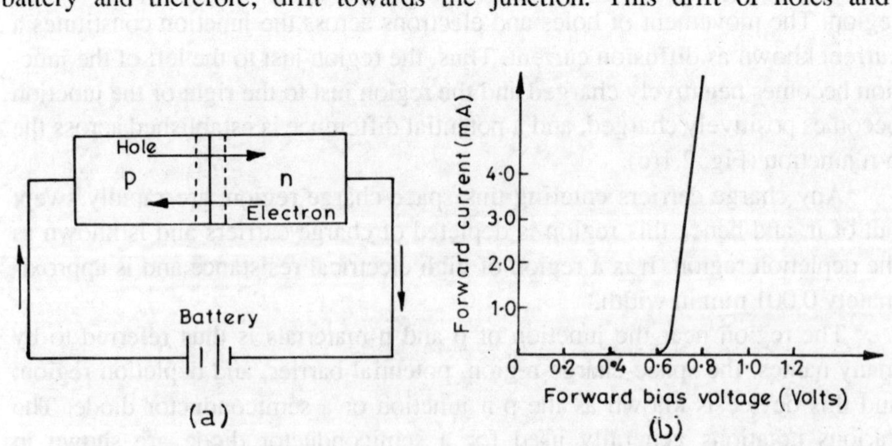

Fig. 2.3. (a) Forward Biasing of a p-n Junction Diode, (b) Current/Voltage characteristic in forward biasing.

electrons towards the junction reduces the width of the depletion region and also the junction potential barrier. This helps the holes of relatively lower energy in the p-region to cross the junction and combine with the electrons in the n-region. Similarly, more free electrons in the n-region cross the junction and combine with the holes in the p-region. The reason for the continuous presence of a large number of holes in the p-region and the electrons in the n-region inspite of diffusion is that the positive and the negative terminals of the battery are considered to inject holes and electrons to the p-side and the n-side of the diode, respectively. In other words, it may be regarded that the holes and the electrons diffusing across the junction are coming from the battery. This diffusion of majority charge carriers constitute a current across the junction and the total current is the sum of the hole and the electron currents. The resultant current enters the p-region as a hole current and leaves the n-region as an electron current. Therefore, it is said that in the forward biasing the current flow is by majority charge carriers.

Now, if the forward voltage applied across the p-n junction is increased, more majority charge carriers cross the junction and more forward current flows across the junction. However, when the forward bias becomes more than the junction potential, the barrier disappears and the current increases rapidly, but it is limited by the bulk resistance of the crystal and the resistance of the ohmic contacts of the p- and n-regions. Thus, in the forward biasing, initially the current increases slowly with the biasing potential but when this becomes slightly more than the junction potential, the current is governed by the ohmic contact resistance and the crystal bulk resistance. Thus, the current/voltage characteristic becomes approximately a straight line, as shown in Fig. 2.3 (b). The diode offers very small resistance in the forward biasing.

2.2.2. Reverse Biasing of a Diode

When the positive terminal of a battery is connected to the n-side and the negative terminal to the p-side of a p-n diode, the diode is said to be reverse biased, as shown in Fig. 2.4 (a). In this condition, the holes in the p-region and the electrons in the n-region of the diode experience attractive forces from the battery and therefore, drift away from the junction. This drift of holes and electrons away from the junction increases the width of the depletion region and also the junction potential barrier. Fewer holes in the p-region and electrons in the n-region now have sufficient energy to overcome the junction potential and therefore, the majority charge carrier current decreases. Since the minority charge carrier current is constant, the resultant current flows across the junction from the n-region to the p-region. When the reverse bias is increased, more majority charge carriers become incompetent to overcome the junction potential and the majority charge carrier current reduces further. Hence the net current from the n-region to the p-region increases. This increase

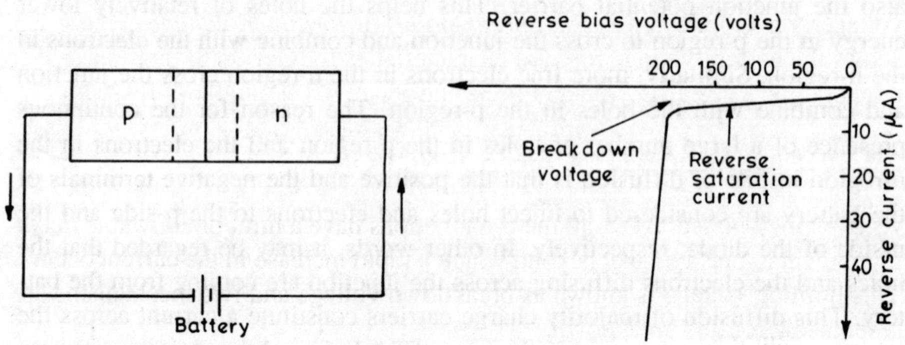

Fig. 2.4. (a) Reverse biasing of a p-n junction diode (b) Current/Voltage characteristic in reverse biasing.

in current with increasing reverse bias voltage continues until a point is reached when no majority charge carrier is able to cross the junction. In this condition, the constant current flowing across the junction is equal to the minority charge carrier current and is known as reverse saturation current. Hence, in the reverse biasing, the diode offers a very high resistance to the flow of current. If the reverse bias is further increased, a critical voltage known as breakdown voltage is reached, at which a rapid increase in current occurs, as shown in Fig. 2.4 (b). The breakdown of a diode is described in Sec. 2.3. The maximum reverse biased voltage that can be maintained across a p-n junction is known as peak inverse voltage (PIV).

2.2.3. The Cut-in Voltage

For every semiconductor diode, there exists a cut-in voltage V_γ, below which

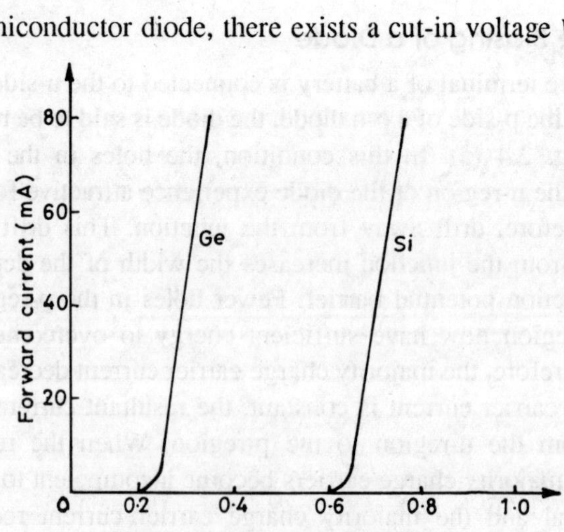

Fig. 2.5. Cut-in voltages in Ge and Si diodes.

the current is very small and beyond which the current rises very rapidly. This voltage is also known as *offset*, *break-point*, or *threshold voltage*. Figure 2.5 shows the current/voltage forward characteristics for Ge and Si diodes at room temperature. The cut-in voltage for Ge and Si are 0.2 V and 0.6 V respectively.

2.3. THE BREAKDOWN OF A p-n JUNCTION

As explained in Sec. 2.2.2, all the p-n junctions have a limit of allowable maximum reverse-bias voltage, beyond which a rapid increase in the current occurs. This limiting voltage is known as breakdown voltage and two mechanisms are responsible for this breakdown.

2.3.1. Zener Breakdown

When the applied reverse voltage produces a sufficient large electric field across the junction, some of the covalent bonds are ruptured owing to this electric field, and a large number of holes and electrons is produced in the crystal. These holes and electrons are responsible for the rapid increase in the saturation current. This phenomenon is known as Zener breakdown.

2.3.2. Avalanche Breakdown

When the applied reverse voltage is large enough to accelerate the minority charge carriers to gain sufficient energy, some of the covalent bonds are broken owing to the collision of these accelerated charge carriers with the atoms. The large number of holes and electrons thus produced is responsible for the rapid increase in the saturation current. This type of breakdown is known as avalanche breakdown.

2.4 IDEAL DIODE

As discussed in Sec. 2.2, a semiconductor diode has a very high resistance in

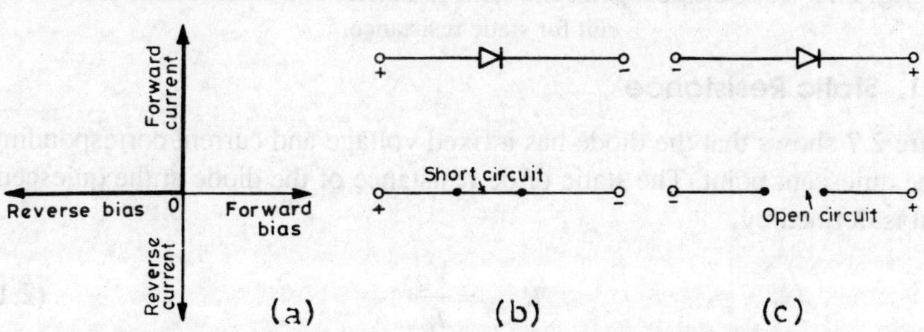

Fig. 2.6. (a) Voltage/Current characteristic of an ideal diode,
(b) Forward biased diode showing short circuit,
(c) Reverse biased diode showing open circuit.

the reverse biasing and a very low resistance in the forward biasing. However, in the case of an ideal diode, the forward resistance of the diode is zero, while the reverse resistance is infinite. The voltage/current characteristic for an ideal diode is shown in Fig 2.6 (a). It is clear from the figure that the forward resistance is zero and the diode behaves as a short circuit [Fig. 2.6 (b)]. In the reverse biasing, the diode behaves as an open circuit [Fig. 2.6(c)] showing infinite resistance.

2.5. STATIC AND DYNAMIC RESISTANCES

Figure 2.7 shows voltage/current characteristics along with the load line for a semiconductor diode. The intersection of the load line and the diode characteristic is known as the *quiescent point* or the *point of operation* for the diode. The voltage across and the current through the diode can be very easily found by drawing vertical and horizontal lines respectively.

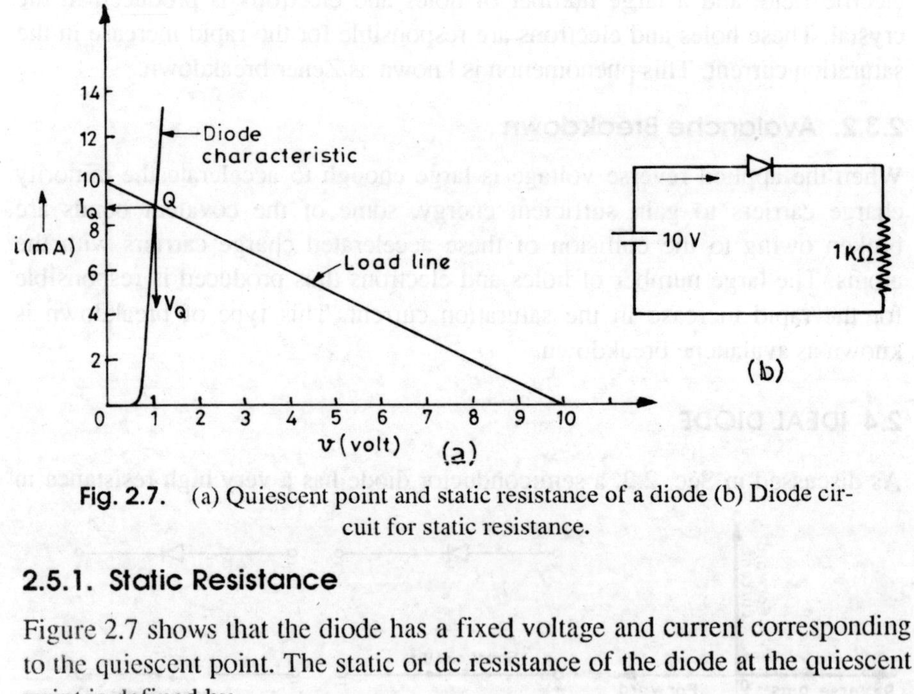

Fig. 2.7. (a) Quiescent point and static resistance of a diode (b) Diode circuit for static resistance.

2.5.1. Static Resistance

Figure 2.7 shows that the diode has a fixed voltage and current corresponding to the quiescent point. The static or dc resistance of the diode at the quiescent point is defined by,

$$R_{\text{static}} = \frac{V_Q}{I_Q} \tag{2.1}$$

The advantage of finding the dc resistance is that the diode can be replaced by a simple resistance equal to the dc resistance, while drawing an equivalent circuit.

2.5.2. Dynamic Resistance

Figure 2.8 shows that a sinusoidal signal on a dc level is applied to the diode circuit. With this additional ac signal, two additional load lines are obtained.

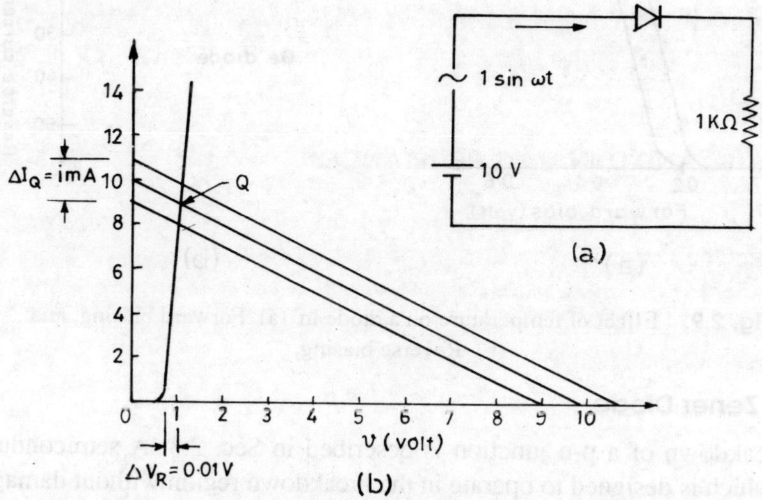

Fig. 2.8. (a) Quiescent point and dynamic resistance of a diode
(b) Diode circuit for dynamic resistance.

The dynamic or ac resistance is obtained by drawing a tangent on the characteristic at the quiescent point and using the relation.

$$R_{\text{dynamic}} = \frac{\Delta V_Q}{\Delta I_Q} \tag{2.2}$$

2.6. EFFECT OF TEMPERATURE ON A DIODE

The decrease with barrier potential in a forward biased *p-n* junction with increase in the temperature is approximately given by

$$\Delta V \cong -2.5 \text{ mV/°C} \tag{2.3}$$

Because of this change in the barrier potential, the cut-in voltage decreases with increase in temperature, as shown in Fig. 2.9. In reverse biasing the available leakage current depends upon the temperature. For a Ge diode, the leakage current becomes double for every 10°C rise in temperature. For a Si diode, the leakage current becomes double for every 6°C rise in temperature.

2.7. TYPES OF DIODE

The various types of diodes used in all the branches of modern electronic circuitry and radio engineering are described in this section.

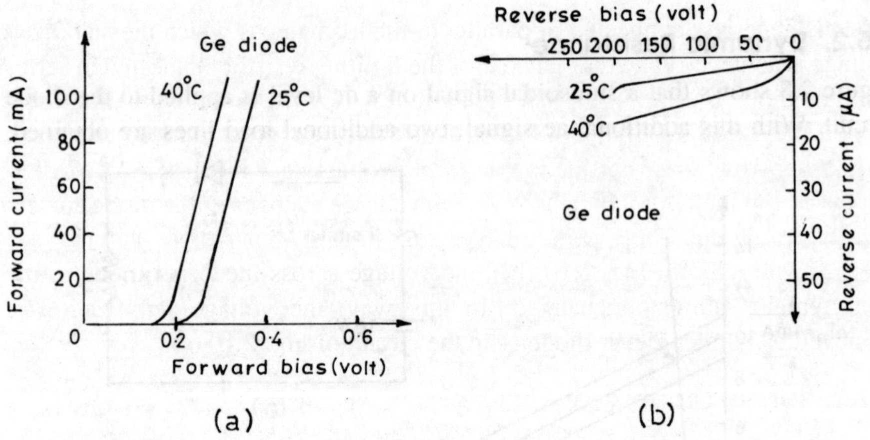

Fig. 2.9. Effect of temperature on a diode in (a) Forward biasing, and (b) Reverse biasing.

2.7.1. Zener Diode

The breakdown of a p-n junction is described in Sec. 2.3. A semiconductor diode which is designed to operate in the breakdown region without damage is known as a Zener, avalanche, or breakdown diode. The current is restricted to a safe value by using an external resistance so that the diode is not damaged.

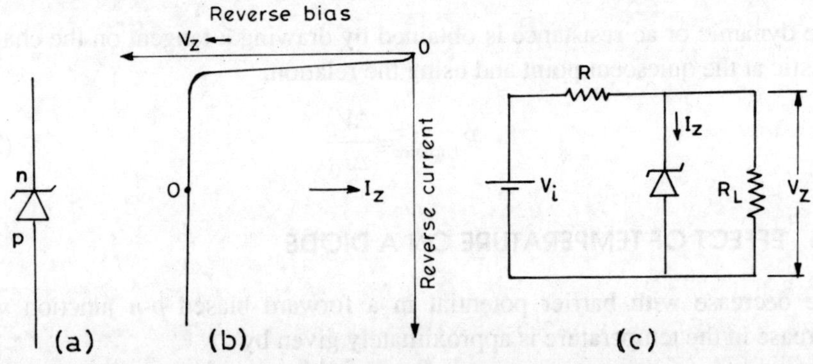

Fig. 2.10. (a) Symbol for a Zener Diode, (b) Voltage/current characteristic of a Zener diode, (c) Zener diode circuit for voltage stabilization.

Figure 2.10 (a) shows the symbol and the circuit of a Zener diode, and its voltage/current characteristic. The location of the Zener potential V_Z, also known as reference voltage can be adjusted by varying the doping levels, as Zener potential decreases by increasing the doping.

The Zener diodes are used for stabilizing voltage and therefore, are also called *stabilizer diodes* or *stabilitrons*. The voltage/current characteristic of a Zener diode shows that in the breakdown region, an extremely minor change in voltage causes very large variations in the current across the diode. The Zener diode is generally connected in parallel to the load across which the stabilized voltage is achieved. The resistance R of the limiting resistor connected in series with the diode is so adjusted that the current I_Z flowing through the diode corresponds to the middle point of the breakdown region [point O in Fig. 2.10 (c)]. Suppose there is some change in the supply voltage (say increase). As a result, the current in the circuit increases, but since large changes in the diode current are able to produce only very small changes in the Zener diode voltage [see breakdown region in Fig. 2.10 (b)], the voltage across the Zener diode or the load remains almost unchanged. In this way, the voltage stabilization is achieved by using a Zener diode as in the circuit of Fig. 2.10 (c).

2.7.2. Tunnel Diode

The tunnel diode was first manufactured by Dr Leo Esaki in 1958. The symbol of tunnel diode is shown in Fig. 2.11(a). It was found that if the depletion layer in a p-n junction was very thin, the electrons could penetrate

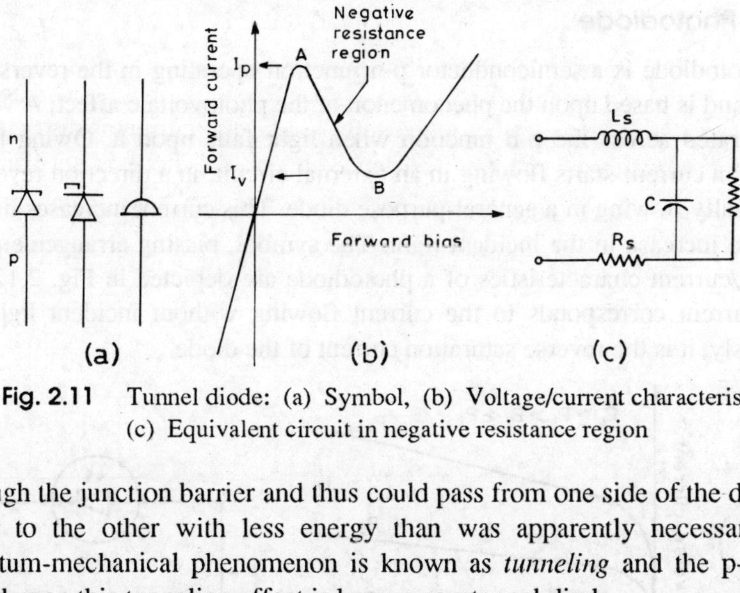

Fig. 2.11 Tunnel diode: (a) Symbol, (b) Voltage/current characteristic, (c) Equivalent circuit in negative resistance region

through the junction barrier and thus could pass from one side of the depletion layer to the other with less energy than was apparently necessary. This quantum-mechanical phenomenon is known as *tunneling* and the p-n diode based upon this tunneling effect is known as a tunnel diode.

In a tunnel diode, the width of the depletion layer is reduced by increasing the concentration of impurity atoms, say, 1 part in 10^3 (in ordinary diode, it is 1 part in 10^8). With this impurity concentration, the width of the depletion layer is reduced to less than 100°A which is one-fiftieth the wavelength of light. For potential barriers of this thickness, the probability of an electron to penetrate this barrier (as calculated quantum-mechanically using Schrödinger equation) becomes very large, and the voltage/current characteristic of this high impurity density diode is completely changed, as shown in Fig. 2.11 (b). The tunnel diode shows a negative resistance for part of its characteristic [A to B in Fig. 2.11 (b)] and therefore, is also called a negative resistance device. For currents between I_P (peak current) and I_V (valley current) the characteristic is triple valued because the same current can be achieved for three different applied voltages. The multivalued feature of the tunnel diode makes it a useful device in the high speed switching circuits. The equivalent circuit of a tunnel diode is shown in Fig. 2.11(c).

2.7.3. Power Diode

The power diode is a p-n junction diode specially designed to be used in the applications of high power and at high temperature. The power diode is constructed to have higher forward current, as well as higher peak inverse voltage. The typical value of PIV is somewhere between 50 V and 1000 V, with a maximum current of 30A. The forward resistance of a power diode is also very low, in the order of one ohm. In many cases, the heat sinks are also used along with a power diode to draw away the heat due to heavy current.

2.7.4. Photodiode

The photodiode is a semiconductor p-n junction operating in the reverse-bias region and is based upon the phenomenon of the photovoltaic affect. A voltage is generated across the p-n junction when light falls upon it. Owing to this voltage, a current starts flowing in an external circuit, in a direction reverse to that usually flowing in a general-purpose diode. This current increases linearly with the increase in the incident light. The symbol, biasing arrangement, and voltage/current characteristics of a photodiode are depicted in Fig. 2.12. The dark current corresponds to the current flowing without incident light and obviously, it is the reverse saturation current of the diode.

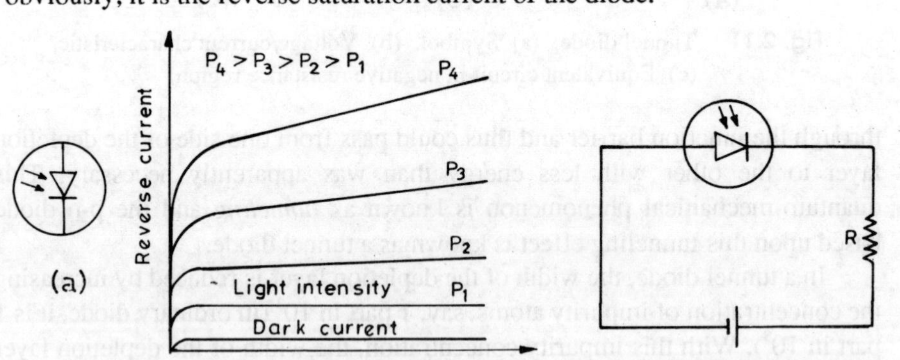

Fig. 2.12. Photodiode: (a) Symbol, (b) Voltage/Current characteristic, (c) Biasing arrangement.

As discussed in Sec. 2.2.2, the current in a reverse-biased p-n junction is owing to diffusing minority charge carriers. As the incident light produces electron-hole pairs in the reverse-biased diode, the electrons in the p-region and the holes in the n-region become additional minority charge carriers and the current increases. If the incident light falls at a distance from the p-n junction, the minority carriers produced due to light may recombine before diffusing across the junction and a small current will flow. This current will increase if the light falls near the junction because the probability of recombination is less. Hence, the photocurrent is a function of the distance from the junction at which the light falls, as shown in Fig. 2.13. As the diffusion lengths of minority charge carriers in the p- and the n-regions are different, the curve is not symmetrical on both sides.

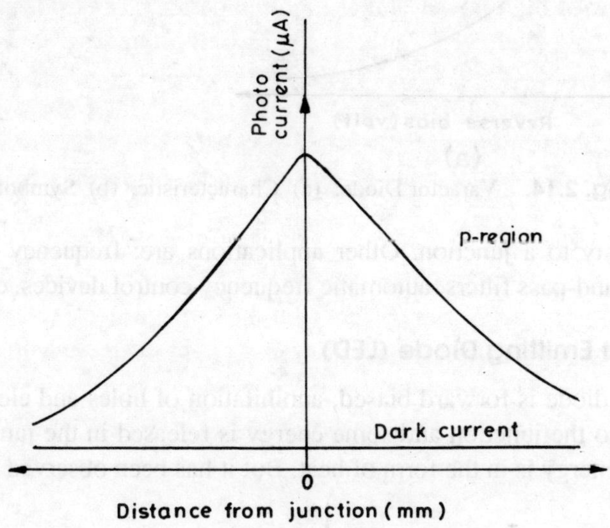

Fig. 2.13. Variation of photocurrent with distance from the junction at which light falls.

2.7.5. Varactor Diode

The varactor diode is a semiconductor variable voltage capacitor. It is also called *varicap diode* or *capacitor diode*. It has been discussed in Sec. 2.2.2 that in the reverse-biased diode, the depletion region is depleted of charge carriers. So, the junction area can be used as a capacitor because the depletion region acts as a dielectric medium, and the ions on each side of the depletion region act as conductors. The capacitance of this capacitor, known as transition region capacitance, decreases as the reverse-bias voltage is increased, as shown in Fig. 2.14. This is because the depletion region becomes wider as the reverse bias is increased. This process is analogous to separating the plates of parallel plate capacitor, resulting in the fall of capacitance. The normal range of reverse bais for VVC diodes is limited to about 20 V. The Figure 2.14 also shows the symbols used for the varactor diode. These diodes are used in the field of parametric amplification, in which a bias voltage and a signal are applied

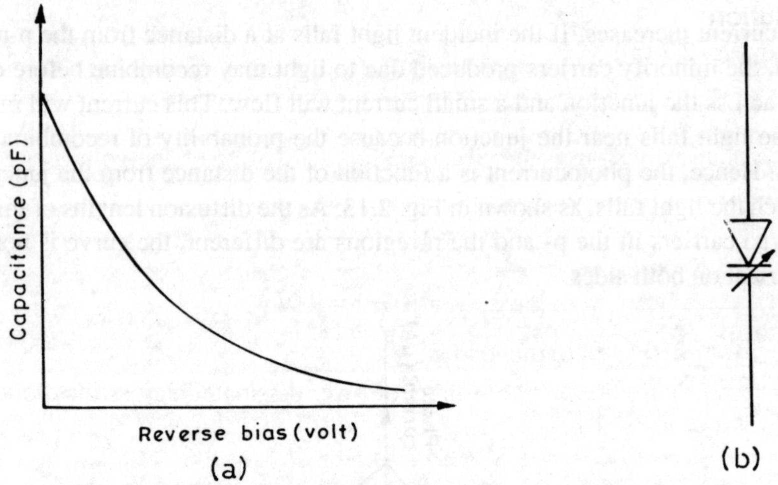

Fig. 2.14. Varactor Diode: (a) Characteristic, (b) Symbol.

simultaneously to a junction. Other applications are: frequency modulators, adjustable band-pass filters, automatic frequency control devices, etc.

2.7.6. Light Emitting Diode (LED)

When a p-n diode is forward biased, annihilation of holes and electrons takes place close to the junction and some energy is released in the junction. In Ge and Si, this energy is in the form of heat. But it has been observed that in some

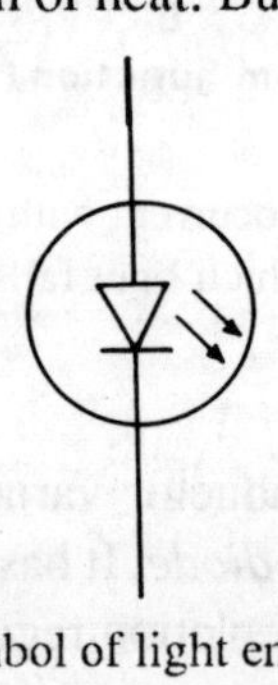

Fig. 2.15. Symbol of light emitting diode (LED).

other semiconductors, e.g. gallium arsenide, the energy is released in the form of radiation. This phenomenon is known as *electro-luminiscence* and the diode based upon this phenomenon is called a light emitting diode. The LED's are used in pilot lamps, display devices, etc.

2.8. SOLVED PROBLEMS

Problem 1

Calculate the voltage across the load and the power delivered to the load for the circuit of Fig. 2.7.

Solution

From Fig. 2.7 (a),

$$V_Q = 1\text{V} \quad \text{and} \quad I_Q = 9\,\text{mA}$$

The voltage division in the circuit of Fig. 2.7 (b) gives the voltage across the load as

$$V_R = V - V_D = V - V_Q$$

$$= 10 - 1 = 9\text{ V}$$

The power delivered to the load is

P_R = power supplied by the source – Power dissipated across the diode.

$$= V \cdot I_Q - V_Q \cdot I_Q$$

$$= 10 \times 9 \times 10^{-3} - 1 \times 9 \times 10^{-3}$$

$$= 81\text{ m W}.$$

Problem 2

Find the static or dc resistance of the diode of the circuit of Fig. 2.7.

Solution

Using Eq. 2.1, the static resistance of the diode is given by,

$$R_{\text{static}} = \frac{V_Q}{I_Q}$$

$$= \frac{1}{9 \times 10^{-3}} = 111.1\ \Omega$$

Problem 3

Find the dynamic resistance for the diode of circuit given in Fig. 2.8.

Solution

Using Eq. 2.2, the dynamic resistance of the diode is given by,

$$R_{\text{dynamic}} = \frac{\Delta V_Q}{\Delta I_Q}$$

From Fig. 2.8(a), $\Delta V_Q = 0.01\text{V}$; $\Delta I_Q = 2$ mA

$$\therefore \qquad R_{\text{dynamic}} = \frac{0.01}{2 \times 10^{-3}} = 5\ \Omega.$$

Problem 4

Find the voltage drop and the current across the diode for the input signal shown in Fig. 2.16. Given: the diode resistance = 4 Ω.

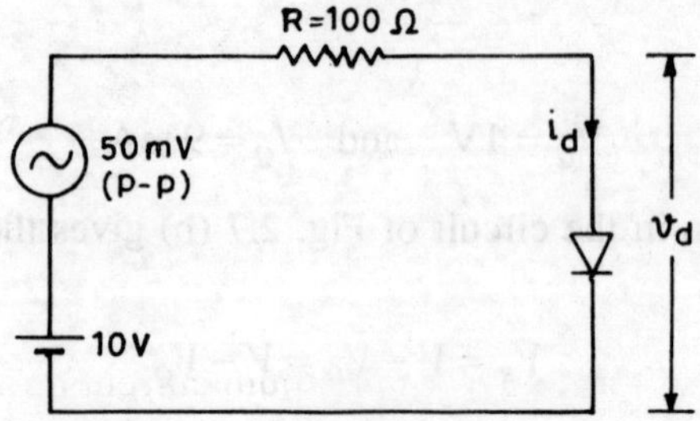

Fig. 2.16. Circuit for Problem 4.

Solution

In the a c equivalent circuit for the circuit shown in Fig. 2.16, the d c supply is ignored and the a c voltage drop across the diode is given by,

$$v_d = \frac{4 \times 50 \times 10^{-3}}{4 + 100} \cong 2\ \text{mV}(p - p)$$

and the diode current owing to the ac signal is

$$i_d = \frac{v_d}{4} = \frac{2\ \text{m V}}{4\ \Omega}$$

$$= 0.5\ \text{mA}\ (p - p).$$

Problem 5

In the circuit of Fig. 2.10, $R = 500\ \Omega$, $V_i = 25$ V, $V_Z = 10$ V, and $P_{d\,max} = 250$ m W. Find the minimum and maximum load currents.

Solution

When the diode is ON, R_L will be minimum. Therefore, from the circuit,

$$V_z = 10 = \frac{R_{L\,min} \times 25}{R_{L\,min} + 500}$$

or, $\quad 10\, R_{L\,min} + 5000 = 25\, R_{L\,min}$

or $\quad R_{L\,min} = 334\ \Omega.$

The minimum R_L allows maximum current through R_L.

$$\therefore \quad I_{L\,max} = \frac{V_Z}{R_{L\,min}} = \frac{10\text{V}}{334\ \Omega}$$

$$= 30\ \text{mA}$$

The maximum power capability limits the maximum current through the diode, i.e.

$$I_{Z\ max} = \frac{250\ \text{m V}}{10\text{V}} = 25\ \text{mA}$$

when diode is ON, the voltage across R is 25 V – 10 V = 15 V and the current through R is $\frac{15\text{V}}{500\ \Omega} = 30$ m A. Hence, the current division dictates that the current through the load is

$$I_{L\ min} = \text{Current through } R - \text{maximum current through diode}$$

$$= 30\ \text{mA} - 25\ \text{mA}$$

$$= 5\ \text{mA}$$

This current yields the maximum value of load resistance, i.e.

$$R_{L\ max} = \frac{V_Z}{I_{L\ min}} = \frac{10\text{V}}{5\ \text{mA}}$$

$$= 2\ \text{k}\ \Omega.$$

Questions and Problems

1. Describe the characteristics of an ideal diode.
2. What is the meaning of forward bias and reverse bias in the case of a diode?
3. How does a diode work as a switch?
4. Which diode works on the principle of breakdown and how?
5. What are the barrier potentials for germanium and silicon diodes?
6. What does the space charge at a p-n junction consist of ?
7. Describe various types of diode.
8. What do you understand by diffusion current in a p-n diode?
9. What is a depletion region and how is it formed?
10. What is the meaning of breakdown of a p-n junction?
11. Describe static and dynamic resistances of a p-n diode.
12. On what principle does a tunnel diode work?
13. Discuss the effect of biasing on the width of the depletion region at a p-n junction.
14. Why cannot the p-n junction barrier potential be measured by connecting a voltmeter across the diode terminals?
15. Can a p-n diode be used as a capacitor in an ac circuit? Give examples.
16. What is the importance of defining the peak inverse voltage for a p-n diode?
17. Calculate the voltage across the load and the power delivered to the load for the circuit shown in Fig. 2.7, if R_L is changed to 1.5 kΩ.

18. Find the dynamic resistance of the diode using the circuit given in Fig. 2.8 (b), if $V = 15$ V.
19. Calculate the power dissipated by a diode of internal resistance 20 Ω and barrier potential 0.7 V, when a current of 400 m A is flowing through it.
20. Find the minimum and maximum values of the load resistance for a 10 V Zener diode, if $R = 500\ \Omega$, $V_i = 20$ V, and $P_{d\ max} = 500$ m W.
21. Find the voltage drop and the current across the diode for the input signal of 100 m V in Fig. 2.16, if the battery potential is changed to 15 V.
22. If $R = 1\ k\ \Omega$, $V_i = 20$ V, $V_Z = 10$ V, and $P_{d\ max} = 250$ mW, find the minimum and maximum load currents in the circuit shown in Fig. 2.10.

3 Diode Circuits

In this Chapter some practical examples of diode application in various electronic circuits are discussed.

3.1. CLIPPERS

The clipping circuits are used to clip–off a portion of the input signal. In the output of a clipping circuit, only that part of the input waveform appears which lies above or below some reference level. The clippers are also called *voltage* or *current limiters, amplitude selectors*, or *slicers*. A clipping circuit involves a diode, a resistor and in some circuits, a d c supply too. Some basic clipping circuits are discussed in the following sections. The input waveform is taken to be sinusoidal in all these circuits.

3.1.1. Simple Series Positive Clippers

The relevant circuit diagram along with input and output waveforms is shown in Fig. 3.1. For the positive half-cycle of the input, the diode is reverse-biased and no current flows through R. Therefore, no output voltage appears across R. But for the negative half-cycle of the input, the diode is forward-biased and the

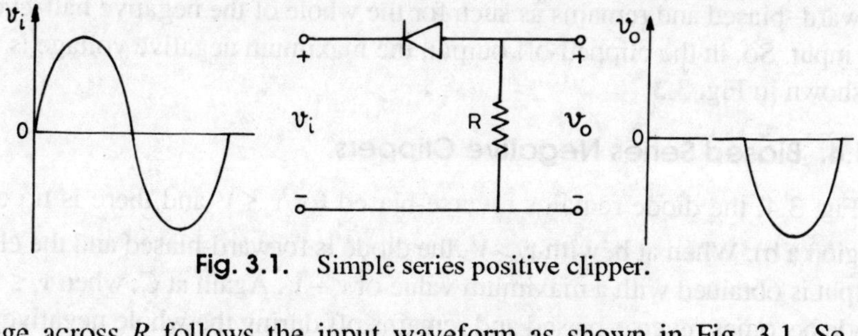

Fig. 3.1. Simple series positive clipper.

voltage across R follows the input waveform as shown in Fig. 3.1. So, the positive portion of the input waveform is clipped–off.

3.1.2. Simple Series Negative Clippers

The relevant circuit diagram along with input and output waveforms are shown in Fig. 3.2. In this case, the diode is forward-biased for the positive cycle and reverse-biased for the negative cycle of the input waveform. Therefore, the negative portion of the input waveform is clipped–off.

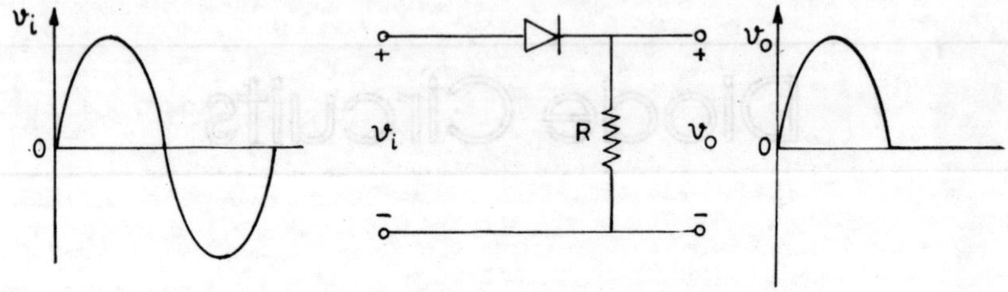

Fig. 3.2. Simple series negative clipper.

3.1.3. Biased Series Positive Clippers

Figure 3.3 shows the relevant circuit along with the input and the output waveforms. When the input in the positive cycle is less than the bias voltage, i.e. $v_i < V$, the diode is forward-biased (short-circuit). A decreasing negative

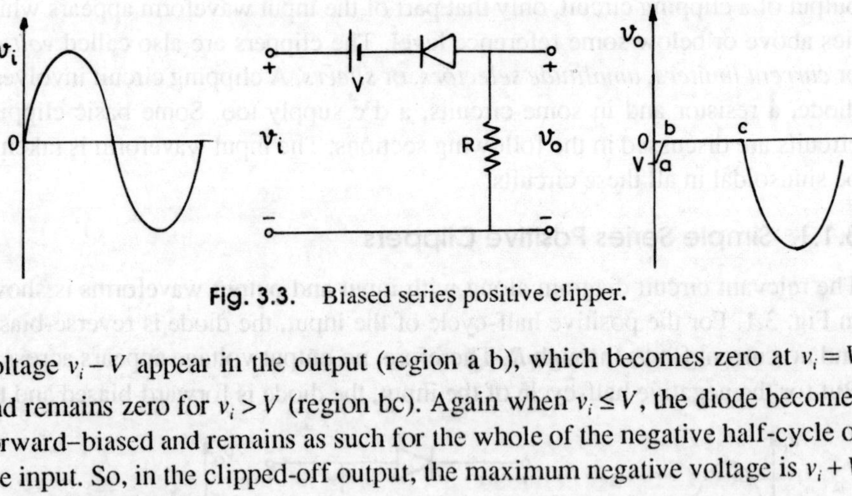

Fig. 3.3. Biased series positive clipper.

voltage $v_i - V$ appear in the output (region a b), which becomes zero at $v_i = V$ and remains zero for $v_i > V$ (region bc). Again when $v_i \leq V$, the diode becomes forward–biased and remains as such for the whole of the negative half-cycle of the input. So, in the clipped-off output, the maximum negative voltage is $v_i + V$ as shown in Fig. 3.3.

3.1.4. Biased Series Negative Clippers

In Fig. 3.4, the diode remains reverse-biased for $v_i \leq V$ and there is no output (region a b). When at b, with $v_i > V$, the diode is forward-biased and the clipped output is obtained with a maximum value of $v_i - V$. Again at C, when $v_i \leq V$, the diode becomes reverse-biased and remains off during the whole negative cycle of the input and no output is obtained.

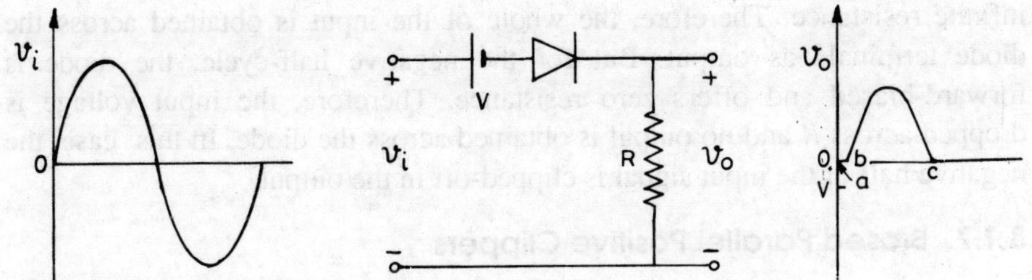

Fig. 3.4. Biased series negative clipper.

3.1.5. Simple Parallel Positive Clippers

In Fig. 3.5, the diode is forward-biased for the positive half-cycle of the input and offers no resistance. Therefore, the whole of the input voltage is dropped

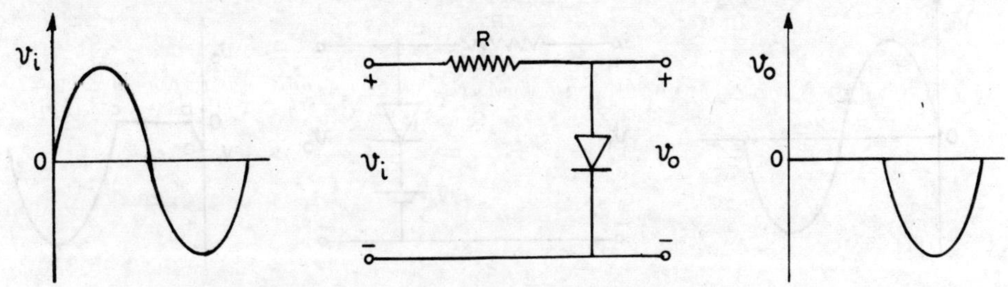

Fig. 3.5. Simple parallel positive clipper.

across *R* and no output is generated. For the negative half-cycle of the input, the diode is reverse-biased and offers infinite resistance. Since there is no current through *R*, the output voltage is equal to the input voltage. Thus, the positive half-portion of the input signal is clipped-off in the output.

3.1.6. Simple Parallel Negative Clippers

In Fig. 3.6, for the positive half-cycle, the diode is reverse-biased and offers

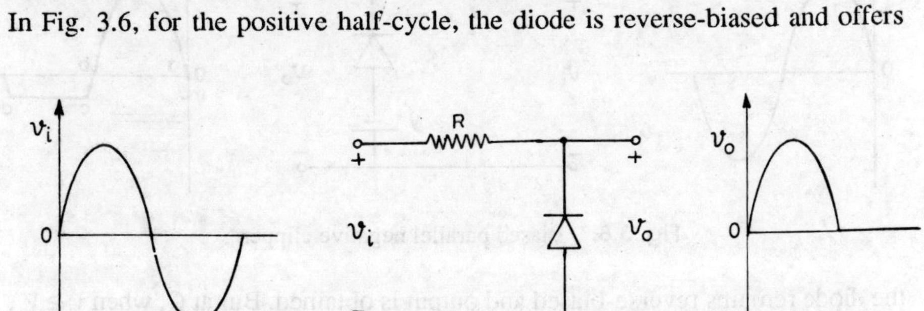

Fig. 3.6. Simple parallel negative clipper.

infinite resistance. Therefore, the whole of the input is obtained across the diode terminals as output. But for the negative half-cycle, the diode is forward-biased and offers zero resistance. Therefore, the input voltage is dropped across R and no output is obtained across the diode. In this case, the negative half of the input signal is clipped-off in the output.

3.1.7. Biased Parallel Positive Clippers

In Fig. 3.7, the diode is reverse-biased owing to the battery V. When the positive half-cycle of the input signal is applied, the diode remains reverse-biased up to $v_i \leq V$ (region a b), and the output is obtained. When at $v_i > V$, the diode becomes forward-biased and no output is obtained. Again, when at C, $v_i \leq V$, the diode becomes reverse-biased and remains so for the whole of the negative cycle also, and the clipped output is obtained as shown in Fig. 3.7.

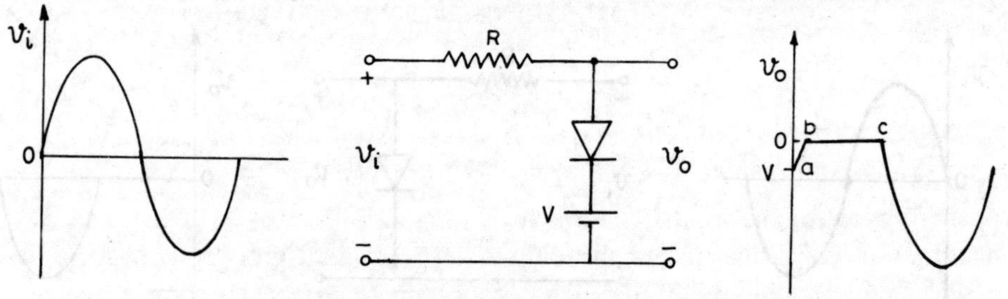

Fig. 3.7. Biased parallel positive clipper.

3.1.8. Biased Parallel Negative Clippers

In Fig. 3.8, the diode is made reverse-biased by a potential V. For the whole of the positive cycle and up to $v_i < V$ in the negative cycle of the input waveform,

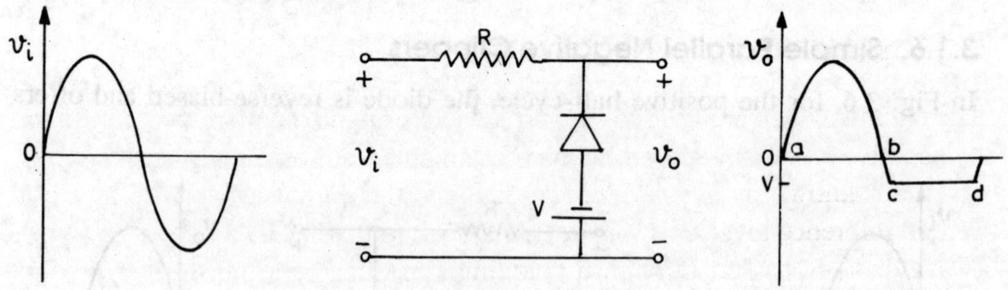

Fig. 3.8. Biased parallel negative clipper.

the diode remains reverse-biased and output is obtained. But at C, when $v_i \geq V$, in the negative cycle of the input the diode becomes forward-biased and no output is obtained. Again at d, when $v_i \leq V$, the diode becomes reverse-biased and the clipped output is obtained as shown in Fig. 3.8.

3.1.9. Double-ended Clippers

Two diodes are used to achieve double-ended clipping at independent levels. This circuit is used to convert a sinusoidal waveform into a square wave. The circuit for a double-ended clipper along with the input and output waveforms are shown in Fig. 3.9. The two diodes D_1 and D_2 are reverse-biased by V_1 and V_2 respectively, where $V_1 > V_2$. In the positive half-cycle, the diode D_1 remains

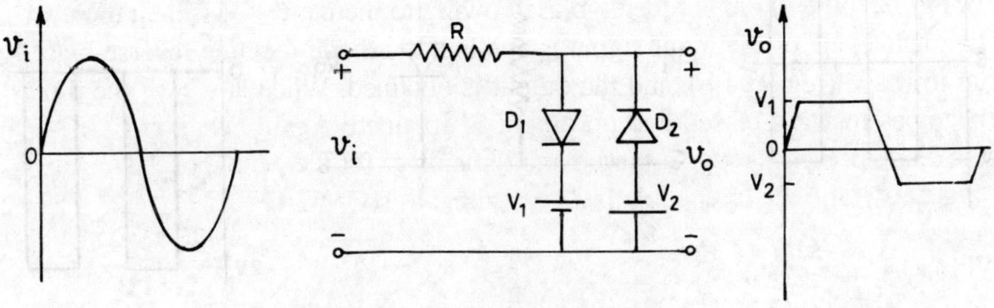

Fig. 3.9. Double ended clipper.

reverse-biased until $V_i \leq V_1$ and D_2 remains reverse-biased for the whole of the positive cycle of the input. The output is obtained but at $v_i = V_1$, the diode D_1 becomes forward-biased and the output drops to zero and remains so until $v_i \leq V_1$ again. At this point, the output is obtained, since D_1 becomes reverse-biased again. In the negative half-cycle, the diode D_2 remains reverse-biased until $v_i \geq V_2$ (D_1 remains reverse-biased in the whole of the negative cycle). At $v_i = V_2$, D_2 becomes forward-biased and the output becomes zero. Hence, the output is clipped at V_1 in the positive cycle and at V_2 in the negative cycle, as shown in Fig. 3.9. In order to get symmetrical square waveform, V_1 must be equal to V_2.

3.2. CLAMPERS

The circuits which are used to convert a waveform into another whose positive or negative– or both–extremities are maintained at some constant reference level, are known as *clamping circuits*. Since the output cannot rise above a certain reference level, it is said to be clamped to that level. Whenever a signal (waveform) is passed through a coupling capacitor or a transformer, its d c component is lost because neither the capacitor nor the transformer allows d c to pass through it. The clamping circuits are used to restore the lost d c or introduce a new one to such signals. Therefore, the clampers are also called d c restorers or d c reinserters. A diode, a capacitor, and a resistor are the three essential elements required for a clamping circuit. The values of the resistor and the capacitor are selected so that the time constant RC is very large in comparison to the time period of the input signal. Under this condition, the voltage across the capacitor may be assumed to remain fairly constant during the input signal, when the capacitor is not charging. The description of a simple clamping circuit, shown along with input and output waveforms in Fig. 3.10 is given below.

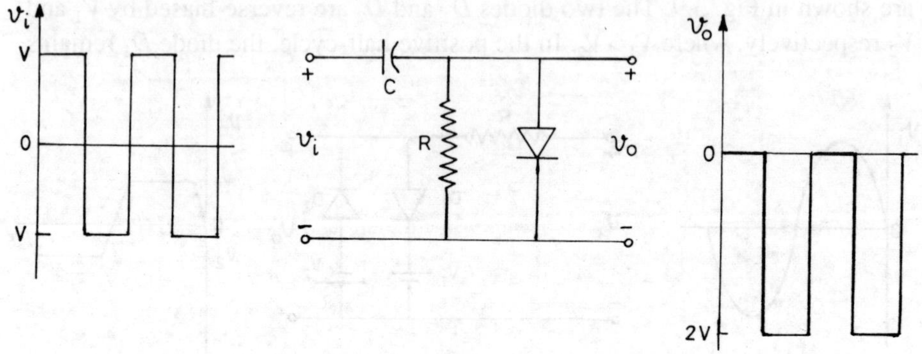

Fig. 3.10. Clamping Circuit along with Input and output waveforms.

When the input is $+V$, the diode is forward-biased and the output taken across the diode is zero. The capacitor charges to $+V$ immediately as the time constant $RC = 0$ (forward-biased diode has zero resistance). Now, when the input switches to $-V$, the diode becomes reverse-biased, showing infinite resistance. The voltage across the capacitor does not change appreciably during this interval of time and the capacitor acts like a fixed voltage source. The output is obtained as the sum of the input signal voltage $(-V)$ and the capacitor voltage $(-V)$, i.e. $(-V) + (-V) = -2\,V$.

Thus, the output is clamped to the negative region at $2\,V$ and will repeat itself at the frequency of the input signal. An important point here is that the swing of the input $(2V)$ and of the output $(2V)$ is the same, and it is true for all the clamping circuits.

3.3. COMPARATORS

As its name indicates, the comparator circuit is used to compare two waveforms. If one waveform is taken as the reference level, the comparator is used to mark the instant when another waveform attains this reference level. Thus, a comparator output may be a sharp pulse which appears when the increasing signal becomes equal to the reference. A comparator circuit along with the input and the output waveforms is shown in Fig. 3.11. The input signal is taken as a ramp. The diode remains cut off until $v_i \le V_R$ and the output remains quiescent at $v_0 = V_R$. When $v_i \ge V_R$, the diode is forward-biased and the output rises with the input signal. In this way, the comparator output marks the instant t at which the input becomes V_R, the reference voltage.

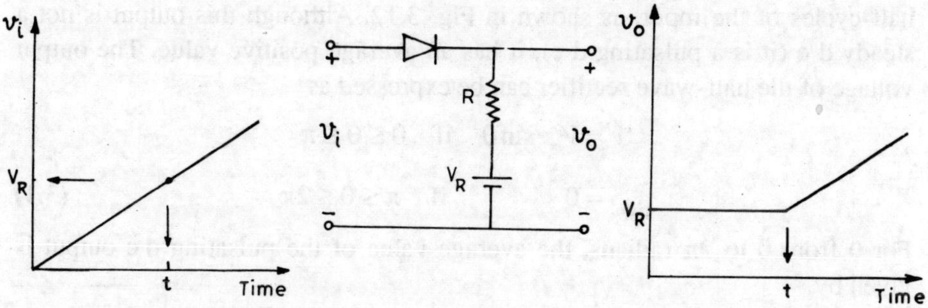

Fig. 3.11. Comparator circuit along with input and output waveforms.

3.4. RECTIFIERS

An electronic circuit which is used to convert an alternating waveform (voltage or current) into a pulsating d.c or unidirectional waveform is known as a *rectifying circuit*, or simply a rectifier. The *p-n* junction diode is well suited for this purpose because it conducts heavily in the forward biasing and only slightly in the reverse biasing. There are three most common rectifying circuits: half-wave, full-wave, and bridge rectifiers.

3.4.1. Half-wave Rectifier

Figure 3.12 shows a circuit for a half-wave rectifier with its input and output waveforms. A half-wave rectifier use only alternate half-cycles (positive or

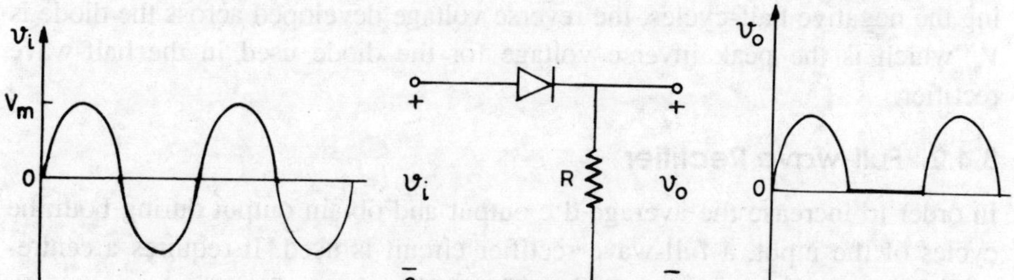

Fig. 3.12. Half wave rectifier circuit along with input and output waveforms.

negative) of the a c input waveform to produce a pulsating d c output. For the other half-cycles of the input, there is no output.

For the positive half-cycles of the input signal, the diode becomes forward-biased (diode resistance becomes zero) and this input waveform appears as the output across the resistor *R*. For the negative half-cycles of the input, the diode becomes reverse-biased (diode resistance is infinite), and no output appears across *R*. Thus, the output is available only for the positive half-cycles of the input, as shown in Fig. 3.12. Although this output is not a steady d c (it is a pulsating d c), it has an average positive value. The output voltage of the half-wave rectifier can be expressed as

$$v = V_m \sin\theta \quad \text{if} \quad 0 \le \theta \le \pi$$

$$= 0 \qquad \text{if} \quad \pi \le \theta \le 2\pi \tag{3.1}$$

For θ from 0 to 2π radians, the average value of the pulsating d c output is given by,

$$V_{dc} = \frac{1}{2\pi}\int_0^{2\pi} V_m \sin\theta \, d\theta$$

$$= \frac{V_m}{2\pi}\int_0^{\pi} \sin\theta \, d\theta \qquad \text{(using Eq.3.1)}$$

$$= \frac{V_m}{2\pi}[-\cos\pi + \cos 0]$$

$$= \frac{V_m}{\pi}$$

$$= 0.318\, V_m \tag{3.2}$$

where V_m is the peak value of the input and V_{dc} is the average d c output. During the negative half-cycles, the reverse voltage developed across the diode is V_m which is the peak inverse voltage for the diode used in the half-wave rectifier.

3.4.2. Full-wave Rectifier

In order to increase the average d c output and obtain output during both the cycles of the input, a full-wave rectifier circuit is used. It requires a centre-tapped transformer and two diodes. The peak voltage developed across each half of the transformer during the positive cycles is V_m and during the negative cycles is $-V_m$. During the alternate positive half-cycles of the input, the upper half of the transformer makes the diode D_1 forward-biased and the lower half of the transformer makes the diode D_2 reverse-biased. Hence, with the diode D_1 conducting, a positive half-cycle of the input waveform, with peak value V_m, is developed across *R*. Similarly, during the alternate negative half-cycles of the input, the diode D_2 is conducting and the diode D_1 is nonconducting. In this condition, the output voltage with peak value V_m developed across R is of the same polarity as for the positive half-cycles because the current through R

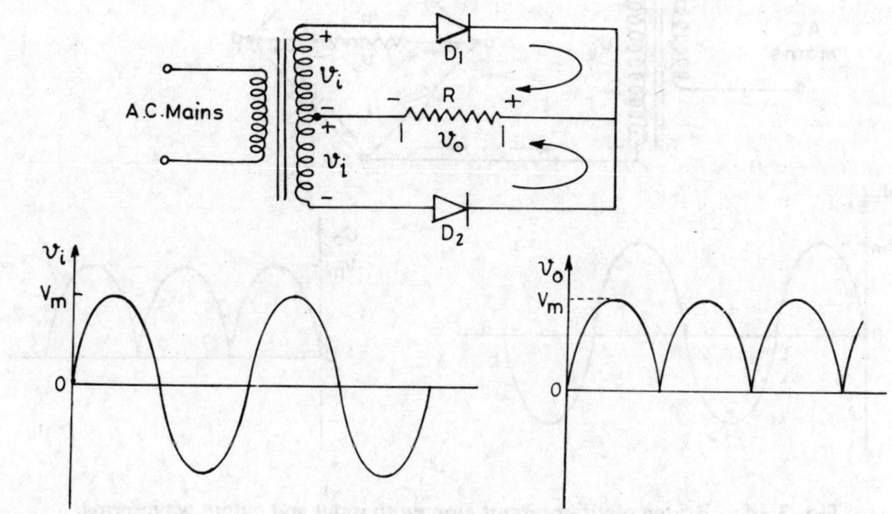

Fig. 3.13. Full wave rectifier circuit along with input and output waveforms.

flows in the same direction in both the cycles, as shown in Fig. 3.13. Hence, the output of a full-wave rectifier is two positive-going half cycles. Therefore, the average value for the dc output of a full-wave rectifier is twice the average value obtained in the case of a half-wave rectifier, i.e.

$$V_{dc} = 2(0.318\ V_m)$$

$$= 0.636\ V_m \tag{3.3}$$

For the positive half-cycles (or negative half-cycles), the reverse voltage developed across D_2 (or D_1) is $2\ V_m$, i.e. the sum of the voltage V_m across the lower half (or the upper half) of the transformer and the voltage V_m across the load resistor R. This is the peak inverse voltage for the diodes used in the full-wave rectifier, which is twice the peak value of the input signal.

3.4.3. Bridge Rectifier

The circuit for a bridge rectifier is shown in Fig. 3.14. It is another form of a full-wave rectifier, in which a centre-tapped transformer is not required. Also, the diodes used have peak inverse voltage rating of V_m, rather than $2V_m$.

During the positive half-cycles of the input signal, the didoes D_1 and D_3 become forward-biased (D_2 and D_4 being reverse-biased) and a current flows along the path $a\ b\ d\ c$. A voltage from the transformer V_m appears across R.

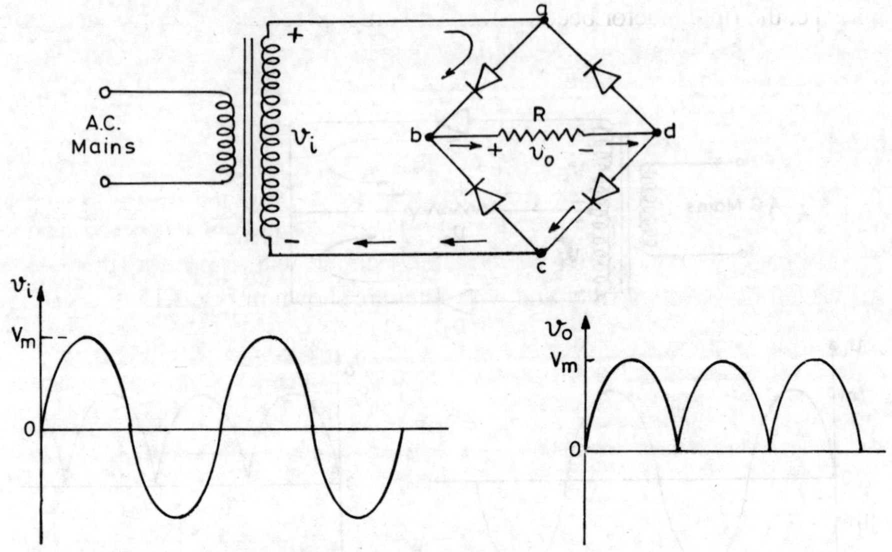

Fig. 3.14. Bridge rectifier circuit alongwith input and output waveforms.

Similarly, during the negative half-cycles of the input signal, the diodes D_2 and D_4 become forward-biased (D_1 and D_3 being reverse-biased) and a current flows along the path *c b d a*. Since this current direction through *R* is the same as in the case of the positive half-cycles, the voltage from the transformer V_m appears across *R* in the same polarity as in the positive half-cycles. Thus, the output of a bridge rectifier is also positive-going half cycles.

During the negative half-cycles, the reverse-biasing across D_2 and D_4 is the sum of the transformer voltage V_m and the voltage across the load, V_m owing to the conduction of D_1 and D_3, i.e. $2V_m$. Since there are two diodes in the path, the reverse voltage across each diode is V_m. Similarly, the reverse voltage across either of D_1 and D_3 during the positive half-cycles is also V_m. Hence, the peak inverse voltage rating for diodes used in the bridge rectifier is V_m, rather than $2V_m$.

3.5. FILTER CIRCUITS

The output from a rectifier is obtained as pulsating d c signals, whose frequency is the same as the input frequency for the half-wave rectifier, or twice the input frequency for the full-wave rectifier. Hence, a rectifier is unable to give a steady d c; it has ripple. The ripple is defined as the departure of the output waveform of a rectifier from pure d c. The ripple in the output of a rectifier circuit is measured in the form of the ripple factor. This factor is defined as the ratio of the r.m.s. value of the a.c. component to the average d c value. In pure d c, the ripple factor becomes zero. In order to reduce the ripple factor to a minimum value (ideally zero), or to obtain pure d c, filtering or smoothing circuits are used.

3.5.1. Capacitor Filter

If a simple capacitor is connected across the rectifier output, the ripple factor is considerably reduced and the d c output voltage is available across the capacitor. Figure 3.15 shows a full-wave rectifier circuit with capacitor filter. Also, the waveforms without filter and with filter are shown in Fig. 3.15.

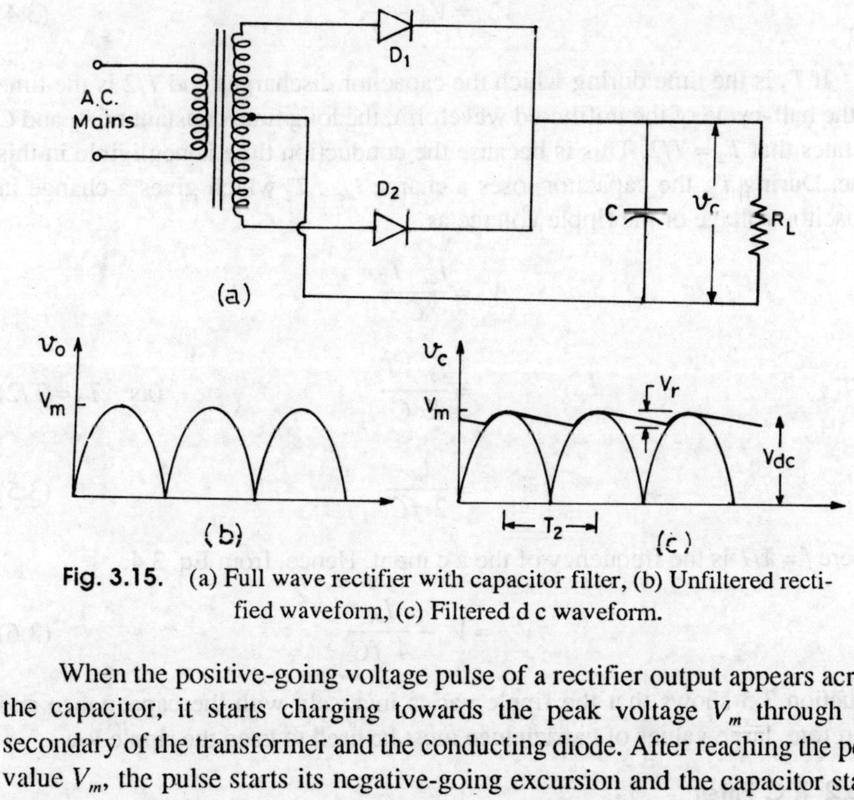

Fig. 3.15. (a) Full wave rectifier with capacitor filter, (b) Unfiltered rectified waveform, (c) Filtered d c waveform.

When the positive-going voltage pulse of a rectifier output appears across the capacitor, it starts charging towards the peak voltage V_m through the secondary of the transformer and the conducting diode. After reaching the peak value V_m, the pulse starts its negative-going excursion and the capacitor starts discharging through R_L, thereby holding the voltage across R_L at some higher value than its unfiltered value. Before the capacitor can fully discharge, the positive-going excursion of the next half-cycle of the unfiltered voltage is nearing its peak value, recharging the capacitor again. Now the pulse, after reaching the peak value V_m again starts its negative-going excursion and the capacitor starts discharging once again. Before a complete discharging of the capacitor, the positive-going excursion of the next half-cycle comes in and the capacitor recharges again. This action of charging and recharging of the capacitor continues as long as the circuit is in operation, as shown in Fig. 3.15. In this way, the time during which the current passes through the load R_L is prolonged and the ripple factor is reduced.

If the average value of d c output is to be very close to V_m, the capacitor must discharge only slightly. In order to fulfil this condition, the time constant of R_L and C is chosen to be very large in comparison to the time period of the unfiltered waveform. Because of the long time constant, the discharge of the capacitor is taken to be approximately linear and the ripple, as shown in Fig. 3.15, is similar to a sawtooth waveform. The average value of the output voltage is given by.

$$V_{dc} = V_m - \frac{V_r}{2} \tag{3.4}$$

If T_2 is the time during which the capacitor discharges and $T/2$ is the time of the half-cycle of the unfiltered waveform, the long time constant of R_L and C dictates that $T_2 = T/2$. This is because the conduction time is negligible in this case. During T_2, the capacitor loses a charge $I_{dc} \cdot T_2$ which gives a change in capacitor voltage or the ripple voltage as

$$V_r = \frac{I_{dc} \cdot T_2}{C}$$

$$= \frac{I_{dc} \cdot T}{2\,C} \qquad (\text{as} \quad T_2 = T/2)$$

$$= \frac{I_{dc}}{2\,fC} \tag{3.5}$$

where $f = 1/T$ is the frequency of the a c input. Hence, from Eq. 3.4,

$$V_{dc} = V_m - \frac{I_{dc}}{4\,fC} \tag{3.6}$$

Equation 3.5 shows that the ripple varies inversely with the capacitance and therefore, large values of capacitance must be used to keep the ripple low.

3.5.2 R C Filter

In order to further reduce the ripple present in the d c obtained after capacitor filter (Sec. 3.5.1), an additional *RC* filtering circuit is used, as shown in Fig. 3.16 (a). Though the *RC* filter is good to attenuate the ripple voltage, it has the drawback of lowering the d c voltage also, owing to a voltage drop across *R*. The final values of the d c and the ripple voltages are described below.

Figures 3.16(b) and (c) show the equivalent circuits to calculate the d c and the ripple voltages after *RC* filtering. V_{dc} and V_r are the d c and the ripple voltages before the *RC* filter circuit, i.e. across the capacitor C_1. As the

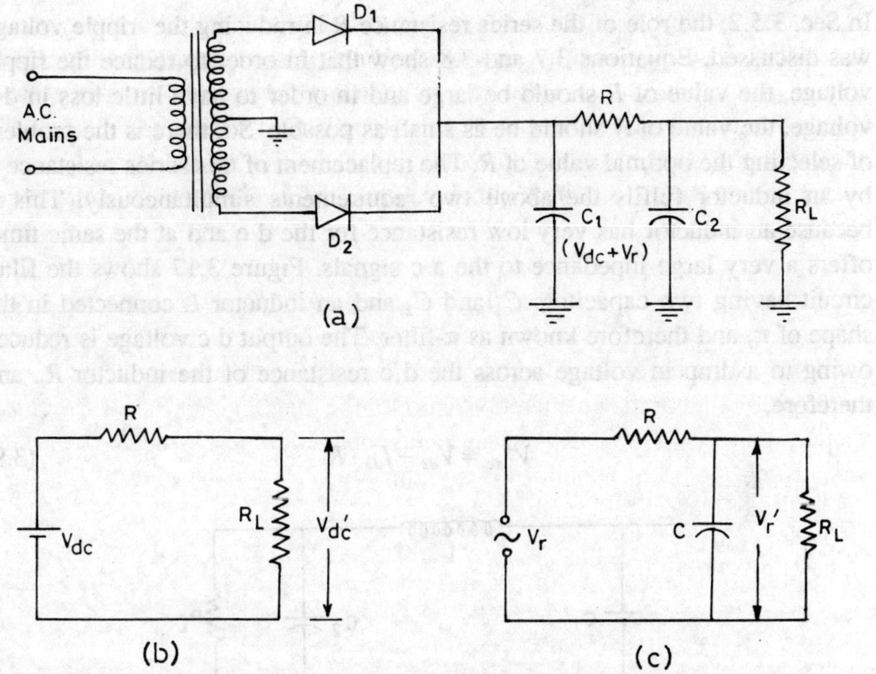

Fig. 3.16. (a) Full wave rectifier with *RC* filter (b) Equivalent circuit for d c operation (c) Equivalent circuit for ac or ripple.

reactance of a capacitor for d c is infinite, the capacitors C_1 and C_2 in the parallel combination are removed in the equivalent circuit of Fig. 3.16 (b). Hence, the final d c voltage across the load R_L is

$$V'_{dc} = \left[\frac{R_L}{R + R_L}\right] V_{dc} \tag{3.7}$$

In the circuit of Fig. 3.16 (c), the resultant impedance of the parallel combination of C_2 and R_L is of the order of C_2 alone and is much less than the load resistance R_L (if $C_2 = 10\ \mu F$ and $R_L = 2$ K ohms, typically). Therefore, the load resistance can be ignored in comparison to C_2 and the ripple voltage after the *RC* filter is

$$V'_r = \frac{1/\omega C_2}{\sqrt{R^2 + (1/\omega C_2)^2}}\, V_r \tag{3.8}$$

using simple voltage division rule. Equation 3.7 and 3.8 show that the *RC* filter is able to reduce the ripple voltage by a greater amount than it decreases the d c voltage, providing better filter operation.

3.5.3. π–Filter

In Sec. 3.5.2, the role of the series resistance R in reducing the ripple voltage was discussed. Equations 3.7 and 3.8 show that in order to reduce the ripple voltage, the value of R should be large and in order to have little loss in d c voltage, the value of R should be as small as possble. So, there is the problem of selecting the optimal value of R. The replacement of this series resistance R by an inductor fulfils the above two requirements simultaneously. This is because an inductor has very low resistance for the d c and at the same time, offers a very large inpedance to the a c signals. Figure 3.17 shows the filter circuit having two capacitors C_1 and C_2 and an inductor L connected in the shape of π, and therefore known as π-filter. The output d c voltage is reduced owing to a drop in voltage across the d c resistance of the inductor R_l, and therefore,

$$V'_{dc} = V_{dc} - I_{dc} \cdot R_l \tag{3.9}$$

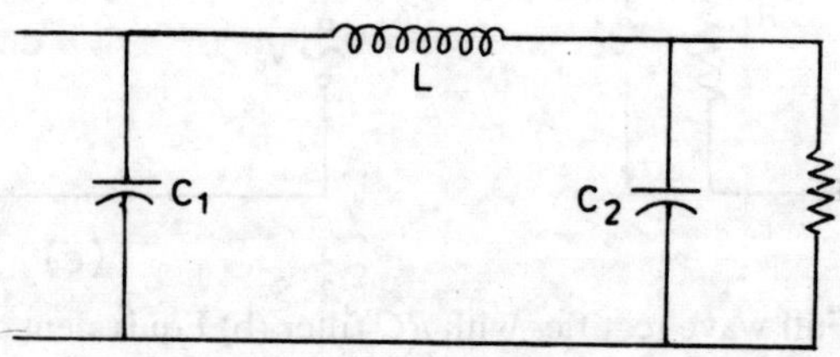

Fig. 3.17. π–Filter circuit.

If R is replaced by a c impedance of the inductor ωL, the expression for the ripple voltage from Eq. 3.8 is

$$V'_r = \frac{1/\omega\, C_2}{(\omega L - (1/\omega\, C_2))} \cdot V_r \tag{3.10}$$

3.6. VOLTAGE MULTIPLYING CIRCUITS

A number of circuits of diodes and capacitors can be used to get d c output voltage that is double, triple, or many times the peak value of the a c input voltage. These circuits are known as *voltage multiplying circuits: doublers, triplers, quadruplers*, etc. All these circuits use the charge stored in the capacitors to increase the output voltage. The input is a c and the output is d c multiple.

3.6.1. Half-wave Voltage Doubler

Figure 3.18 shows the circuit for a half-wave voltage doubler, which gives a d c voltage twice that obtained from the half-wave rectifier circuit. For the positive half-cycles of the input, the top of the transformer secondary is positive and the capacitor C_1, charges through the conducting diode D_1 to approximately the peak value of the secondary voltage. At this time, the diode D_2 is nonconducting and there is no output. During the negative cycle of the input, the top of the transformer secondary becomes negative and the diode D_1 becomes cut off. As the voltage across C_1 is in series with this applied a c input, the two voltages are added to give a forward-biasing of $V_m + V_m = 2V_m$ to the diode D_2, and the capacitor C_2 is charged to this voltage, i.e. $2V_m$. The output waveform across the capacitor C_2 is double that to a half-wave rectifier filtered by a capacitor filter, and the peak inverse voltage across each diode is $2V_m$. In this way, a d c output approximately double the peak value of the a c input is obtained.

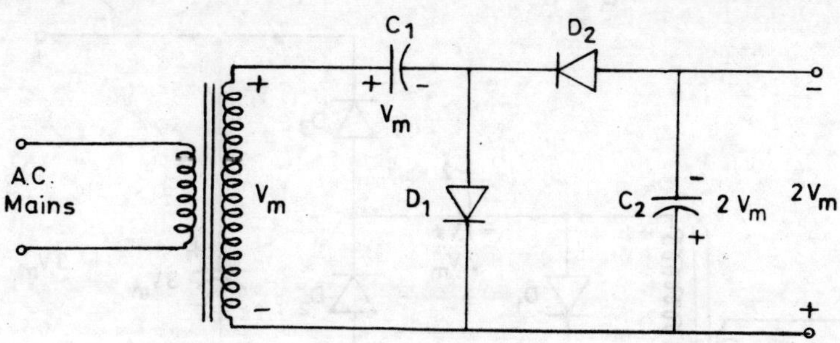

Fig. 3.18. Half-wave voltage doubler circuit.

3.6.2. Full-wave Voltage Doubler

Figure 3.19 shows the circuit of a full-wave voltage doubler. For the positive half-cycle of the a c input, the top of the transformer secondary is positive

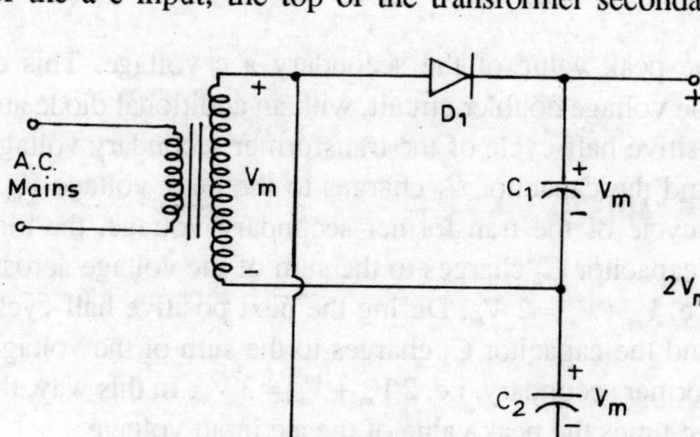

Fig. 3.19. Full-wave voltage doubler circuit.

and the diode D_1 conducts. The capacitor C_1 charges through D_1 and the transformer secondary to the peak value of the secondary voltage V_m. At this time, the diode D_2 is nonconducting. During the negative half-cycle of the a c input, the diode D_2 conducts. The capacitor C_2 charges through D_2 and the transformer secondary to the peak value of the secondary voltage V_m. At this time, the diode D_1 is nonconducting. Thus, the voltage across C_1 and C_2 is $2\ V_m$ because both are in a series combination. The peak inverse voltage for each diode is $2\ V_m$. In this case, the filtering action is poor as the resultant capacitor is less than the capacitance of either C_1 or C_2, as both are in series.

3.6.3. Voltage Tripler and Quadrupler

Figure 3.20 shows a voltage tripler circuit which provides a d c output that is

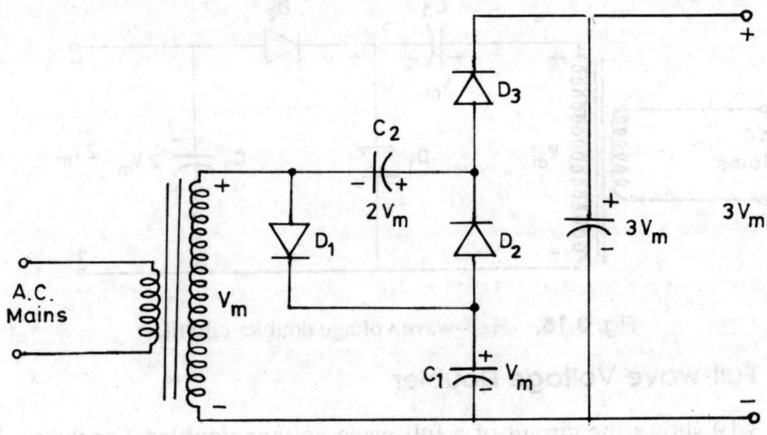

Fig. 3.20. Voltage tripler circuit.

three times the peak value of the secondary a c voltage. This circuit is an extension of the voltage doubler circuit, with an additional diode and capacitor. During the positive half-cycle of the transformer secondary voltage, the diode D_1 conducts and the capacitor C_1 charges to the peak voltage V_m. During the negative half-cycle of the transformer secondary voltage, the diode D_2 conducts and the capacitor C_2 charges to the sum of the voltage across C_1 and the transformer, i.e. $V_m + V_m = 2\ V_m$. During the next positive half-cycle, the diode D_3 conducts and the capacitor C_3 charges to the sum of the voltages across C_2 and the transformer secondary, i.e. $2\ V_m + V_m = 3\ V_m$. In this way, the d c output voltage is three times the peak value of the a c input voltage.

Figure 3.21 shows a voltage quadrupler circuit, which gives a d c output voltage four times the peak value of the secondary a c voltage. The circuit behaviour is on the same lines as explained for the tripler circuit above. This circuit may be further extended to obtain very large voltages.

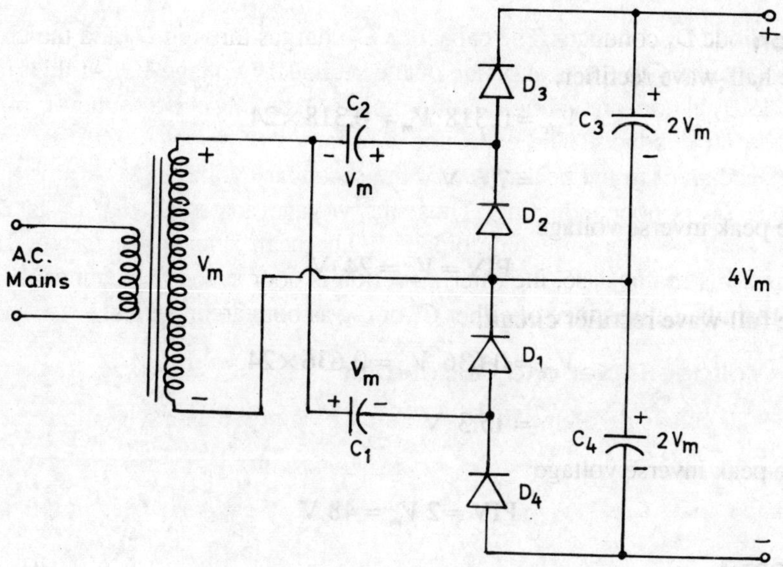

Fig. 3.21. Voltage quadrupler circuit.

3.7. SOLVED PROBLEMS

Problem 1

Calculate the ripple voltage under no-load and full-load conditions in the case of capacitor filter with $C = 2000\ \mu$ F, $I_{dc} = 250$ m A, and $V_{ac} = 9$ V.

Solution

The ripple voltage using a simple capacitor is given by

$$V_r = \frac{I_{dc}}{2fC} = \frac{I_{dc}}{100\ C} \quad \therefore \quad f = 50 \text{ Hz}$$

Under no-load condition, $I_{dc} = 0$

$$\therefore \qquad V_r = 0 \text{ V}$$

Under full-load condition, $I_{dc} = 250$ mA

$$\therefore \qquad V_r = \frac{250 \times 10^{-3}}{100 \times 2000 \times 10^{-6}}$$

$$= 2.5 \text{ V}.$$

Problem 2

Find the d c voltage obtained from a half-wave and a full-wave rectifier circuits, with peak rectified voltage of 24 V. Also, calculate the peak inverse voltage developed in the two cases.

Solution

For the half-wave rectifier,

$$V_{dc} = 0.318\ V_m = 0.318 \times 24$$

$$= 7.6\ \text{V}$$

and the peak inverse voltage

$$\text{PIV} = V_m = 24\ \text{V}$$

For the full-wave rectifier circuit,

$$V_{dc} = 0.636\ V_m = 0.636 \times 24$$

$$= 15.3\ \text{V}$$

and the peak inverse voltage

$$\text{PIV} = 2\ V_m = 48\ \text{V}$$

Problem 3

Find the peak inverse voltage for the diodes used in a bridge rectifier if V_{dc} = 63.6 V.

Solution

For a bridge rectifier,

$$V_{dc} = 0.636\ V_m$$

$$\therefore \quad V_m = \frac{V_{dc}}{0.636} = 100\ \text{V}$$

and the peak inverse voltage

$$\text{PIV} = V_m = 100\ \text{V}$$

Problem 4

What will be the output d c voltage for a π-filter if the input voltage is 30 V, the filter inductor is 5 henry with 200 Ω resistance and the output current is 50 mA?

Solution

Across the π-filter,

$$V'_{dc} = V_{dc} - I_{dc} \cdot R_L$$

$$= 30 - 50 \times 10^{-3} \times 200$$

$$= 20\ \text{V}$$

Problem 5

Draw the output waveforms for the input applied to the circuit of Fig. 3.22.

Solution

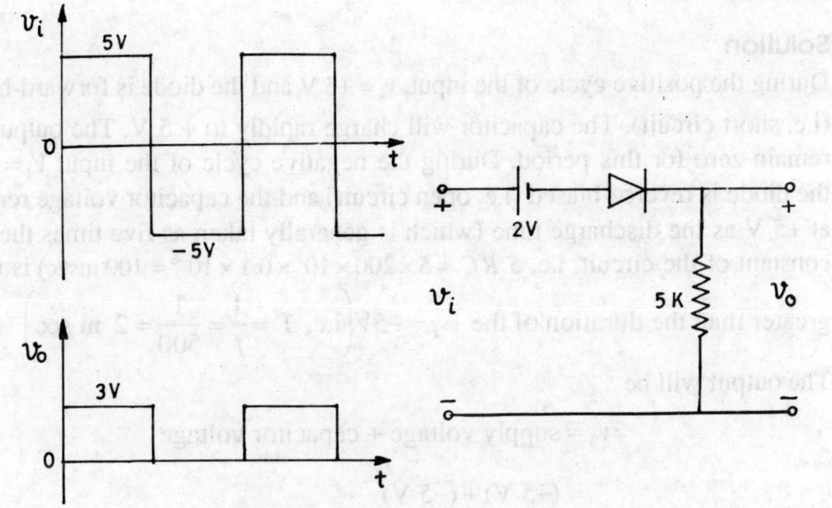

Fig. 3.22. Clipping circuit for problem 5.

For values $v_i > 2$ V, the diode is forward-biased and $V_o = 5 - 2 = 3$ V.

For values $v_i < 2$ V, the diode is reverse-biased and $v_0 = 0$. The resultant output waveforms are shown in Fig. 3.22 (c).

Problem 6

Draw the output waveforms for the input applied to the circuit of Fig. 3.23.

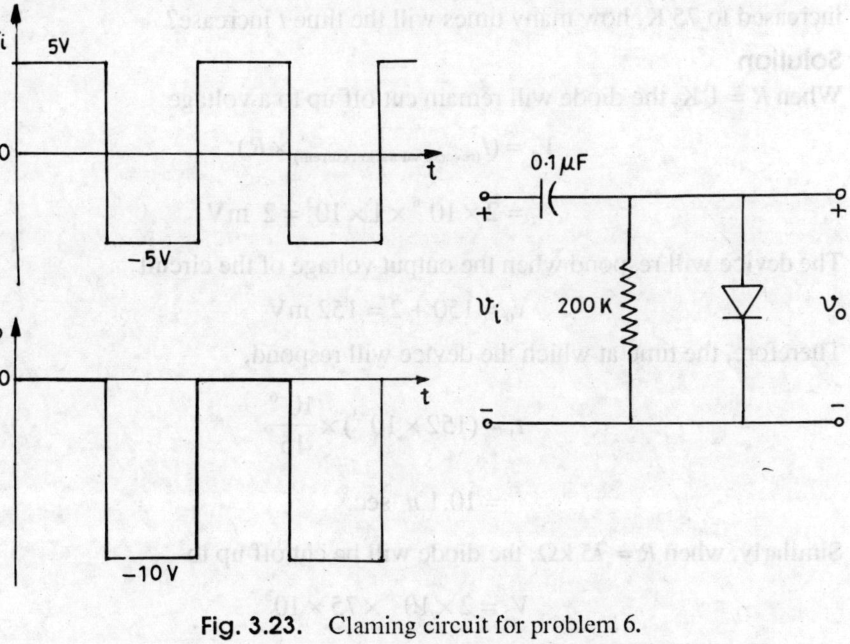

Fig. 3.23. Claming circuit for problem 6.

Solution

During the positive cycle of the input, $v_i = +5$ V and the diode is forward-biased (i.e. short circuit). The capacitor will charge rapidly to + 5 V. The output will remain zero for this period. During the negative cycle of the input $V_i = -5$ V, the diode is reverse-biased (i.e. open circuit) and the capacitor voltage remains at +5 V as the discharge time (which is generally taken as five times the time constant of the circuit, i.e. 5 $RC = 5\times200\times10^3\times0.1\times10^{-6} = 100$ msec) is much greater than the duration of the $v_i = -5V\left(\text{i.e. } T = \frac{1}{f} = \frac{1}{500} = 2 \text{ m sec}\right)$

The output will be

$$v_0 = \text{supply voltage} + \text{capacitor voltage}$$

$$= (-5\text{ V}) + (-5\text{ V})$$

$$= -10\text{ V}$$

The output waveforms are shown in Fig. 3.23 by which it is clear that the voltage swing is the same for the input as well as the output waveforms.

Problem 7

In the circuit of Fig. 3.11, if $V_R = 0$, $R = 1$ K,v_i is a ramp rising at the rate of 15 V/μ sec, and the diode used has a reverse saturation current of 2 μA, find the time t at which a device operative at 150 mV will respond. If the resistance R is increased to 75 K, how many times will the time t increase?

Solution

When $R = 1$ K, the diode will remain cut off up to a voltage

$$V_1 = (I_{\text{(reverse saturation current)}} \times R)$$

$$= 2\times10^{-6}\times1\times10^3 = 2\text{ mV}$$

The device will respond when the output voltage of the circuit

$$v_0 = 150 + 2 = 152\text{ mV}$$

Therefore, the time at which the device will respond,

$$t_1 = (152\times10^{-3})\times\frac{10^{-6}}{15}$$

$$= 10.1\ n\ \text{sec.}$$

Similarly, when $R = 75$ kΩ, the diode will be cut off up to

$$V_2 = 2\times10^{-6}\times75\times10^3$$

$$= 150\text{ mV}$$

and the device will respond at

$$v_0 = 150 + 150 = 300 \text{ mV}$$

∴ The time at which the device will respond,

$$t_2 = (300 \times 10^{-3}) \times \frac{10^{-6}}{15}$$

$$= 20\ n \text{ sec}$$

Hence, the time increases by two times only even if the resistance is increased 75 times.

Questions and Problems

1. What do you mean by a clipping circuit? Discuss the various clipping circuits.
2. Draw and discuss the circuit of a clipper which is used to convert a sinusoidal waveform to a square wave.
3. Describe a clamping circuit.
4. What do comparators mean?
   For what purpose are these used?
5. Describe a full-wave rectifier circuit.
6. Derive and compare the d c outputs in the half-wave and full-wave rectifiers.
7. What is the advantage of a bridge rectifier over a full-wave rectifier?
8. Compare the peak inverse voltages in the case of the half-wave, full-wave and bridge rectifiers.
9. Why are filtering circuits used in power supplies?
10. Describe the various types of the filters.
11. What is the meaning of the ripple factor? Calculate and compare it in various filtering circuits.
12. Why is π-filter better than other filters?
13. Describe voltage doubler and tripler circuits.
14. On what principle the voltage multiplying circuits work? Describe a quadrupler circuit.
15. Find the ripple voltage in capacitor filter with C = 1000 μF, I_{dc} = 200 mA, and V_{ac} = 10 V.
16. If the peak rectified voltage is 20 V, find the d c voltage obtained in case of half-wave and full-wave rectifiers.
17. What will be the peak inverse voltages in problem 16?
18. Find the PIV for the diodes of bridge rectifier with V_{dc} = 95 V.

19. What will be the output d c voltage for 24 V input if π-filter is used? Given: $L = 5$ henry, $R = 200\ \Omega$, and output current = 40 mA.
20. Draw the output waveforms for the input applied to the circuit of Fig. 3.22, if battery potential is changed to 3 V.

4

Transistors, Integrated Circuits and their Fabrication Technology

A transistor is a device which converts a small change into a large change in current, voltage, or power. Basically, these are the functions of an amplifier and before the advent of the transistor, a vacuum triode was used for these functions. The first transistor was constructed by Willium Schokley, John Bardeen, and Walter Brattain in 1947, and was a three-terminal solid state device. The various transistors are the bipolar junction transistor (BJT), the junction field effect transistor (JFET), and the insulated gate field effect transistor or the metal oxide semiconductor field effect transistor (MOSFET). In this chapter, the various transistors are discussed.

4.1. BIPOLAR JUNCTION TRANSISTOR (BJT)

The bipolar junction transistor is also called, simply, the transistor. There are two fundamental types of bipolar junction transistors. These are the n-p-n and the p-n-p three-layer semiconductor devices. If a layer of p-type silicon (or germanium) crystal is sandwiched between two layers of n-type silicon, the device is known as an n-p-n transistor. Alternatively, if a layer of n-type silicon is sandwiched between two layers of p-type silicon, the device is known as a p-n-p transistor. These three layers of alternately doped silicon (or germanium) are called the emitter, the base, and the collector. Figure 4.1 shows the two types of transistors. The base region is made extremely thin compared to the emitter and the collector regions. Also, the surface area of the emitter-base junction is made less than that of the collector-base junction. The arrowhead in the emitter lead indicates the direction of the flow of the holes.

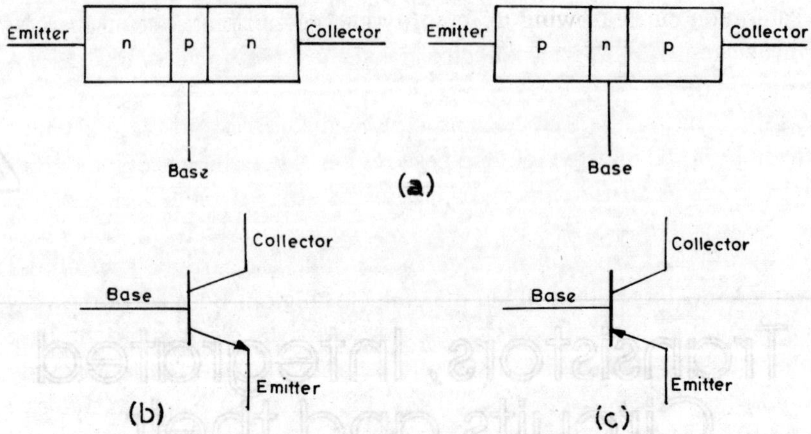

Fig. 4.1. (a) n-p-n and p-n-p transistors (b) Circuit symbols for n-p-n and p-n-p transistors

4.2. TRANSISTOR OPERATING MODES

A transistor consists of two p-n junctions, an emitter-base junction and a collector-base junction. Since a p-n junction can be operated in two types of biasing modes, the forward-biasing and the reverse-biasing, therefore, a transistor can be operated in four types of biasing modes. They are shown in Table 4.1.

Table 4.1. Transistor Operating Modes

	Emitter-base junction	Collector-base junction	Operating mode
1.	Forward-biased	Forward-biased	Saturation
2.	Forward-biased	Reverse-biased	Active
3.	Reverse-biased	Forward-biased	Inverse
4.	Reverse-biased	Reverse-biased	Cut-off

The transistors are not generally operated in the inverse mode. The saturation and the cut-off modes are used in the switching action of a transistor and are discussed in later chapters of this book. In the active mode, a transistor works as an amplifier and this operation is discussed in this chapter.

4.3. TRANSISTOR ACTION

Figure 4.2 shows a p-n-p transistor in the active mode of operation with the emitter junction forward-biased and the collector junction reverse-biased. For understanding the transistor action, initially the emitter base bias V_{EB} is assumed to be zero. In this condition, the majority and the minority charge carrier currents are equal and opposite and therefore, the emitter-base junction current is zero. Since the collector-base junction is reverse-biased by V_{CB}, the reverse saturation current owing to minority charge carriers flows in the collector. This current is known as collector leakage current I_{CO} when the emitter-base junction is open.

Now, the emitter-base junction is slowly made forward-biased, which causes the holes to be injected into the base region, where they become minority charge carriers (the base is n-type). The reverse-biased collector base junction behaves as forward-biased for these injected holes, which are then swept into the collector and cause an increase in the collector current. These holes then recombine with the electrons in the collector region. Because of the very thin base, a small number of holes recombines with the electrons in the base region while passing through it. Thus, the internal transistor current mainly consists of holes flowing from the emitter to the collector. The two batteries V_{EB} and V_{CB} furnish holes to the emitter and electrons to the collector and the base regions so as to compensate for the loss of these charge carriers in the three respective regions. In other words, the holes moving from the emitter to the collector and the electrons in the collector and the base recombining with the holes flowing from the emitter to the collector are continuously being supplied from the batteries. These hole and electron flows from the battery terminals to the emitter, the collector, and the base contribute to the external emitter, collector, and base currents I_E, I_C, and I_B respectively, as shown in Fig. 4.2.

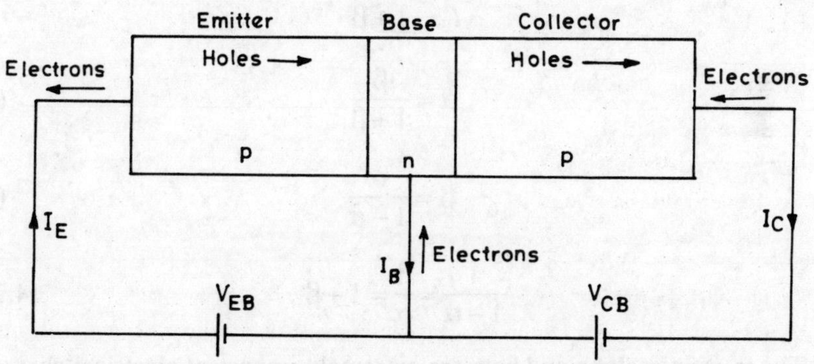

Fig. 4.2. Action of a transistor.

In a transistor, the current enters via the emitter terminal and leaves via both the base and the collector terminals. As the current entering the transistor must be equal to the current leaving the transistor, this can be expressed as

$$I_E = I_B + I_C \tag{4.1}$$

The ratio of the collector current and the base current at a particular point of operation is known as the forward current gain and is denoted by β, which is also represented as h_{FE} in the form of h-parameters. Here, F indicates the forward current gain and E is for the common emitter connection. The details of h-parameters are discussed in Chapter 6. Hence,

$$\beta = \frac{I_C}{I_B} \tag{4.2}$$

The typical value of β is 50 but it may vary between ten and 600. The ratio of the collector current and the emitter current at a particular point of operation is known as the large signal current gain and is denoted by α. Hence,

$$\alpha = \frac{I_C}{I_E} \tag{4.3}$$

The typical value of α lies between 0.95 and 0.995. From Eqs. 4.1 and 4.3,

$$(1-\alpha)\, I_E = I_B \tag{4.4}$$

From Eqs 4.1 and 4.2,

$$I_E = (1+\beta)\, I_B \tag{4.5}$$

The relationship between α and β can be obtained by dividing Eq. 4.2 by Eq. 4.5 as,

$$\frac{I_C}{I_E} = \frac{\beta}{1+\beta}$$

or

$$\alpha = \frac{\beta}{1+\beta} \tag{4.6}$$

and

$$\beta = \frac{\alpha}{1-\alpha} \tag{4.7}$$

or

$$\frac{1}{1-\alpha} = \frac{\beta}{\alpha} = 1+\beta \tag{4.7(a)}$$

The transistor alpha and beta are extremely important electrical characteristics of the BJT. As α is a ratio of two nearly equal quantities and is very nearly equal to one, it is difficult to measure it with precision. However, β may be measured easily and therefore, the values of β are generally given on the manufacturer's data.

4.4. TRANSISTOR BIASING CONFIGURATIONS AND CHARACTERISTICS

There are three biasing configurations in which a transistor can be connected in a circuit. In each configuration, one of the electrodes (emitter, base, collector) is made common to both the input and the output.

4.4.1. Common Base Configuration

Figure 4.3 shows a p-n-p transistor connected in common base configuration. In this case, the emitter-base are the input terminals and the collector-base are the output terminals. The input and the output characteristics of a transistor in common base are shown in Fig. 4.4. The input or the emitter characteristics

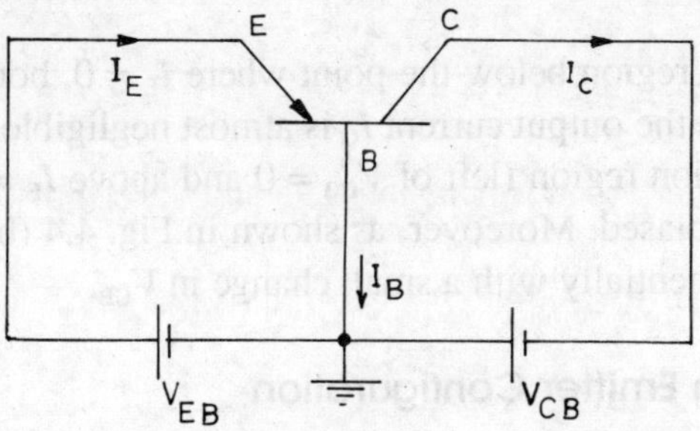

Fig. 4.3. Transistor in common base configuration.

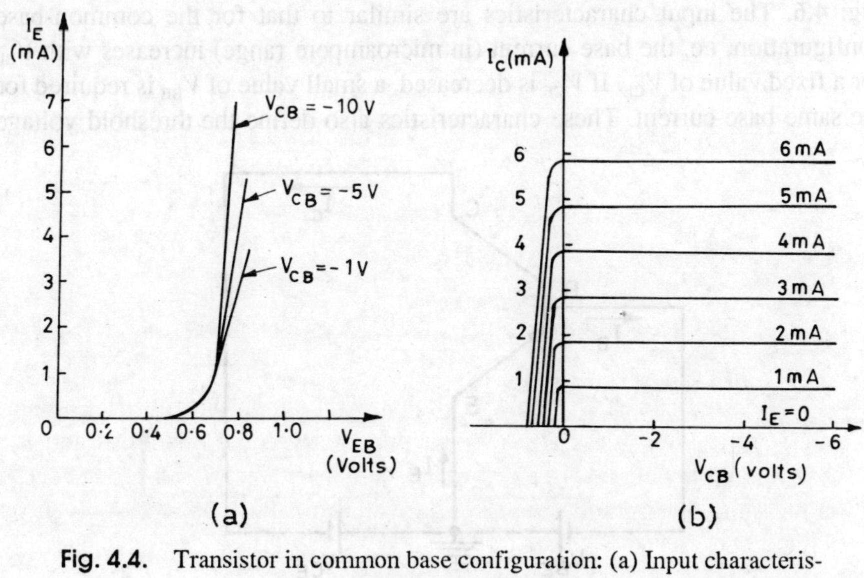

Fig. 4.4. Transistor in common base configuration: (a) Input characteristics, (b) Output characteristics.

[Fig. 4.4 (a)] show that for a fixed value of V_{CB}, the emitter current I_E increases with V_{EB}. If V_{CB} is increased to some other value, the smaller value of V_{EB} is required for the same current. These characteristics also show a cut-in, offset, or threshold voltage V_γ below which the emitter current is very small. The output or collector characteristics [Fig. 4.4 (b)] show three regions of interest: the active, cut-off, and saturation regions.

In the active region, the input junction is forward-biased and the output junction is reverse-biased. When $I_E = 0$, the output current I_C is equal to the collector leakage current I_{CO}, which is very small and can be ignored. When the emitter current I_E increases beyond the zero level, the collector current I_C increases to a magnitude slightly less than I_E. This is because $I_C = \alpha\, I_E$ and $\alpha \cong 0.995$. It is also observed that the collector current I_C is almost independent of V_{CB}.

In the cut-off region below the point where $I_E = 0$, both the junctions are reverse-biased and the output current I_C is almost negligible.

In the saturation region (left of $V_{CB} = 0$ and above $I_E = 0$), both the junctions are forward-biased. Moreover, as shown in Fig. 4.4 (b) the collector current changes exponentially with a small change in V_{CB}.

4.4.2. Common Emitter Configuration

Figure 4.5 shows a p-n-p transistor connected in the common emitter configuration, in which the input terminals are the emitter-base and the output terminals are the collector-emitter. The input and output characteristics are shown in Fig. 4.6. The input characteristics are similar to that for the common-base configuration, i.e. the base current (in microampere range) increases with V_{BE} for a fixed value of V_{CE}. If V_{CE} is decreased, a small value of V_{BE} is required for the same base current. These characteristics also define the threshold voltage V_γ.

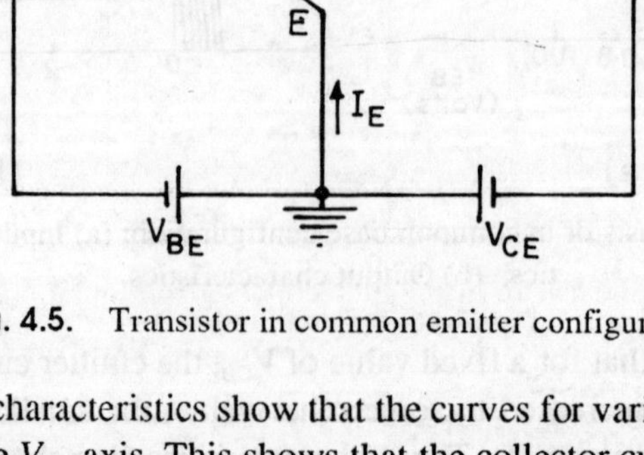

Fig. 4.5. Transistor in common emitter configuration.

The output characteristics show that the curves for various values of I_B are not parallel to the V_{CE} axis. This shows that the collector current is affected by

V_{CE}, which was not the case for the common-base configuration. Also, there is some output current even for the zero-base current. This current is known as common emitter leakage current I_{CEO}.

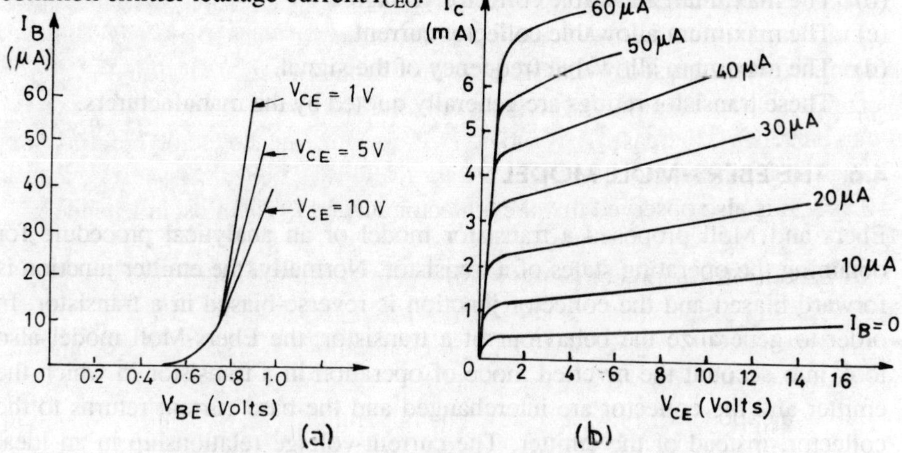

Fig. 4.6. Transistor in common emitter configuration: (a) Input characteristics, (b) Output characteristics.

4.4.3. Common Collector Configuration

Figure 4.7 shows a transistor connected in the common collector configuration. The collector-base terminals are the input terminals and the collector-emitter terminals are the output terminals. The input and the output characteristics for this configuration are similar to that for the common emitter configuration. This circuit is also known as the emitter follower circuit.

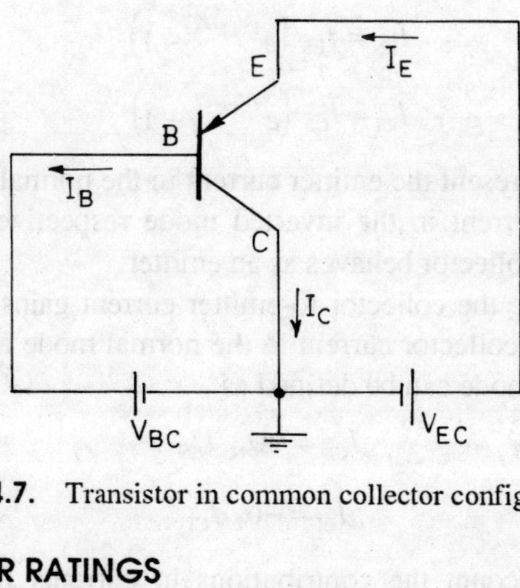

Fig. 4.7. Transistor in common collector configuration.

4.5. TRANSISTOR RATINGS

There are certain parameters which should not be exceeded while operating a transistor. They are:

(a) The maximum allowable power dissipation.
(b) The maximum allowable collector voltage.
(c) The maximum allowable collector current.
(d) The maximum allowable frequency of the signal.

These transistor ratings are generally quoted by the manufacturers.

4.6. THE EBERS-MOLL MODEL

Ebers and Moll proposed a transistor model or an analytical procedure for obtaining the operating states of a transistor. Normally, the emitter junction is forward-biased and the collector junction is reverse-biased in a transistor. In order to generalize the behaviour of a transistor, the Ebers-Moll model also took into account the inverted mode of operation in a transistor in which the emitter and the collector are interchanged and the base current returns to the collector, instead of the emitter. The current-voltage relationship in an ideal p-n junction diode is represented by

$$I = I_S\ (e^{qV/kT} - 1) \tag{4.8}$$

where I_S = saturation current
V = applied junction voltage
q = electron charge
k = Boltzmann's constant
T = temperature in Kelvin.

Equation 4.8 can be rewritten for normal and inverted modes of operation as

$$I_{EN} = I_{ES}\left(e^{qV_{EB}/kT} - 1\right) \tag{4.9}$$

and

$$I_{CI} = I_{CS}\left(e^{qV_{CB}/kT} - 1\right) \tag{4.10}$$

where I_{EN} and I_{CI} represent the emitter current in the normal mode of operation and the collector current in the inverted mode respectively, because in the inverted mode, the collector behaves as an emitter.

If α_N and α_I are the collector-to-emitter current gains in the normal and inverted modes, the collector current in the normal mode and the emitter current in the inverted mode can be defined as

$$I_{CN} = -\alpha_N\ I_{EN} \tag{4.11}$$

and

$$I_{EI} = -\alpha_I\ I_{CI} \tag{4.12}$$

Taking into account the contributions in currents for the normal and inverted modes, the emitter and the collector currents can be represented as,

$$I_E = I_{EN} + I_{EI} \quad (4.13)$$

and
$$I_C = I_{CN} + I_{CI} \quad (4.14)$$

Using Eqs 4.9 to 4.12, these can be rewritten as,

$$I_E = I_{ES}\left(e^{qV_{EB}/kT} - 1\right) - \alpha_I I_{CS}\left(e^{qV_{CB}/kT} - 1\right) \quad (4.15)$$

and
$$I_C = -\alpha_N I_{ES}\left(e^{qV_{EB}/kT} - 1\right) + I_{CS}\left(e^{qV_{CB}/kT} - 1\right) \quad (4.16)$$

Equation 4.15 shows that the emitter current in a transistor has two parts, one is the usual forward-biased diode current controlled by the emitter-base voltage, and the other is the reverse current generator $\alpha_I I_{CI}$. This can be represented by a parallel combination of these two components as shown in Fig. 4.8 (a).

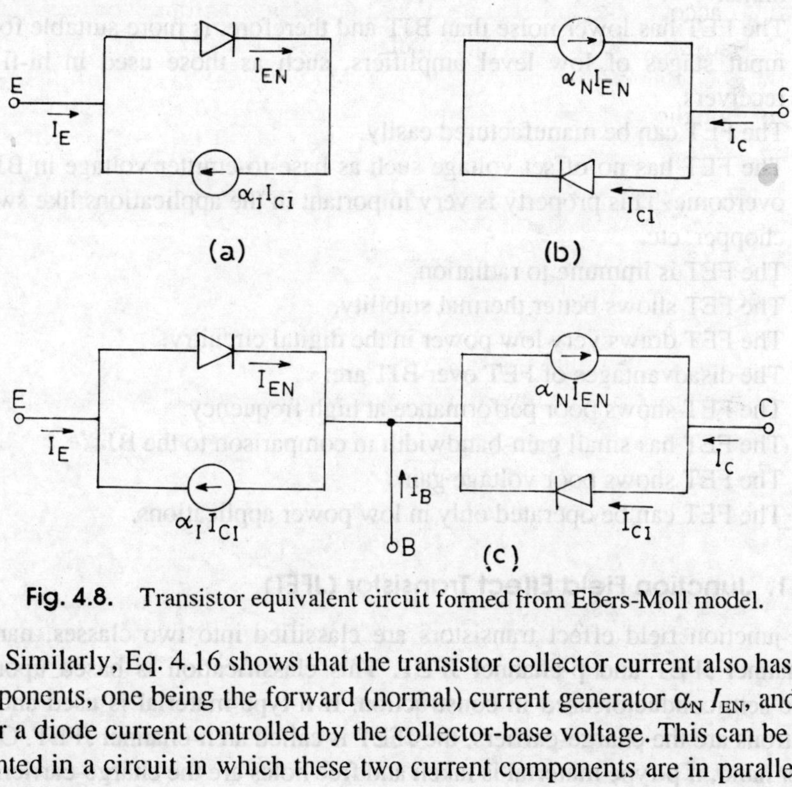

Fig. 4.8. Transistor equivalent circuit formed from Ebers-Moll model.

Similarly, Eq. 4.16 shows that the transistor collector current also has two components, one being the forward (normal) current generator $\alpha_N I_{EN}$, and the other a diode current controlled by the collector-base voltage. This can be represented in a circuit in which these two current components are in parallel, as shown in Fig. 4.8 (b).

Figure 4.8 (c) represents a complete transistor circuit model based upon the Ebers-Moll model defined by Eqs. 4.15 and 4.16. It is clear from these equations and the figure that a transistor is not a combination of two diodes only. This model is very useful because it gives a lot of information about the d c characteristic curves of a transistor.

4.7. FIELD EFFECT TRANSISTORS

The field effect transistor (FET) is another type of solid state device which has become very popular in recent years in linear and digital circuit applications. The current flow through a FET is controlled by an electrostatic field and therefore, it is called a field effect transistor. As the operation of FET depends only on the motion of one type of charge carriers, either holes or electrons, but not of both, it is also called an unipolar transistor. There are two categories of field effect transistor: (a) the junction field effect transistor (JFET) and (b) the insulated gate field effect transistor (IGFET), which is also known as metal oxide semiconductor field effect transistor (MOSFET).

The advantages of FET over BJT are:

(a) The FET has an extremely high input impedance, typically, of 100 meg-ohms.
(b) The FET has lower noise than BJT and therefore, is more suitable for the input stages of low level amplifiers, such as those used in hi-fi FM receivers.
(c) The FET can be manufactured easily.
(d) The FET has no offset voltage such as base-to-emitter voltage in BJT to overcome. This property is very important in the applications like switch, chopper, etc.
(e) The FET is immune to radiation.
(f) The FET shows better thermal stability.
(g) The FET draws very low power in the digital circuitry.

The disadvantages of FET over BJT are:

(a) The FET shows poor performance at high frequency.
(b) The FET has small gain-bandwidth in comparison to the BJT.
(c) The FET shows poor voltage gain.
(d) The FET can be operated only in low power applications.

4.7.1. Junction Field Effect Transistor (JFET)

The junction field effect transistors are classified into two classes, namely, n-channel JFET and p-channel JFET. This classification is based upon the basic semiconductor used in construction. If n-type material is used and free electrons are the charge-carriers, the JFET is called an n-channel JFET. On the other hand, if p-type material is taken and free holes are the charge-carriers, the JFET is called a p-channel JFET. Figure 4.9 shows the physical structure of these two types of JFET's along with their circuit symbols.

In an n-channel JFET, a bar of n-type semiconductor is taken and ohmic contacts are made at each end of the bar. One terminal is called the source S, and the other is called the drain D. On either side of the bar, heavy doping of the p-material is done to construct p-n junctions. These p-regions are called

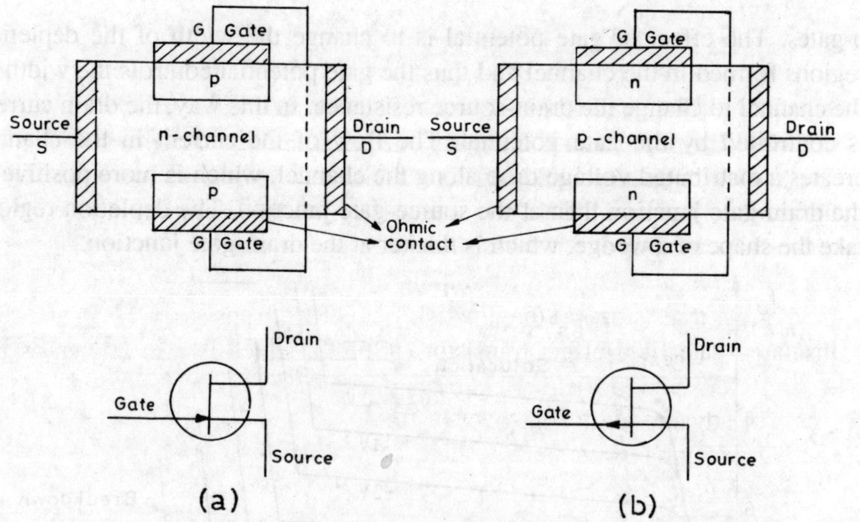

Fig. 4.9. Physical structure and circuit symbol for: (a) n-channel JFET, (b) p-channel JFET.

gates (G). The thin region between these gates is called the channel. In the operation of JFET, these p-n junctions are not given forward-biasing and thus, have no offset voltage. In fact, the p-n junctions are given reverse-biasing and therefore, the input resistance of the JFET is very high. The current in the channel is controlled by the potential applied to the gates.

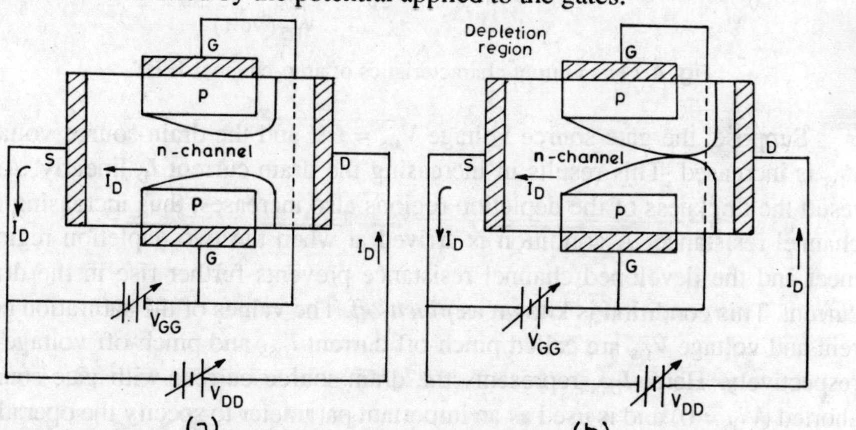

Fig. 4.10. (a) Biasing of n-channel JFET, (b) Pinch-off condition when two depletion regions close the channel.

Figure 4.10 shows the biasing of an n-channel JFET. The JFET output characteristics are shown in Fig. 4.11. The voltage across the drain-source V_{DS} results in a current I_D in the channel from the drain to the source (electrons in the n-channel move from the source to the drain). This drain current passes through a channel in between two reverse-biased p-n junctions formed by p-gates. The effect of gate potential is to change the width of the depletion regions formed in the channel and thus the gate potential controls the width of the channel to change the drain-source resistance. In this way, the drain current is controlled by the gate potential. The flow of the current in the channel creates a distributed voltage drop along the channel, which is more positive at the drain-gate junction than at the source-gate junction. The depletion regions take the shape of a wedge, which is thicker at the drain-gate junction.

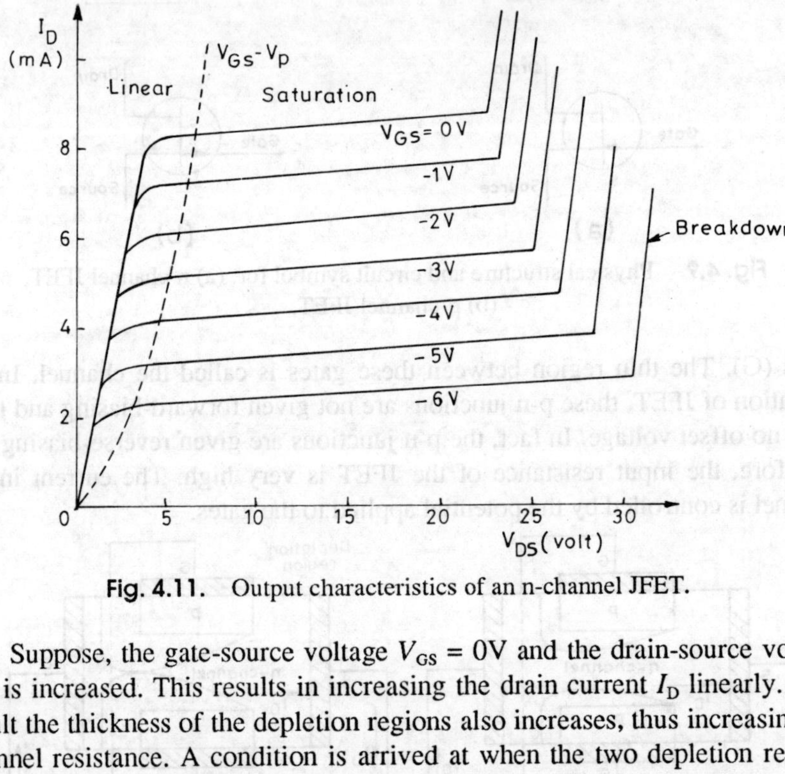

Fig. 4.11. Output characteristics of an n-channel JFET.

Suppose, the gate-source voltage V_{GS} = 0V and the drain-source voltage V_{DS} is increased. This results in increasing the drain current I_D linearly. As a result the thickness of the depletion regions also increases, thus increasing the channel resistance. A condition is arrived at when the two depletion regions meet and the developed channel resistance prevents further rise in the drain current. This condition is known as *pinch-off*. The values of the saturation current and voltage V_{DS} are called pinch-off current I_{DSS} and pinch-off voltage V_P respectively. Here, I_{DSS} represents the drain-source current with gate-source shorted ($V_{GS} = 0$) and is used as an important parameter to specify the operation of the JFET. The further rise in V_{DS} breaks down the junctions and large current flows in the JFET. This Zener breakdown is shown in Fig. 4.11.

Now, if V_{GS} is increased (more negative in the n-channel JFET), the pinch-off condition is arrived at a lower value of drain current (Fig. 4.11). Thus, the gate potential acts as a control for drain current at a particular value of V_{DS}.

Another type of JFET characteristics are the transfer characteristics. This is a curve between the drain current and the gate-source voltage for a constant value of drain-source voltage, as shown in Fig. 4.12. There are two important points in a transfer characteristic, viz. the pinch-off current I_{DSS} (when $V_{GS} = 0$) and the pinch-off voltage V_P.

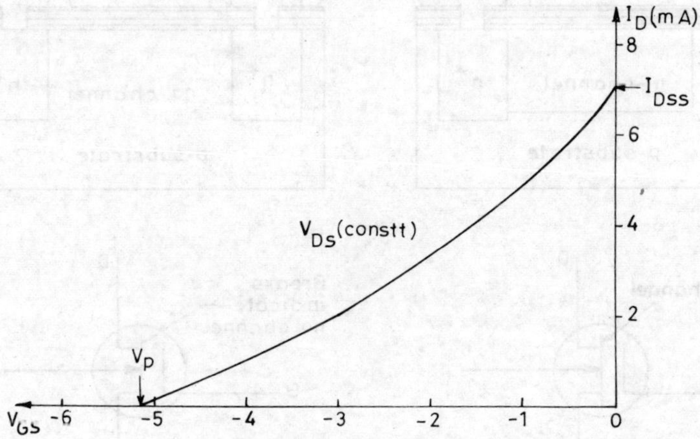

Fig. 4.12. Transfer characteristic for an n-channel JFET.

4.7.2. Metal Oxide Semiconductor Field Effect Transistor (MOSFET)

A second and more useful FET is the metal oxide semiconductor field effect transistor (MOSFET). In this FET, the channel resistance is controlled by an electric field owing to a capacitor unlike a p-n junction in a JFET. There are basically two types of MOSFET. The first type, in which the channel is physically constructed between the source and the drain, is known as depletion MOSFET. The other type, in which no channel is constructed at the time of manufacturing but it is developed only when voltage is applied to the gate, is known as enhancement MOSFET. The p-type semiconductor is a foundation block of both the MOSFET. Figure 4.13 shows the two types of MOSFET with an n-channel, along with their circuit symbols.

The n-channel MOSFET is formed on a p-substrate (p-doped semiconductor). The source and the drain regions are formed by diffusing heavily doped n-type material (N^+) into the substrate. These regions are connected to the source and the drain terminals by metal ohmic contacts. An n-channel is diffused between the source and the drain onto the substrate in the case of a depletion mode MOSFET. However, in the case of an enhancement mode MOSFET, no channel is diffused. An insulating layer of SiO_2 (silicon dioxide) is deposited above the n-channel. A metal layer is deposited above this SiO_2 layer and is connected to the gate terminal. This metal layer functions as one plate of the capacitor. The SiO_2 acts as the dielectric of the capacitor. The n-channel semiconductor material in depletion mode and the p-type substrate in the enhancement mode MOSFET acts as the other plate of the capacitor. The

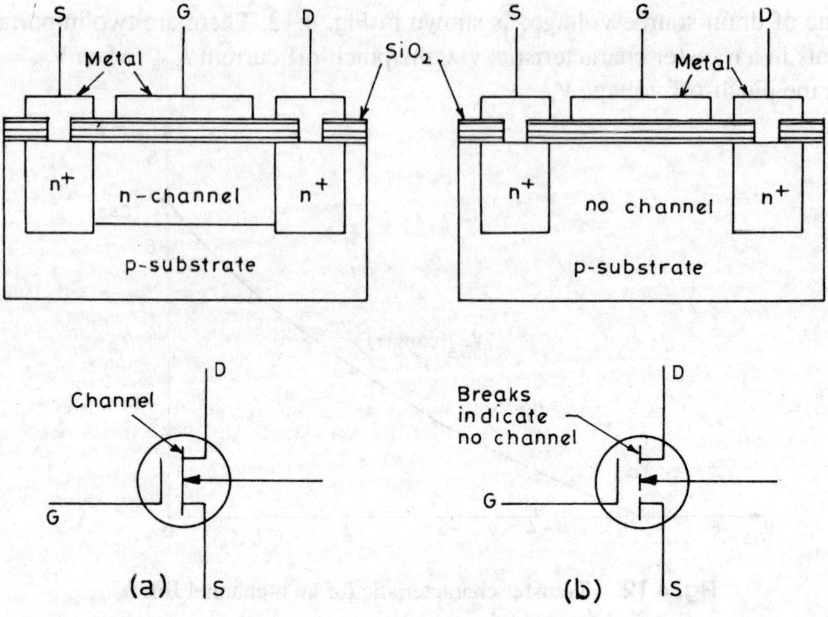

Fig. 4.13. Physical structure and circuit symbol for: (a) n-channel depletion mode MOSFET, (b) n-channel enhancement mode MOSFET.

presence of this gate capacitor makes the gate input resistance of a MOSFET very large, because leakage current in a capacitor is infinitesimally small in comparison to the junction leakage current in a JFET.

Figure 4.14 shows the characteristics of an n-channel depletion mode MOSFET. The negative gate-source voltages V_{GS} push electrons out of the channel to deplete it and therefore, large enough negative V_{GS} values pinch off the channel. On the other hand, positive V_{GS} values result in an increase in the channel size because the positive values of V_{GS} push away the p-type carriers. This results in more charge carriers and larger channel current. The transfer characteristics in this MOSFET continue for positive values of V_{GS} also. This MOSFET can be operated with either polarity of V_{GS} because the gate is isolated from the channel for both the negative and the positive values of V_{GS}—no gate current flows in either case.

Figure 4.15 shows the characteristics of an n-channel enhancement mode MOSFET. In this case, there is no channel between the source and the drain. When a positive V_{GS} is applied, the holes are repelled from the p-substrate under the gate region and thus, this region is depleted. When V_{GS} is increased further, the electrons are attracted into this depletion region, making it like an n-channel between the source and the drain. There is no drain current for smaller values of V_{GS}. The drain current is available only when V_{GS} is sufficient

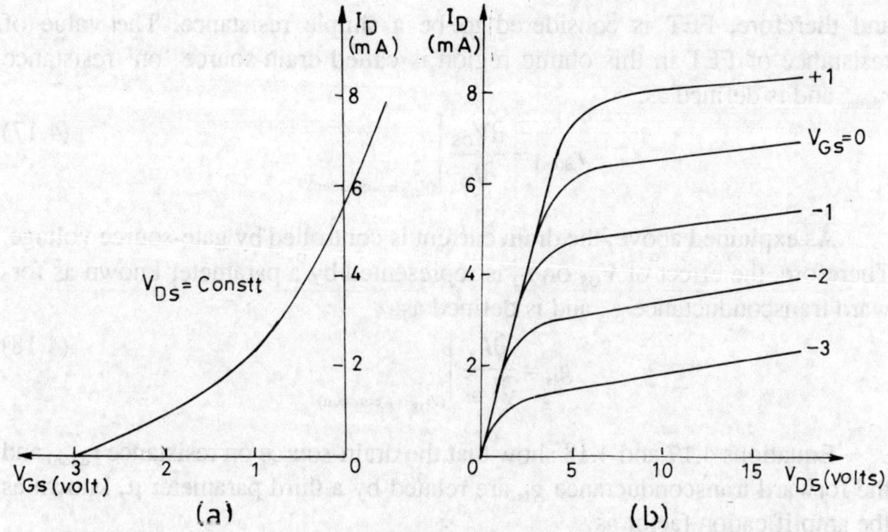

Fig. 4.14. Depletion mode MOSFET characteristics: (a) Transfer curves, (b) Output curves.

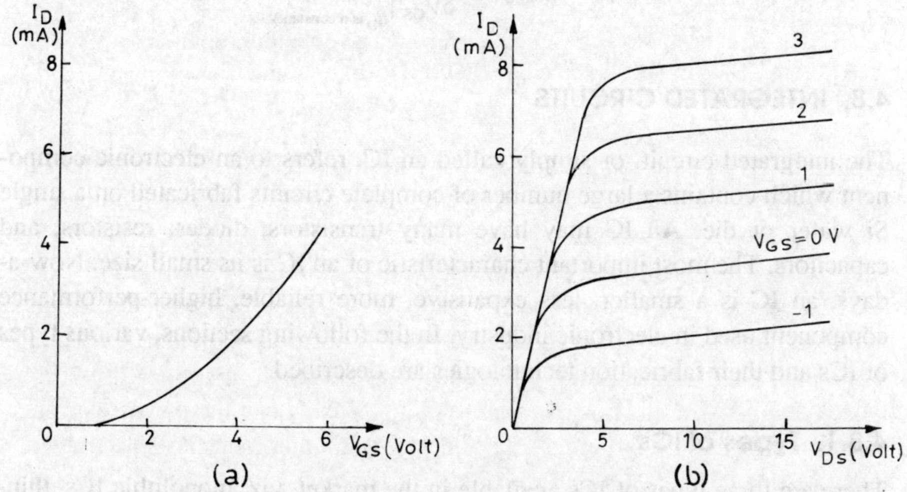

Fig. 4.15. Enhancement mode MOSFET characteristics: (a) Transfer curves, (b) Output curves.

enough to create a channel. As there is no drain current for $V_{GS} = 0$, no value of I_{DSS} can be associated with the enhancement mode MOSFET.

4.7.3. FET Parameters

In the low region of V_{DS}, the output characteristics (V_{DS} *vs* I_D) are quite linear and therefore, FET is considered to be a simple resistance. The value of resistance of FET in this ohmic region is called drain-source 'on' resistance $r_{d\,(on)}$ and is defined as,

$$r_{d(on)} = \left.\frac{\partial V_{DS}}{\partial I_D}\right|_{(V_{GS}\text{ is a constant})} \tag{4.17}$$

As explained above, the drain current is controlled by gate-source voltage. Therefore, the effect of V_{GS} on I_D is represented by a parameter known as forward transconductance g_{fs} and is defined as,

$$g_{fs} = \left.\frac{\partial I_D}{\partial V_{GS}}\right|_{(V_{DS}\text{ is a constant})} \tag{4.18}$$

Equations 4.17 and 4.18 show that the drain-source on resistance $r_{d\,(on)}$ and the forward transconductance g_{fs} are related by a third parameter μ, known as the amplification factor as,

$$\mu = r_{d(on)}\, g_{fs} = \left.\frac{\delta V_{DS}}{\delta V_{GS}}\right|_{(I_D\text{ is a constant})} \tag{4.19}$$

4.8. INTEGRATED CIRCUITS

The integrated circuit, or simply called an IC, refers to an electronic component which contains a large number of complete circuits fabricated on a single Si wafer or die. An IC may have many transistors, diodes, resistors, and capacitors. The most important characteristic of an IC is its small size. Now-a-days, an IC is a smaller, less expansive, more reliable, higher-performance component used in electronic industry. In the following sections, various types of ICs and their fabrication technologies are described.

4.8.1. Types of ICs

There are three types of ICs available in the market, viz. monolithic ICs, thin- and thick-film ICs, and hybrid ICs. The word 'monolithic is a Greek word meaning single stone. The ICs fabricated within a single Si wafer are called monolithic ICs. In monolithic ICs, both active and passive circuit components can be fabricated. The thin- and thick-film ICs are not fabricated on a single die but on the surface of an insulating substrate, which may be of glass or of some ceramic material. In film ICs, only passive circuit components, such as resistors and capacitors can be fabricated on the insulating substrate. The active components fabricated by the monolithic technique are joined as discrete components on the surface of the film ICs. In film ICs, the passive components have a wider range of values in comparison to the monolithic ICs. As the active components are used as discrete components, more and more circuits may be formed in film ICs. These are more costly than the monolithic ICs.

The hybrid ICs are those integrated circuits in which the various components or circuits are fabricated by employing either the monolithic or the film technique. Sometimes, the active components are first fabricated within a silicon wafer by the monolithic technique, and then an insulating substrate of SiO_2 is formed, on which the passive circuit components are fabricated by the film technique. The interconnections between the active and passive components are made through holes made in the SiO_2 layer.

4.8.2. Analog and Digital ICs: Classification

According to the types of signal processing, the integrated circuits are sometimes classified as analog or digital ICs. In order to carry information, two types of electrical signals are used, viz. analog and digital. If the characteristics of input and output are directly related, the system is called an analog system. For example, the modulation and demodulation belong to an analog system because the amplitude or frequency of a modulated wave conform to the characteristics of the original signals. On the other hand, if the number, time duration, or frequency of waves is used to carry the information, it is called a digital system, e.g., as in telegraphy. The analog system is a linear system, whereas a digital system is a nonlinear one.

The semiconductor devices show both linear and nonlinear behaviour, and as the integrated circuits involve a large number of semiconductor devices, the transfer characteristic of an IC shows both linear and nonlinear behaviour. An analog or linear IC is designed for use in the linear part of its characteristics. The operational amplifiers, video amplifiers, and R.F. amplifiers are examples of an analog IC. A digital IC is designed to work in the nonlinear part of its transfer characteristics. The gates, flip-flops, and other switching circuits are examples of digital ICs, where the output is not required to be proportional to the input.

4.8.3. Semiconductor Fabrication

The first step towards producing integrated circuits is to get a semiconductor material of the highest purity. The semiconductor (Ge or Si) must have impurity of less than one part in one billion. In the early period of technology, the Ge was used extensively because it could be refined and processed easily, being less reactive than Si. The mobility of charge carriers is also more than that in Si. But the smaller bandgap of Ge causes the intrinsic charge carriers to reach the dopant atoms at a lower temperature than in Si. Also, it is not possible to dope a small selected area of a piece of Ge and as a result, the size of the devices formed is bigger, whereas a selective doping is possible in Si. The silicon also exhibits semiconducting as well as good insulating properties.

Therefore, Si has replaced Ge as a semiconductor used in device technology. In this section, the fundamental fabrication steps for integrated circuits are described.

(a) Single Crystal Growth

The raw material of a semiconductor is first purified by various chemical processes and the zone-refining technique. This yields a polycrystalline crystal of desired purity. In the zone-refining technique, a bar of semiconductor is placed in a boat of quartz, which is put inside a quartz container filled with some inert gas. The quartz container is covered with radio frequency induction coils at one end. These coils are movable along the container. The setup of the zone-refining process is shown in Fig. 4.16 (a). The radio frequency signal is applied to produce heat and melt the semiconductor material surrounded by the coil. Now, the r.f. coil is moved towards the right to melt the neighbouring part of the rod. This causes the more fluidic impurities to follow the molten region and hence, most impurities reach the right end of the rod when the r.f. coil reaches this end. Now, this end part of the rod is cut off and the whole process is repeated again and again to get the desired value of purity. After getting the pure polycrystalline sample, the *Czochralski technique*, shown in Fig. 4.16 (b),

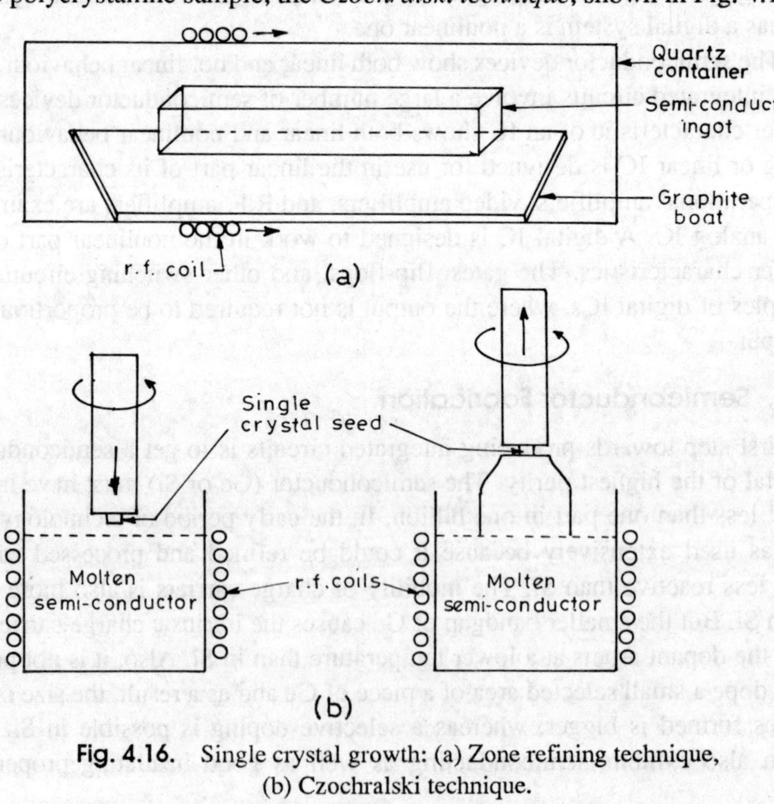

Fig. 4.16. Single crystal growth: (a) Zone refining technique, (b) Czochralski technique.

is used to get the single crystal in which a single crystal seed of desired impurity is immersed in the molten of the polycrystalline semiconductor placed in a crucible of graphite and then withdrawn slowly, keeping the seed rotating. In this process, a single crystal grows on the seed, which can be up to 25 cms in length and 8 cms in diameter.

Another method to produce a single crystal is known as the *floating zone technique*. The apparatus is shown in Fig. 4.17. A small single crystal seed of desired impurity and crystalline orientation is fixed at the lower end of the bar of polycrystalline material and heated by localized heating through r.f. coils to get the molten state. The r.f. coils are moved upwards and in a slow motion, while keeping the bar rotating. In this process, the impurities follow the molten state and as a result, a single crystal of improved impurity level is formed as described above. A very low level of contamination is ensured by this process.

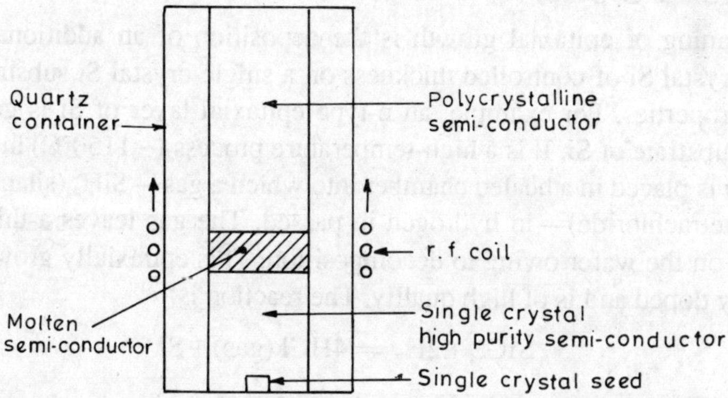

Fig. 4.17. Floating zone technique for single crystal growth.

(b) Wafer Slicing

After getting a single crystal of the semiconductor material, it is cut into thin circular wafers by using a tungsten saw or a rotating cutting knife. These wafers are then passed through the etching process to remove the sawing damage. Then, these wafers are polished to get the mirror-like defect-free surface.

4.8.4. Planar Technology: Fabrication of Monolithic ICs

In the rapid development of microelectronics, the planar technology played a very important role. In this process, various electronic properties are imparted to various regions of Si by a selective introduction of n-type or p-type trace impurities into small areas of Si from one surface of the wafer. As the fabrication of the device is achieved from one surface plane, it is called a *planar* technology. The doping is done by depositing the impurity on the surface and then by diffusing it into the desired regions. In this technology, it is possible to dope several thousand small separate regions and make their interconnections in one wafer simultaneously and therefore, the entire circuit can be fabricated in a single Si chip. This is the key principle of IC fabrication by the planar technology. The fabrication consists of a series of complicated manufacturing procedures, which are described briefly in the following subsections. The purpose of this description is to enable some familiarity with the basic fabrication steps. For detailed expertise, reference may be made to more specialized and advanced texts available on this subject.

(a) Photomask Making

The complete circuit is first designed and the pattern of all the components and the interconnections is specified with the help of pattern generator. A set of working masks is prepared separately for emitters, bases, and interconnections. These masks are made from glass substrates, which are coated by a thin chromium film.

(b) Epitaxial Growth

The meaning of epitaxial growth is the deposition of an additional layer of single-crystal Si of controlled thickness on a single-crystal Si substrate of different properties. For example, an n-type epitaxial layer of Si is grown on a p-type substrate of Si. It is a high-temperature process (~ 1150°C) in which the Si wafer is placed in a heated chamber into which a gas—SiH_4 (silane) or $SiCl_4$ (silcon tetrachloride)—in hydrogen in passed. The gas leaves a thin layer of solid Si on the wafer owing to decomposition. This epitaxially grown Si layer is lightly doped and is of high quality. The reaction is:

$$SiCl_4 + 2H_2 = 4HCl\,(gas) + Si$$

The LPE (liquid phase epitaxy) is used to deposit many layers of GaAs with a different composition, thickness, and doping. The MBE (molecular beam expitaxy) is used to grow extremely thin multilayers of homostructures and heterostructures. Actually, 'epitaxial' is a Greek word in which 'epi' means 'upon' and taxis' means 'arrangement'.

(c) Oxidation

In this process, the Si wafer is covered with a thin layer of amorphous SiO_2, which is a good insulator and has the structure of fused glass. The Si wafer is placed inside a quartz tube placed at a temperature of 900–1200°C and wet or dry oxygen is passed over the wafer. The oxidation takes place to make an SiO_2 layer on the wafer. The block diagram of the setup is shown in Fig. 4.18.

(d) Diffusion

The process of diffusion is used to form p-type, n-type, and all the circuit

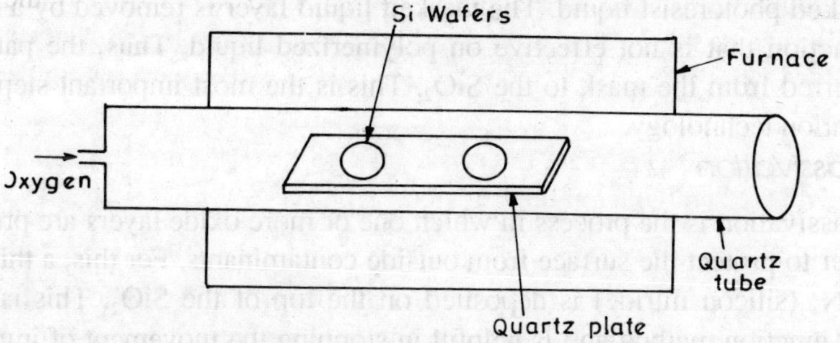

Fig. 4.18. Set-up for oxidation.

elements in an integrated circuit. It is followed in two steps. The dopant impurity is placed on or near the wafer surface by the gas deposition method or the wafer is coated by a layer of impurity. Then by following driving-in-diffusion at a higher temperature (~ 1000°C), the impurities are further moved into the wafer over the desired region and depth.

(e) Ion Implantation

The ion implantation is also used to introduce dopant impurity into the selected regions of Si wafer. Actually, it is used as an alternative process to diffusion. The impurity atoms are ionized and accelerated by electric field and then are allowed to strike the Si wafer surface. These high energy ions then penetrate into the wafer but during this process, the crystal is somewhat damaged and therefore, annealing is done. The ion implantation provides higher purity and a control on the number of dopant atoms.

(f) Metallization for Interconnections

For connecting various doped regions, the desired electrical connections are made through the process of metallization or metal deposition. The SiO_2 is removed from the areas of interconnections and a layer of metal (aluminium or an aluminium alloy) is deposited by the vacuum evaporation technique. The wafer is then put at a low temperature (~ 500°C) to form an effective electrical contact between aluminium and silicon.

(g) Photolithography and Photoengraving

Photolithography is the process in which the circuit patterns are transferred from the photomask to the wafer surface and the layers of oxide or metal are removed in specific areas where etching, diffusion, implantation, etc. are to be done. In this process, a photoresist liquid is applied to the wafer surface rotating at several hundred rotations per minute. This helps to have a very thin layer of photoresist liquid. The actual circuit mask is placed over the surface of the wafer and the ultraviolet light is allowed to fall on it, which polymerizes the unmasked photoresist liquid. The masked liquid layer is removed by a chemical reaction that is not effective on polymerized liquid. Thus, the pattern is transferred from the mask to the SiO_2. This is the most important step in the fabrication technology.

(h) Passivation

The passivation is the process in which one or more oxide layers are produced in order to protect the surface from outside contaminants. For this, a thin layer of Si_3N_4 (silicon nitride) is deposited on the top of the SiO_2. This is called sealed junction method and is helpful in stopping the movement of impurities to the surface of Si.

(i) Packaging and Testing

After the various processes described earlier, the wafer is scribed and broken down into individual chips which are then packaged in various types of packaging, viz. flat package, To-type package, and dual-in-line plastic package. The wires are connected from the leads on the package to the metal on the semiconductor chip. The package is finally sealed, with ceramic, metal cover, or plastic. The last step is the testing of the final products, which is done in many steps.

4.8.5. Monolithic Passive and Active Circuit Components

(a) IC Resistor

The resistance of any material depends upon the length, area, temperature, and resistivity and is given by

$$R = \rho \frac{l}{A} = \rho \frac{l}{b \times t} \quad (4.20)$$

where, ρ = resistivity, l = length of the material, and A = area of the material if t is the thickness and b is the breadth. This has been represented in Fig. 4.19.

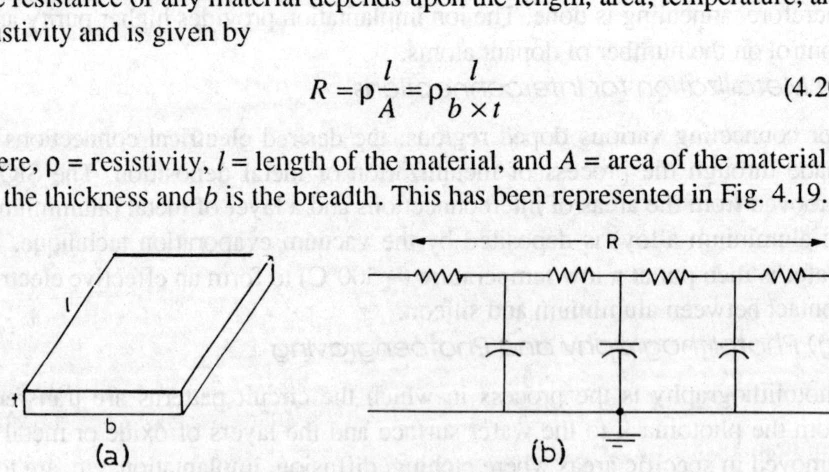

Fig. 4.19. Integrated circuit resistor.

For making an IC resistor, a hole is made in the SiO_2 layer above the uniformly doped Si wafer and the impurity of opposite conductivity is diffused into the wafer. If the voltage is applied between the two ends of the diffused region, a current flows along the diffused area parallel to the surface. Thus, the diffused region works as a resistor.

The diffused resistor differs from the discrete component resistor because it has a distributed capacitance owing to the p-n junction associated with it (Fig. 4.19). This resistance has almost nil inductance. As the IC resistor involves capacitance, it shows frequency response also, as shown in Fig. 4.20. The value of resistance at low frequency is called the d c resistance. For a square sheet ($l = b$), Eq. 4.20 yields

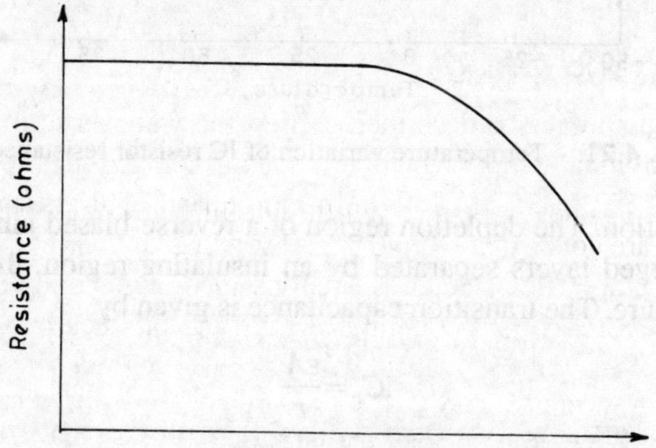

Fig. 4.20. Frequency response for an IC resistor.

$$R_S = \rho \cdot \frac{l}{t \times l} = \frac{\rho}{t} \tag{4.21}$$

where R_s is called the sheet resistance and its units are ohm/cm^2. Equation 4.21 shows that the sheet resistance is independent of the size of the square. Equation 4.20 can be written as,

$$R = R_S \cdot \frac{l}{b} \tag{4.22}$$

The temperature coefficient of an IC resistor is higher than an ordinary discrete resistor. The variation in resistance of an IC resistor with temperature is shown in Fig. 4.21. In an integrated circuit, many resistors are fabricated simultaneously by defining different geometric patterns on the same photomask, thus reducing the manufacturing cost.

(b) IC Capacitor

The IC capacitor is formed by using the transition capacitance of a reverse-

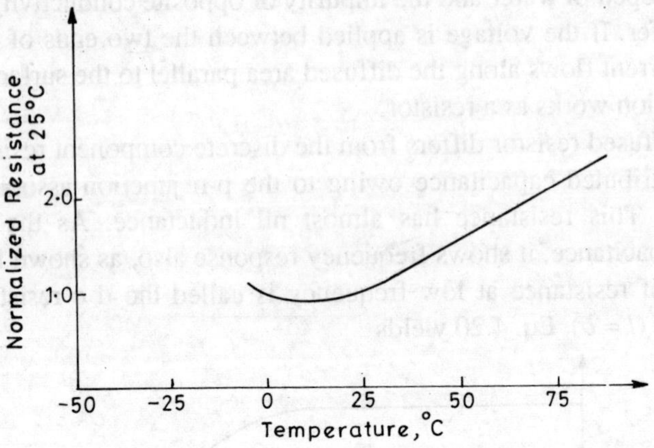

Fig. 4.21. Temperature variation of IC resistor resistance.

biased p-n junction. The depletion region of a reverse-biased junction has two oppositely charged layers separated by an insulating region, thus showing a capacitor structure. The transition capacitance is given by

$$C_t = \frac{\varepsilon A}{t}$$

where, ε = permittivity of the material; A = area of the junction; and t = thickness of the depletion region.

(c) IC Inductor

The inductors are not designed in monolithic ICs, whereas in film ICs, the inductors can be joined as discrete components. But the inductors are generally avoided owing to their large size.

(d) IC Transistor

An IC transistor is shown in Fig. 4.22. On a p-substrate, an n-epitaxial layer is grown. Here, the p-substrate marks only as a support and provides an isolating region. The base, the emitter, and the collector regions are formed during the diffusion processes. An IC transistor has large parasitic collector resistance because the collector contact is taken at the surface of the Si wafer. Owing to the capacitance of the isolation diode, an IC transistor also shows a collector-to-ground parasitic capacitance. The leakage current of this isolation diode is also important at low powers in IC transistors.

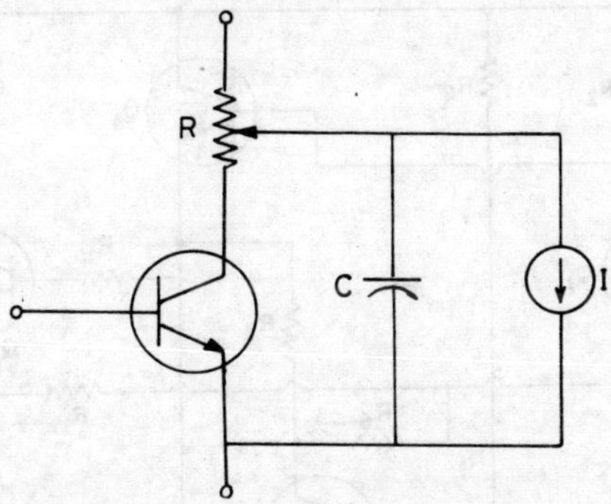

Fig. 4.22. IC transistor model.

(e) IC Diode

The diodes are fabricated by making the three regions of a transistor and then by masking the diode terminals.

4.8.6. Typical IC Low Frequency Amplifier

As the technology of IC fabrication is not capable of fabricating large values of capacitors, the circuit for an amplifier having more than one stage can only be a direct coupled amplifier, or external capacitors are to be used. The actual circuit of the amplifier depends upon the requirements of the gain, frequency response, input and output impedances, and the output swing. A typical circuit for a low frequency IC amplifier is shown in Fig. 4.23. This amplifier has the following features:

(a) It is completely d c coupled and as such, large capacitors for coupling are avoided.
(b) It uses emitter follower circuits for providing isolation.
(c) Gain and temperature stability are achieved by negative feedback, using a transistor stage and other internal feedback components.

4.8.7. New Technology Trends

(a) Fabrication of FETs

In general, the integrated circuits involve bipolar transistors, which depend upon minority charge carriers in the active region, i.e. electrons in the base region of an n-p-n transistor. Currently, other categories of transistors are becoming important. The action of these transistors depend upon the majority carriers in the active region and that is why, they are called *majority carrier devices*. These are of two categories: JFET and MOSFET. These transistors have already been explained in detail earlier in the text. The most significant

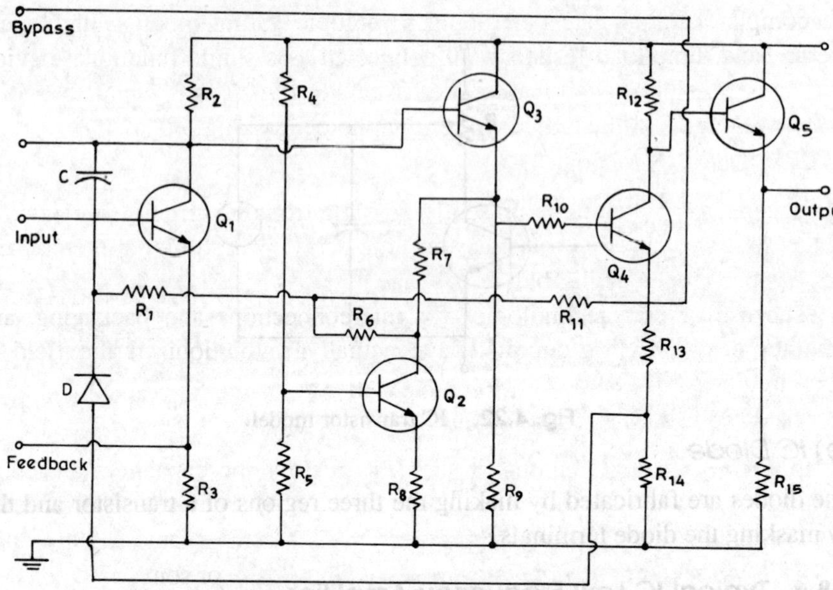

Fig. 4.23. Circuit for a typical IC amplifier.

characteristic of FETs is their high input impedance (~10^{15} Ω) at low frequency, because of which no gate current flows. The JFET is generally used where higher input impedance than that shown by a BJT is required, e.g. a ceramic transducer.

As the threshold voltage for a MOSFET is about 4 V, it has become possible to make dynamic multiphase logic circuits in the integrated circuits. An example of this type of circuit is very large serial shift resistors, in which the information 0 or 1 is stored by charging the input capacitance of a MOSFET. Further details on this topic may be obtained from specialized textbooks on complex silicon integrated circuits.

(b) Geometry and Circuit Complexity

In order to reduce the cost of construction of ICs, the number of circuit functions per unit area of the wafer is increased which, in turn, increases the complexity. This process can be achieved by using small geometry-techniques to put more components on a circuit chip. As the number of components increases, the number of lead points also rises rapidly and hence, the designing ability of more complex circuits is limited by the number of input and output leads. The actual approach towards microcircuits started with the MOSFETs, and now single-chip microcircuits with hundreds of components are being manufactured, because MOSFET circuits do not need isolation regions. A 100-bit shift resistor containing 100–800 MOS devices is also available. BCD counters, dual JK flip-flop, master slave JK flip-flop, etc. are various examples of microcircuits. A microcircuit chip may be of 3 mils from side-to-side, which is a complete digital gate containing a multiple gating input, a high-speed inverter, and three load resistors, all connected as a single functional device. Now, it is possible to fabricate patterns having dimensions equal to two or three wavelengths of the light used for the photoengraving process.

(c) Complexity and Speed

The nanosecond switching is presently possible from the circuits fabricated in a single chip, because the transmission-time delay owing to long interconnections has been made negligible.

The use of new technologies for interconnections and packaging, and computer designing for circuits, have actually revolutionized the field of microcircuitry and ICs.

(d) Thin-film Passive Components

The thin-film passive components such as resistors and capacitors are being formed directly on the insulating surface of the active devices. The isolation for these elements is obtained owing to SiO_2. Tantalum or a nickel-chromium alloy is used for fabricating resistors and tantalum oxide or some oxide of Si is used for fabricating capacitors. Nowadays, high-resistivity sputtered tantalum is also under study as a suitable material for thin-film resistors. This material will reduce the size of these resistors and, in turn, lead to the overall size of the ICs being reduced.

4.9. SOLVED PROBLEMS

Problem 1

Find α and β for a BJT if $I_C = 2$ mA and $I_B = 50$ µA.

Solution

The forward current gain is given by

$$\beta = \frac{I_C}{I_B} = \frac{2 \times 10^{-3}}{50 \times 10^{-6}} = 40$$

and the large signal current gain

$$\alpha = \frac{\beta}{1+\beta} = \frac{40}{1+40} = 0.975 .$$

Problem 2

Calculate the base and collector currents for a BJT if $I_E = 1.6$ m A and $\beta = 39$.

Solution

The relation for currents in a BJT is given by

$$I_E = I_B + I_C$$

$$= I_B + \beta\, I_B$$

$$\therefore \qquad I_B = \frac{I_E}{1+\beta} = \frac{1.6 \text{ mA}}{40} = 40\ \mu A$$

and $$I_C = I_E - I_B = 1.6 - 0.040$$

$$= 1.56 \text{ mA}.$$

Problem 3

In Fig. 4.4, Calculate

(i) I_C if $I_E = 4$ mA and $V_{CB} = -4$ V
(ii) I_C if $V_{CB} = -5$ V and $V_{EB} = 0.8$ V
(iii) V_{EB} if $V_{CB} = -1$V and $I_C = 3$ mA.

Solution

(i) $I_C \equiv 4$ m A, from Fig. 4.4 (b)
(ii) From Fig. 4.4 (a), when $V_{CB} = -5$ V, $V_{EB} = 0.8$ V, $I_E = 4$ mA and from Fig. 4.4(b), $I_C = 4$ mA if $I_E = 4$ mA.
(iii) Fig. 4.4 (b) shows $I_E = 3$ mA for $V_{CB} = -1$V and $I_C = 3$ mA and Fig. 4.4 (a) shows

$$V_{EB} = 0.8 \text{ V} \quad \text{for} \quad I_E = 3 \text{ mA} \quad \text{and} \quad V_{CB} = -1 \text{ V}.$$

Problem 4

Using Fig. 4.6, find the values of I_C if $V_{BE} = 0.8$ V and $V_{CE} = 5$ V.

Solution

In the input characteristics,

$$I_B = 20\ \mu\text{A for } V_{BE} = 0.8 \text{ V and } V_{CE} = 5 \text{ V}$$

and for these values, the output characteristics yield

$$I_C = 2.5 \text{ mA}.$$

Problem 5

Find the FET parameters for an n-channel FET for which the characteristics are drawn in Fig. 4.11.

Solution

From Fig. 4.11, if we choose $V_{DS} = 10$ V and change the value of V_{GS} from 4 V to 6 V, I_D changes from 4.3 mA to 2.5 mA. Hence, the forward transconductance

$$g_{fs} = \left.\frac{\partial I_D}{\partial V_{GS}}\right|_{(V_{DS} \text{ is a Constant})}$$

$$= \frac{1.8 \times 10^{-3}}{2} = 9 \times 10^{-4} \text{ mhos}$$

If we select V_{GS} = 5 V and change the value of V_{DS} from 5 V to 15 V, I_D changes from 3.2 mA to 3.5 mA. Hence, the drain-source resistance is given by

$$r_d = \left.\frac{\partial V_{DS}}{\partial I_D}\right|_{(V_{GS} \text{ is a Constant})}$$

$$= \frac{10}{0.3 \times 10^{-3}} = 33.4 \text{ k}\Omega$$

The amplification factor is given by

$$\mu = r_d \cdot g_{fs}$$

$$= 33.4 \times 10^3 \times 9 \times 10^{-4}$$

$$= 30.$$

Questions and Problems

1. What do you mean by a bipolar junction transistor? How does it work in a circuit?
2. How many types of BJT's are there? Distinguish among them.
3. Describe the various biasing configurations for a BJT.
4. How is a BJT biased normally?
5. Describe the Ebers-Moll model for a BJT.
6. Which are the terminals of a BJT that are used as input terminals for a CE amplifier?
7. Which are the terminals of a BJT that are used as input terminals for an emitter follower?
8. Draw the input and output characteristics for a BJT in the CE configuration.
9. Draw the input and output characteristics for a BJT in the CB configuration.
10. Describe a JFET and draw its characteristics.
11. Explain a depletion mode MOSFET along with its characteristics.
12. Describe the enhancement mode MOSFET and draw its characteristics.
13. Describe various FET parameters.
14. What are the three regions of operation for a BJT? Explain the importance of each of the regions.

15. What do you mean by collector leakage current?
16. Define α and β for a BJT.
17. What is the meaning of transistor ratings?
18. What are the advantages and disadvantages of FET over BJT?
19. Explain Zener breakdown in case of a JFET.
20. What is the role of SiO_2 in a MOSFET?
21. Find α and β of a BJT if $I_C = 2.5$ mA and $I_B = 50$ μA.
22. Find β if $I_C = 1.56$ mA and $I_E = 1.6$ mA.
23. Find α if $I_C = 1.56$ mA and $I_E = 1.6$ mA.
24. What is the advantage of the multiple emitter transistor fabricated in an IC?
25. How does the resistance of an IC resistor change with frequency?
26. Why are junction capacitors not commonly used in ICs?
27. Where may the first silicon seed required in crystal growth have come from?
28. Why does the semiconductoı remain electrically neutral after doping?
29. What happens to the behaviour of a transistor if the base is made very wide?
30. What are the advantages of ICs?
31. Give a broad classification of ICs.
32. Describe the techniques of getting high-purity single semiconductor crystal.
33. Why is the floating zone technique superior to the zone refining technique for crystal growth?
34. What do you mean by planar technology?
35. How is a layer of SiO_2 developed on a Si wafer?
36. Describe in brief the various processes involved in the fabrication of an IC.
37. Which is the most important step in the fabrication of an IC?
38. What do you mean by an IC resistor?
39. What is sheet resistance?
40. Discuss the model of an IC transistor.
41. Describe an IC low-frequency amplifier.
42. What is the meaning of circuit complexity in an IC fabrication?
43. What are the new technology trends in IC fabrication?
44. How is isolation in various components in an IC obtained?

5 Transistor Biasing Techniques

Transistors are used in variety of circuits, e.g. amplifiers, switches, waveform shapers, logic circuits, etc. Depending upon the particular application, the transistors are given proper biasing so that they can operate in the desired region of their characteristics. Several biasing techniques are used for transistors so as to select the dc operating point in a desired operating region, which is required for the particular performance criteria that the circuit should meet.

5.1 OPERATING POINT OF A TRANSISTOR

The operating point of a transistor represents a point on the transistor characteristics, which has the desired values of the current and the voltage. It is also called the *quiescent point* and is selected with the help of biasing technique. There are certain factors which should be taken into account while selecting an operating point. They are as follows:

(a) The transistors have to be operated within the maximum limits for the current, the voltage, and the power. If the transistor is operated outside these maximum ratings, the lifetime of the transistor is reduced and it may be completely damaged. Therefore, the operating point is selected so that the transistor does not cross the maximum ratings while handling the input signal.

(b) The operating point is selected so that the transistor may handle the entire range of the input signal. For example, in an amplifier the positive and the negative parts of the input signal change the voltage and the current of the transistor. The operating point for an amplifier should, therefore, be so selected that the input signal should not drive the transistor into the cut-off or the saturation region.

(c) The operating point must be selected in such a way that in an amplifier, the entire input signal is covered in the linear region of the characteristics so that the gain remains constant for the entire swing of the input signal.

(d) The operating point must be stabilized against the temperature changes because the ac and the dc parameters of a transistor vary with the operating point.

(e) It is also desirable that the biasing circuit deciding the operating point should be practical and economical, and for that, the circuit must involve only one power supply for both the input and the output circuits.

In order to achieve these requirements, a large number of biasing circuits is designed. In the following sections, various biasing circuits are given. The dc currents and the voltages in the base and collector of the transistor are determined in order to understand these circuits.

5.2 CONSTANT-BASE CURRENT CIRCUIT

A constant-base current bias circuit, or a fixed bias circuit, is shown in Fig. 5.1 (a). The input signal is applied to the base of an n-p-n transistor , with the emitter as common terminal. The circuit shows only one power supply which provides both the forward and the reverse bias conditions. The base is at a positive potential with respect to the common emitter (ground) as it is connected to the positive supply voltage through a resistor R_B. Hence, the base-emitter junction is forward biased. The collector-base junction becomes reverse-biased if the collector is made more positive than the base. This is achieved with the help of a resistor R_L, whose value is selected at less than that of R_B. Thus, the voltage drop across R_L is less than that across R_B and as a result, the potential of the collector is more than that of the base. Hence, the single battery provides the forward bias to the input junction and the reverse bias to the output junction, which is the required condition for a transistor to act as an amplifier.

Writing the Kirchoff's voltage equation for the input section of the circuit shown in Fig. 5.1(b),

$$V_{CC} = I_B R_B + V_{BE} \tag{5.1}$$

The base current can be calculated from this equation as

$$I_B = \frac{V_{CC} - V_{BE}}{R_B} \tag{5.2}$$

Since V_{CC} is always large in comparison to V_{BE} ($V_{CC} \gg V_{BE}$), V_{BE} can be

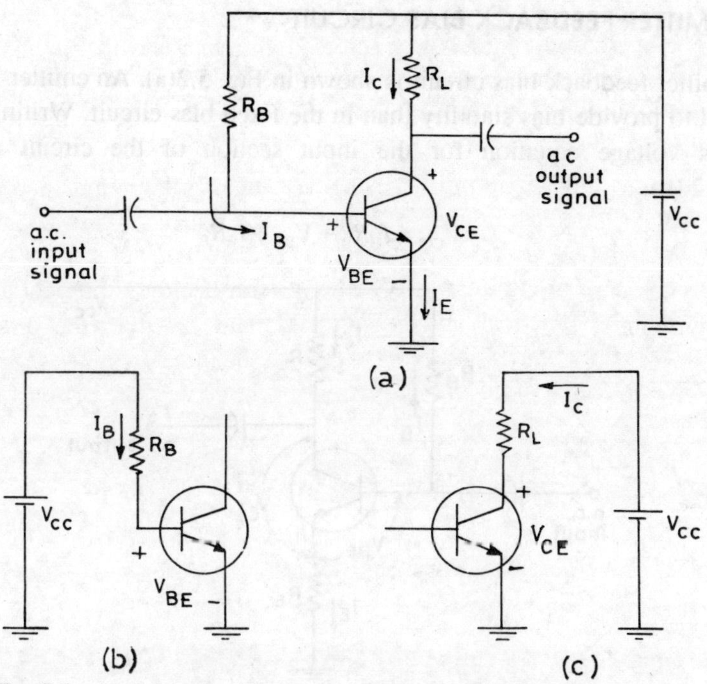

Fig. 5.1. (a) Constant base current bias circuit (b) Input section (c) Output Section.

neglected and

$$I_B \approx \frac{V_{CC}}{R_B} \tag{5.3}$$

Thus, the base current can be held constant by the constant current source V_{CC} and R_B.

Writing the Kirchoff's voltage equation for the output section of the circuit shown in Fig. 5.1(c),

$$V_{CC} = I_C R_L + V_{CE} \tag{5.4}$$

The collector current I_C is defined as

$$I_C = \beta I_B \tag{5.5}$$

Hence, Eq. 5.4 becomes

$$V_{CE} = V_{CC} - \beta I_B R_L \tag{5.6}$$

As the current gain β changes with temperature, the values of V_{CE} and I_B also change even when V_{CC} and R_L are held constant. Hence, the operating point shifts with temperature. This is a disadvantage of the constant-base current circuit. But this shift in operating point may be minimized by using large collector currents with small loads.

5.3 EMITTER FEEDBACK BIAS CIRCUIT

An emitter feedback bias circuit is shown in Fig. 5.2(a). An emitter resistance is used to provide bias stability than in the fixed bias circuit. Writing the Kirchoff's voltage equation for the input section of the circuit shown in Fig. 5.2 (b),

$$V_{CC} = I_B R_B + V_{BE} + I_E R_E \tag{5.7}$$

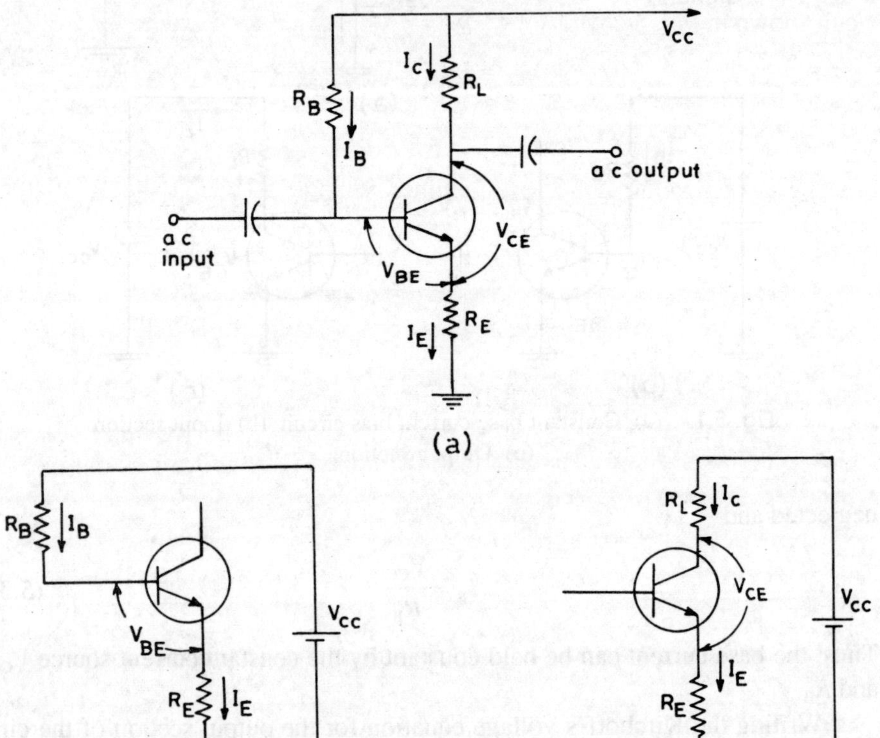

Fig. 5.2. (a) Emitter feedback bias circuit (b) Input section (c) Output section.

Using Eq. 4.5 of Chapter 4, I_E can be replaced by $(1 + \beta)\ I_B$. Eq. 5.7 now becomes

$$V_{CC} = I_B R_B + V_{BE} + (1+\beta) I_B R_E$$

or

$$I_B = \frac{V_{CC} - V_{BE}}{R_B + (1+\beta) R_E}$$

$$\cong \frac{V_{CC}}{R_B + (1+\beta)R_E} \quad (\text{as } V_{CC} \gg V_{BE})$$

$$\cong \frac{V_{CC}}{R_B + \beta R_E} \qquad (\text{as } \beta \gg 1) \tag{5.8}$$

On comparing this equation with Eq. 5.3, it is obvious that the emitter feedback bias circuit provides better stability than a fixed bias circuit.

Writing the Kirchoff's voltage equation for the output section of the circuit shown in Fig. 5.2(c),

$$V_{CC} = I_C R_L + V_{CE} + I_E R_E$$

or

$$V_{CE} = V_{CC} - I_C R_L - I_E R_E \tag{5.9}$$

This gives the collector-to-emitter voltage at which the transistor is biased. The term $V_{CC} - I_C R_L$ represents the collector potential measured between the collector and the ground, whereas $I_E R_E$ is the emitter potential measured between the emitter and the ground. Hence, the difference of the two shows V_{CE}, the potential difference between the collector and the emitter.

5.4 UNIVERSAL BIAS CIRCUIT

Figure 5.3(a) shows a bias circuit which provides better stability and is independent of β also. This circuit is very popular and involves load resistor R_L, emitter resistor R_E and two voltage dividing resistors R_{B_1} and R_{B_2}. As it involves four resistors, the circuit is also called the four resistor biasing circuit. The voltage dividing resistors can be theveninized as shown in Fig. 5.3(b) with

$$R_B = \frac{R_{B_1} R_{B_2}}{R_{B_1} + R_{B_2}} \tag{5.10}$$

and

$$V_{BB} = V_{CC} \cdot \frac{R_{B_2}}{R_{B_1} + R_{B_2}} \tag{5.11}$$

Substituting the value of $R_{B_2}/R_{B_1} + R_{B_2}$ from Eq. 5.11 in Eq. 5.10,

$$R_B = R_{B_1} \cdot \frac{V_{BB}}{V_{CC}}$$

or

$$R_{B_1} = \frac{V_{CC}}{V_{BB}} \cdot R_B \tag{5.12}$$

From Eqs. 5.10 and 5.12,

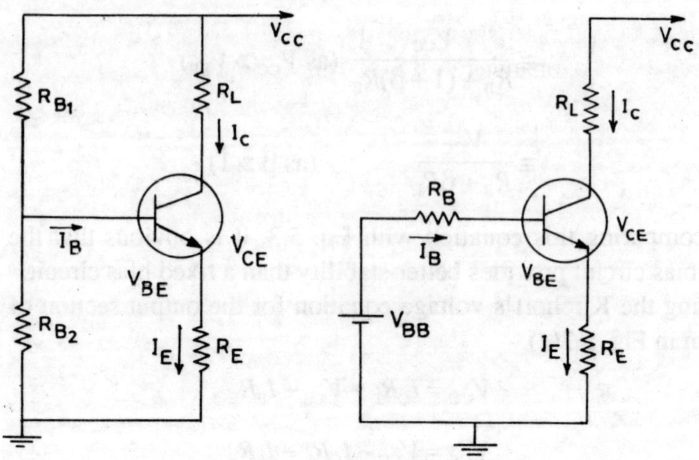

Fig. 5.3. (a) Universal bias circuit (b) Theveninized equivalent circuit.

$$R_{B_2} = \frac{V_{CC}}{V_{CC} - V_{BB}} \cdot R_B \tag{5.13}$$

Writing the Kirchoff's voltage equation for the input section of the circuit,

$$V_{BB} = I_B R_B + V_{BE} + I_E R_E$$

$$= I_B R_B + V_{BE} + (1 + \beta) I_B R_E$$

as $\qquad I_E = (1 + \beta) I_B$

Hence

$$V_{BB} - V_{BE} = I_B [R_B + (1 + \beta) R_E] \tag{5.14}$$

From this equation, I_B can be written as

$$I_B = \frac{V_{BB} - V_{BE}}{R_B + (1 + \beta) R_E}$$

$$= \frac{V_{BB} - V_{BE}}{R_B [1 + (1 + \beta) R_E / R_B]}$$

$$= \frac{V_{BB} - V_{BE}}{R_B} \cdot \frac{1}{1 + \beta (R_E / R_B)} \quad \text{as} \quad \beta \gg 1 \tag{5.15}$$

Hence, I_C can be written as

$$I_C = \beta \cdot I_B$$

$$= \frac{V_{BB} - V_{BE}}{R_B} \cdot \frac{\beta}{1 + \beta (R_E / R_B)} \tag{5.16}$$

Equations 5.15 and 5.16 show that this universal bias circuit provides better stability against the changes in β with temperature, as I_B and I_C include more complex terms in β. Actually, complete independence against the changes in β can be achieved if we adjust

$$1+\beta\cdot\frac{R_E}{R_B}=\beta \quad \text{in Eq. 5.16}$$

or

$$\frac{R_E}{R_B}=\frac{\beta-1}{\beta}=1-\frac{1}{\beta}$$

$$\cong 1$$

or

$$R_E \cong R_B \tag{5.17}$$

In general, $R_B > R_E$ and in order to achieve the above requirement (Eq. 5.17), either of two alternatives can be followed:

(a) The value of R_B can be reduced. However, this results in drawing some extra current from the power supply. Moreover, it reduces the gain because it provides lower resistance shunt path for the input signal.

(b) The value of R_E can be increased. But this again results in reducing the gain as this process provides a negative feedback path. For some ac applications, R_E is decoupled with a capacitor C_E to eliminate negative feedback, causing loss of gain. But this is achieved at the cost of bias stabilization. Therefore, C_E is avoided and the reduction in the gain owing to negative feedback through R_E is accepted because the circuit then shows stabilization against the changes in β.

5.5 COMMON COLLECTOR BIAS (EMITTER FOLLOWER) CIRCUIT

Figure 5.4 shows a common collector bias circuit in which the input is applied to the base circuit and the output is taken from the emitter circuit. This circuit is also known as emitter follower circuit (refer Chapter 7). Writing the Kirchoff's voltage equation for the input section of the circuit,

$$V_{CC}=I_BR_B+V_{BE}+I_ER_E \tag{5.18}$$

The emitter current I_E can be expressed as

$$I_E=(1+\beta)I_B \tag{5.19}$$

Substituting the value of I_E in Eq. 5.18,

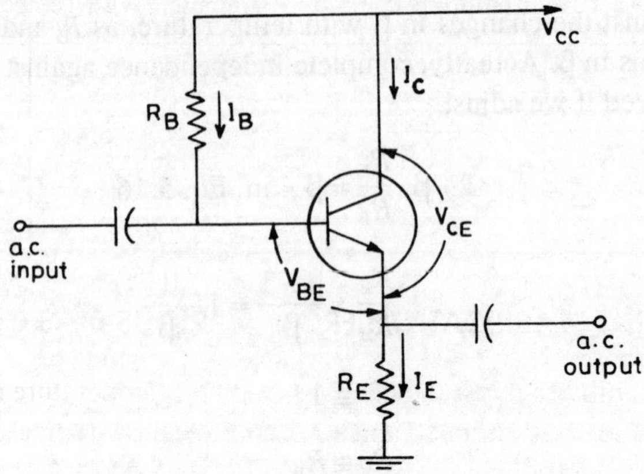

Fig. 5.4. Common collector bias circuit.

$$I_B = \frac{V_{CC} - V_{BE}}{R_B + (1+\beta)R_E}$$

$$\cong \frac{V_{CC} - V_{BE}}{R_B[1 + \beta(R_E/R_B)]} \quad \text{as } \beta \gg 1 \qquad (5.20)$$

Writing the Kirchoff's voltage equation for the output section of the circuit,

$$V_{CC} = V_{CE} + I_E R_E$$

or

$$V_{CE} = V_{CC} - I_E R_E \qquad (5.21)$$

Using Eqs 5.20 and 5.21, various bias conditions can be calculated.

5.6 Q-POINT STABILIZATION

Q-point stabilization is an important requirement of the electronic circuits. Briefly, it means that the bias point must not shift from the values originally set for the circuit. But it is observed that the operating point shifts because of the changes in the temperature and β. The collector current I_C varies with a change in temperature because of the following reasons:

(a) The transistor leakage current I_{CO} becomes double for every 10°C increase in the temperature.

(b) The base emitter voltage V_{BE} decreases by 2.5 mV for every degree centigrade rise in the temperature.

(c) The transistor current gain β or h_{FE} increases with the temperature. Table 5.1 shows typical variations in I_{CO}, V_{BE}, and β with temperature for a silicon transistor.

Table 5.1 Variation in Transistor Parameters with Temperature

T (°C)	I_{co} (nA)	β	V_{BE} (volt)
25	0.1	50	0.65
100	20	80	0.48
175	3.3×10^3	120	0.30

5.7 Q-POINT STABILIZATION FROM VARIOUS BIAS CIRCUITS

Q-point instability against the changes in β and temperature can be reduced by applying proper bias circuit. Tables 5.2 and 5.3 show typical variation in I_C and V_{CE} with β and temperature respectively for the constant-base current biasing, the emitter-feedback biasing, and the universal biasing. The three biasing methods have already been discussed in earlier sections. For varying β, five transistors of the same number have been taken randomly.

Table 5.2 Various bias circuits depicting Q-stabilization against temperature for constant β

	Constant-base Current Biasing		Emitter-feedback Biasing		Universal Biasing	
	25°C	65°C	25°C	65°C	25°C	65°C
I_C (mA)	1.1	2.2	0.95	1.6	1.25	1.5
V_{CE} (V)	5.2	0.5	5.0	1.0	5.5	2.9

Table 5.3 Various bias circuits depicting Q-stabilization against β at constant temperature

β for various transistors of same number	Constant-base Current Biasing		Emitter-feedback Biasing		Universal Biasing	
	I_C (mA)	V_{CE} (V)	I_C (mA)	V_{CE} (V)	I_C (mA)	V_{CE} (V)
β_1	1.1	5.2	0.95	5.0	1.25	5.5
β_2	1.35	3.8	1.15	3.6	1.26	5.4
β_3	2.05	1.3	0.72	6.0	1.27	5.4
β_4	0.78	6.6	0.89	5.4	1.24	5.6
β_5	1.0	5.8	1.15	3.6	1.24	5.6

On comparing the three biasing techniques, Tables 5.2 and 5.3 indicate clearly that there are wide variations in Q-point (defined here by I_C and V_{CE}) in the constant-base biasing and these variations reduce to a great extent in the emitter feedback biasing. It is also clear that the universal biasing is the best, because the variations in I_C and V_{CE} with changes in β and temperature are the least in this case.

5.8 STABILITY FACTORS

The variations in transistor parameters I_{CO}, β, and V_{BE} with temperature have been shown in Table 5.1. The effect of the change in these parameters on the value of I_C can be represented by stability factors. The stability factor S can be defined for each of these parameters as

$$\left.\begin{aligned} S(I_{CO}) &= \left.\frac{\Delta I_C}{\Delta I_{CO}}\right|_{(V_{BE} \text{ and } \beta \text{ are constant})} \\ S(V_{BE}) &= \left.\frac{\Delta I_C}{\Delta V_{BE}}\right|_{(I_{CO} \text{ is constant})} \\ S(\beta) &= \left.\frac{\Delta I_C}{\Delta \beta}\right|_{(I_{CO} \text{ and } V_{BE} \text{ are constant})} \end{aligned}\right\} \tag{5.22}$$

The change in the collector current from the designed Q-point because of the temperature and β changes is then,

$$\Delta I_C = S_{I_{CO}} \cdot \Delta I_{CO} + S_{V_{BE}} \cdot \Delta V_{BE} + S_\beta \cdot \Delta\beta \tag{5.23}$$

The Kirchoff's voltage equation for the input section of the universal bias circuit of Fig. 5.3(b) as given by Eq. 5.14 can be reproduced as

$$V_{BB} = I_B R_B + V_{BE} + I_E R_E \tag{5.24}$$

The collector current for a transistor in CE configuration is given by

$$I_C = \alpha I_E + I_{CEO} \tag{5.25}$$

where I_{CEO} is the transistor leakage current. Using $I_E = I_B + I_C$, Eq. 5.25 becomes

$$I_C = \alpha(I_B + I_C) + I_{CO}$$

$$= \frac{\alpha}{1-\alpha} \cdot I_B + \frac{1}{1-\alpha} \cdot I_{CO} \tag{5.26}$$

In Chapter 4, it is found that $\beta = \dfrac{\alpha}{1-\alpha}$ and $1 + \beta = \dfrac{1}{1-\alpha}$ as in Eq. 4.7. Therefore,

$$I_C = \beta I_B + (1+\beta) I_{CO} \tag{5.27}$$

Using Eqs 5.24, 5.25 and 5.27, one gets

$$I_C = \frac{\beta(V_{BB} - V_{BE}) + I_{CO}(1+\beta)(R_B + R_E)}{R_B + (1+\beta)R_E} \tag{5.28}$$

The stability factors as defined in Eq. 5.22 can be calculated from Eq. 5.28 as

$$S(I_{CO}) = \frac{(1+\beta)(R_B+R_E)}{R_B+(1+\beta)R_E}$$

$$S(V_{BE}) = -\frac{\beta}{R_B+(1+\beta)R_E}$$

$$S(\beta) = \frac{I_C - I_{CO}}{\beta}\left[\frac{R_B+R_E}{R_B+(1+\beta)R_E}\right] \tag{5.29}$$

These stability factors are used to study the comparative effects of various transistor parameters on the Q-stabilization.

5.9 DIODE COMPENSATION BIAS CIRCUIT

A diode compensation bias circuit is shown in Fig. 5.5(a). This biasing technique eliminates the need of an emitter resistor and a bypass capacitor for the Q-point stabilization against temperature changes. The temperature characteristics of the diode D used in this circuit is identical to that of the transistor. Thus, the voltage change across the diode compensates for the change in V_{BE} to maintain the bias value of I_C. In order to achieve perfect identical matching between the diode and the transistor, and identical transistor is used in place of the diode. This transistor used as a diode is connected with its base and collector joined together, as shown in Fig. 5.5(b).

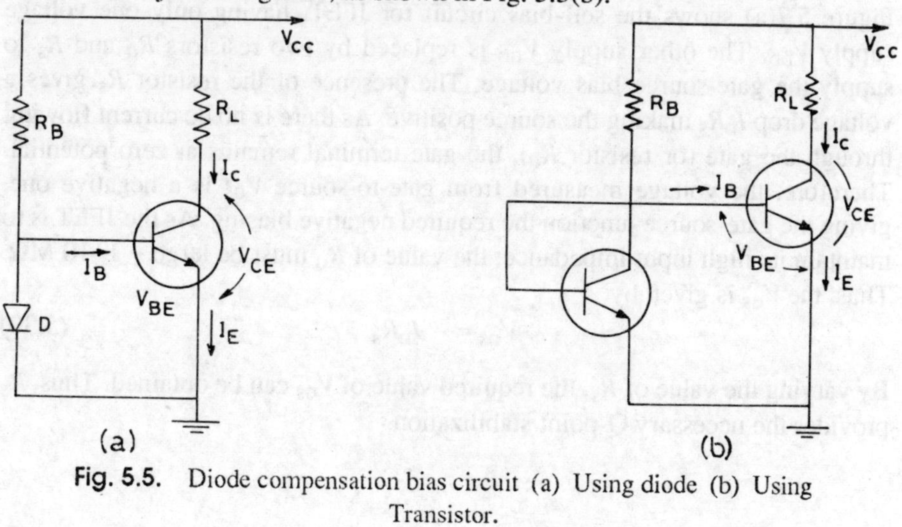

Fig. 5.5. Diode compensation bias circuit (a) Using diode (b) Using Transistor.

5.10 SHUNT FEEDBACK BIAS CIRCUIT

Figure 5.6 shows another biasing technique commonly used. This circuit has the base resistance connected to the collector rather than to the supply voltage. If the collector current I_C increases, say, because of temperature, the voltage drop across R_L increases and as a result V_{CE} decreases. This decrease in V_{CE} results in a decrease in I_B and hence, I_C also decreases to reach the originally set value. Thus, the quiescent point of the transistor is very near its designed Q-point.

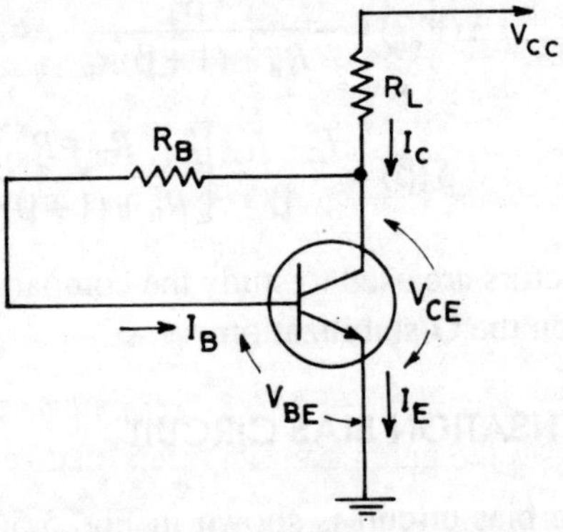

Fig. 5.6. Shunt feedback bias circuit.

5.11 JFET BIASING

There are two commonly used biasing techniques for a junction field effect transistor.

5.11.1 Self-bias Circuit

Figure 5.7(a) shows the self-bias circuit for JFET, having only one voltage supply V_{DD}. The other supply V_{GG} is replaced by two resistors R_G and R_S to supply the gate-source bias voltage. The presence of the resistor R_S gives a voltage drop I_DR_S making the source positive. As there is no dc current flowing through the gate (or resistor R_G), the gate terminal remains at zero potential. Therefore, the voltage measured from gate-to-source V_{GS} is a negative one, giving the gate-source junction the required negative biasing. As the JFET is to maintain its high input impedance, the value of R_G must be large ~ 1–10 MΩ. Thus, the V_{GS} is given by

$$V_{GS} = - I_D R_S \tag{5.30}$$

By varying the value of R_S, the required value of V_{GS} can be obtained. Thus, R_S provides the necessary Q-point stabilization.

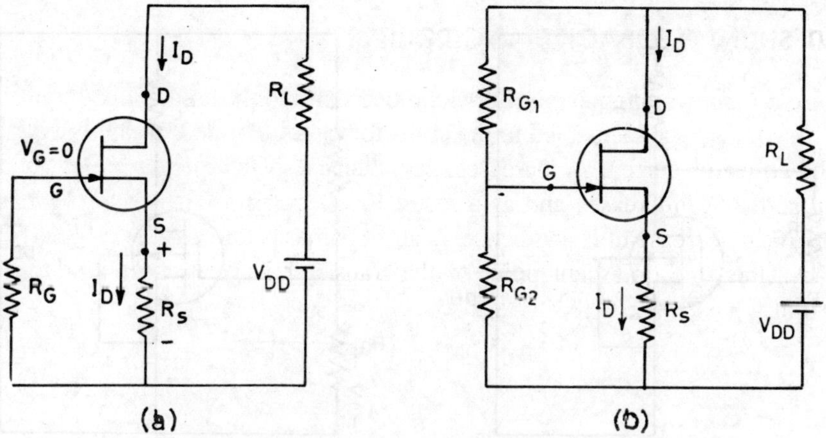

Fig. 5.7. JFET biasing technique (a) Self bias circuit (b) Voltage divider gate bias circuit.

5.11.2 Voltage Divider Gate Bias (Fixed Bias) Circuit

Figure 5.7(b) shows another form of JFET bias circuit using voltage divider resistors R_{G_1} and R_{G_2} to set the gate potential other than zero. The value of V_G is given by

$$V_G = \frac{R_{G_2}}{R_{G_1} + R_{G_2}} \cdot V_{DD} \tag{5.31}$$

and the gate-to-source voltage

$$V_{GS} = V_G - I_D R_S \tag{5.32}$$

This circuit provides a greater dc bias stability.

5.12 MOSFET Biasing

There are different biasing techniques used in the enhancement mode and the depletion mode MOSFET.

5.12.1 Feedback Biasing for Enhancement Mode MOSFET

The circuit diagram for feedback biasing of the enhancement mode MOSFET is shown in Fig. 5.8(a). The drain is connected to the gate through a feedback bias resistor R_G. As there is no gate current, there is no voltage drop across R_G and V_{DS} = V_{GS}. Hence, the drain and the gate potentials are the same with respect to the ground source and any value of R_G allows V_{GS} to be equal to V_{DS}. The bias point can only be changed by varying the value of R_L.

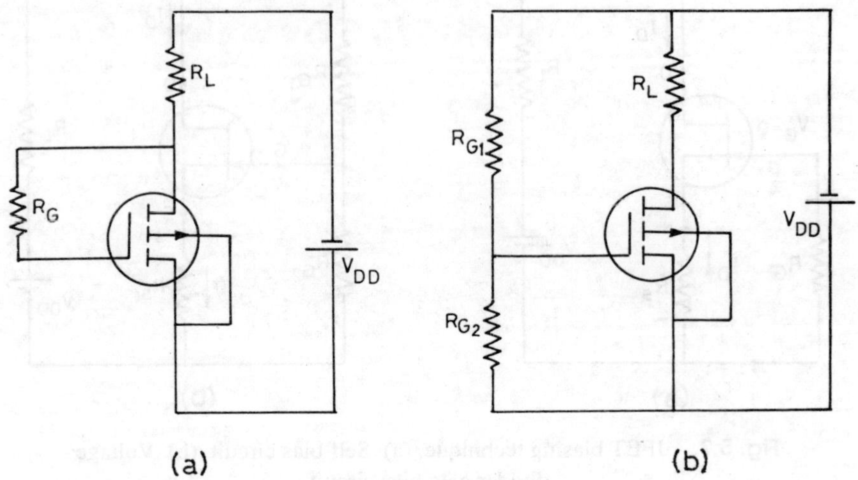

Fig. 5.8. MOSFET biasing in enhancement mode (a) Feedback biasing (b) Fixed biasing.

5.12.2 Fixed Biasing for Enhancement Mode MOSFET

The circuit for fixed biasing of enhancement mode MOSFET is shown in Fig. 5.8(b). Two resistors R_{G_1} and R_{G_2} are used as a voltage divider for V_{DD} and the gate-to-source voltage V_{GS} is held fixed at a value

$$V_{GS} = \frac{R_{G_2}}{R_{G_1} + R_{G_2}} V_{DD} \tag{5.33}$$

5.12.3 Self-biasing for Depletion Mode MOSFET

Figure 5.9(a) shows the self-biasing for depletion mode MOSFET. This circuit is almost identical to that of JFET discussed in Sec. 5.11.1. The only difference is that R_G is larger in the case of MOSFET. As in the MOSFET, the gate-source voltage can be positive also, it is possible to bias the device at only slightly negative V_{GS}.

5.12.4 Fixed Biasing for Depletion Mode MOSFET

Figure 5.9(b) shows the fixed biasing circuit for the depletion mode MOSFET. The fixed V_{GS} can be obtained as

$$V_{GS} = \frac{R_{G_2}}{R_{G_1} + R_{G_2}} \cdot V_{DD} \tag{5.34}$$

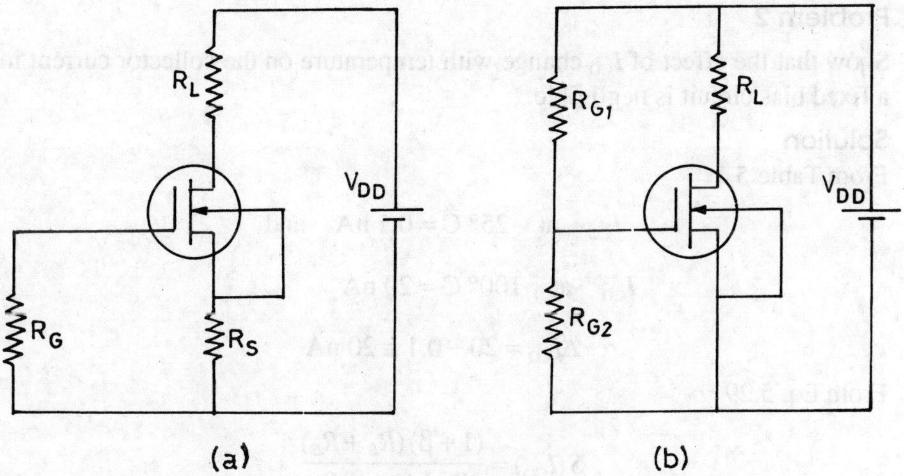

Fig. 5.9. Depletion mode MOSFET biasing (a) Self biasing (b) Fixed Biasing.

5.13 SOLVED PROBLEMS

Problem 1

Calculate the voltage and current for the quiescent point for a constant base current bias circuit if

$R_B = 470\ \text{k}\Omega,\ \ R_L = 5\ \text{k}\Omega,\ \ \beta = 50,$
$V_{BE} = 0.6\ \text{V},\ \text{and}\ \ V_{CC} = 10\ \text{V}$

Solution

In a constant-base current bias circuit as shown in Fig. 5.1(a), the base current for the quiescent point is given by Eq. 5.2 as

$$I_B = \frac{V_{CC} - V_{BE}}{R_B} = \frac{10 - 0.6}{470 \times 10^3}$$

$$= 20\ \mu\text{A}$$

and the collector-emitter potential for the quiescent point is given by Eq. 5.6 as

$$V_{CE} = V_{CC} - \beta \cdot I_B R_L$$

$$= 10 - 50 \times 20 \times 10^{-6} \times 5 \times 10^3$$

$$= 5\ \text{V}$$

and the collector current

$$I_C = \beta I_B = 50 \times 20 \times 10^{-6}$$

$$= 1\ \text{mA}$$

Problem 2

Show that the effect of I_{CO} change with temperature on the collector current in a fixed bias circuit is negligible.

Solution

From Table 5.1

$$I_{CO} \text{ at } 25° \text{ C} = 0.1 \text{ nA} \quad \text{and}$$

$$I_{CO} \text{ at } 100° \text{ C} = 20 \text{ nA}$$

$$\therefore \quad \Delta I_{CO} = 20 - 0.1 \cong 20 \text{ nA}$$

From Eq. 5.29

$$S(I_{CO}) = \frac{(1+\beta)(R_B + R_E)}{R_B + (1+\beta)R_E}$$

but for a fixed-bias circuit

$$R_E = 0$$

$$\therefore \quad S(I_{CO}) = 1 + \beta = 1 + 50$$

$$= 51$$

because $\beta = 50$ at 25°C from Table 5.1. Hence from Eq. 5.22,

$$\Delta I_C = S.\Delta I_{CO}$$

$$= 51 \times 20 \times 10^{-9}$$

$$= 1 \ \mu\text{A}$$

If the collector current is set at 2.5 mA, the change is 1 μA, which is insignificant.

Problem 3

Calculate the quiescent point for an emitter feedback bias circuit with the following circuit parameters:

$$V_{CC} = 9 \text{ V}, \quad R_B = 250 \text{ k}\Omega, \quad R_L = 2 \text{ k}\Omega,$$
$$\beta = 50, \quad \text{and} \quad R_E = 1 \text{ k}\Omega.$$

Solution

In an emitter feedback bias circuit, as shown in Fig. 5.2, the base current is given by Eq. 5.8 as

$$I_B = \frac{V_{CC}}{R_B + \beta R_E} = \frac{9}{250\times 10^3 + 50\times 10^3}$$

$$= 30\ \mu A$$

$\therefore$
$$I_C = \beta . I_B = 50\times 30\ \mu A$$

$$= 1.5\ mA$$

The emitter current $I_E = I_B + I_C = 1.53$ mA

The collector emitter potential for the Q-point is given by Eq. 5.9 as

$$V_{CE} = V_{CC} - I_C R_L - I_E R_E$$

$$= 9 - 1.5\times 10^{-3}\times 2\times 10^3 - 1.53\times 10^{-3}\times 1\times 10^3$$

$$= 4.47\ V$$

Problem 4

Design a fixed bias circuit for an operating point $I_C = 1$ mA and $V_{CE} = 5$ V. The transistor used has $\beta = 50$ and the supply voltage is 15 V.

Solution

The load resistance R_L is given by Eq. 5.6 as

$$R_L = \frac{V_{CC} - V_{CE}}{\beta . I_B}$$

$$= \frac{V_{CC} - V_{CE}}{I_C} = \frac{15 - 5}{1\times 10^{-3}}$$

$$= 10\ k\Omega$$

and
$$\beta = \frac{I_C}{I_B}$$

or
$$I_B = \frac{10^{-3}}{50} = 20\ \mu A$$

The base resistance is given by Eq. 5.3 as

$$R_B = \frac{V_{CC}}{I_B} = \frac{15}{20\times 10^{-6}}$$

$$= 750\ k\Omega$$

Hence using the values of $R_L = 10$ kΩ, $R_B = 750$ kΩ, and $V_{CC} = 15$ V, a fixed bias circuit as shown in Fig. 5.1 can be designed.

Problem 5

In the universal bias circuit of Fig. 5.3, find the operating point and the dc resistance if,
$R_{B_1} = 75\ k\Omega$, $R_{B_2} = 25\ k\Omega$, $R_L = 2\ k\Omega$, $R_E = 1\ k\Omega$, $\beta = 50$, and $V_{CC} = 15$ V.

Solution

In the universal bias circuit described in Sec. 5.4, the base resistance and the base bias are given by Eqs. 5.10 and 5.11 as

$$R_B = \frac{R_{B_1} R_{B_2}}{R_{B_1} + R_{B_2}} = \frac{75 \times 25}{100} = 18.75\ k\Omega$$

and
$$V_{BB} = V_{CC} \cdot \frac{R_{B_2}}{R_{B_1} + R_{B_2}} = 15 \times \frac{25}{100} = 3.75\ V$$

Equation 5.15 gives the base current

$$I_B = \frac{V_{BB} - V_{BE}}{R_B[1 + (1+\beta)R_E/R_B]}$$

$$= \frac{3.75 - 0.7}{18.75\left[1 + (1+50)\frac{1}{18.75}\right]}$$

$$= 43.7\ \mu A$$

$$\therefore \qquad I_C = \beta . I_B = 50 \times 43.7 = 2.185\ mA$$

From the circuit,

$$V_{CE} = V_{CC} - I_C R_L - I_E R_E$$

$$= 15 - 2.185 \times 10^{-3} \times 2 \times 10^3 - 2.229 \times 10^{-3} \times 1 \times 10^3$$

$$= 15 - 4.37 - 2.23$$

$$= 8.4\ V$$

The dc input resistance is given by

$$R_{dc} = R_B 11 (1+\beta) R_E$$

$$= \frac{18.75 \times 51}{18.75 + 51} = 13.7\ k\Omega$$

Questions and Problems

1. What is the meaning of transistor biasing?
2. What do you mean by Q-point and how is it selected?
3. Describe a constant-base current circuit. Why is this circuit not suitable against temperature variations?
4. Draw an emitter feed back bias circuit. Find the operating point in this circuit.
5. Explain how the emitter feed back biasing is better than constant-base current biasing?
6. Describe a four-resistor or universal bias circuit. How does this circuit help in stabilizing the Q-point against the β changes?
7. What is the role of emitter resistance in a universal bias circuit?
8. Find the Q-point in a universal bias circuit.
9. Draw and describe a common collector bias circuit. Fix the Q-point for this circuit.
10. What do you mean by Q-point stabilization? What are the reasons for Q-point instability?
11. Compare various bias circuits keeping in view the Q-point stabilization.
12. Define the stability index. Derive the expressions for various stability factors.
13. Describe a bias circuit which shows Q-point stabilization against temperature changes even when emitter resistor is removed from the four-resistor bias circuit.
14. What is the role of an additional transistor connected in a diode compensation bias circuit?
15. Describe a shunt feed back bias circuit.
16. Describe the biasing techniques for JFET.
17. Describe the various biasing techniques for MOSFET connected in the depletion and enhancement modes.
18. What is the role of R_S, the source resistance in a self-bias circuit for JFET?
19. Find the quiescent point for a constant-base current bias circuit if, $R_B = 500\ k\Omega$, $R_L = 4.7\ k\Omega$, $\beta = 75$, $V_{BE} = 0.7\ V$, and $V_{CC} = 9\ V$.
20. Find the voltage and current for the quiescent point for an emitter feed back bias circuit if, $\beta = 75$, $V_{CC} = 9\ V$, $R_E = 1\ k\Omega$, $R_B = 200\ k\Omega$, and $R_L = 2\ k\Omega$.
21. Calculate the values of load and base resistances for a fixed-bias circuit if other circuit components are:
$V_{CC} = 9\ V$, $\beta = 75$, $I_C = 1.5\ mA$, and $V_{CE} = 4\ V$.
22. Find the quiescent point for a universal bias circuit if,
$R_{B_1} = R_{B_2} = 50\ k\Omega$, $R_L = 5\ k\Omega$, $R_E = 2\ k\Omega$, $\beta = 75$, and $V_{CC} = 9\ V$

6

Two-Port Network Theory and Circuit Models

For an understanding of the response of transistor circuits to a sinusoidal ac signal applied to its input section, the equivalent circuits are to be developed, because it is impossible to reach inside the device to measure its various parameters. An equivalent circuit is defined as the combination of circuit elements, properly chosen, which gives approximately the same response as that of the device in a particular region of operation. Once the equivalent circuit is developed, the symbol of the device used in the circuit is replaced by this equivalent circuit and the different methods of the ac circuit analysis can be used to measure its various parameters. Thus, by the use of an equivalent circuit, the behaviour of active or passive devices in terms of their input and output voltages and currents can be explained. Various circuit models suggesting equivalent circuits are discussed in this chapter.

6.1 TWO-PORT NETWORK ANALYSIS

Figure 6.1 shows an electronic device inside a black box, also known as a

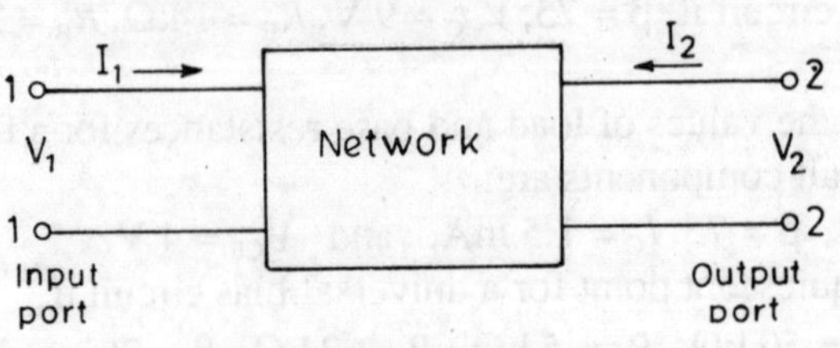

Fig. 6.1. Two port network.

network, with two pairs of terminals (1,1), and (2,2). Each pair of terminals is known as a port. The input voltage and the current, V_1 and I_1 respectively, are measured at the input port (1,1). Similarly, the output voltage and the current V_2 and I_2 are measured at the output port (2,2). These four parameters V_1, I_1, V_2, and I_2 are related to one another and can be expressed as functions of one another. There are several ways of defining their relationships. If two variables are taken as independent variables and the other two as dependent variables, the relevant sets of equations can be obtained. For example, if I_1 and V_2, I_1 and I_2, and V_1 and V_2 are taken as three sets of independent variables, three sets of equations can be formed as

$$\left.\begin{array}{l} V_1 = f_1(I_1, V_2) \\ I_2 = f_1(I_1, V_2) \end{array}\right\} \tag{6.1}$$

$$\left.\begin{array}{l} V_1 = f_2(I_1, I_2) \\ V_2 = f_2(I_1, I_2) \end{array}\right\} \tag{6.2}$$

$$\left.\begin{array}{l} I_1 = f_3(V_1, V_2) \\ I_2 = f_3(V_1, V_2) \end{array}\right\} \tag{6.3}$$

Based upon these equations, three types of circuit parameters are defined:

6.1.1 Hybrid Parameters

Equation 6.1 is used to define hybrid parameters. In order to understand the response of the network for ac signals, the Eq. 6.1 is differentiated as

$$dV_1 = \frac{\partial V_1}{\partial I_1} dI_1 + \frac{\partial V_1}{\partial V_2} dV_2$$

$$dI_2 = \frac{\partial I_2}{\partial I_1} dI_1 + \frac{\partial I_2}{\partial V_2} dV_2 \tag{6.4}$$

When the device is operated in the linear region having constant slope, these equations can be rewritten as

$$V_1 = h_i I_1 + h_r V_2$$

$$I_2 = h_f I_1 + h_o V_2 \tag{6.5}$$

where partial derivatives of Eq. 6.4 are expressed by the letters h as

$$h_i = \frac{\partial V_1}{\partial I_1} \quad \text{(with unit-ohm)}$$

$$h_r = \frac{\partial V_1}{\partial V_2} \quad \text{(with no unit)}$$

$$h_f = \frac{\partial I_2}{\partial I_1} \quad \text{(with no unit)}$$

$$h_0 = \frac{\partial I_2}{\partial V_2} \quad \text{(with unit - mho)} \tag{6.6}$$

These h-parameters are called hybrid parameters because of the mixture of units. Equation 6.5 can be used to understand what the h-parameters represent and to calculate these parameters. When $V_2 = 0$, i.e. the output port is short-circuited

$$h_i = \frac{V_1}{I_1} = \text{(short circuit input impedance)}$$

$$h_f = \frac{I_2}{I_1} = \text{(short circuit forward transfer current gain)} \tag{6.7}$$

when $I_1 = 0$, i.e. input port is open-circuited,

$$h_r = \frac{V_1}{V_2} = \text{(open circuit reverse transfer voltage gain)}$$

$$h_0 = \frac{I_2}{V_2} = \text{(open circuit output admittance)} \tag{6.8}$$

6.1.2 Impedance Parameters

Equation 6.2 is used to define impedance or Z-parameters. This equation is differentiated in a way similar to that explained in Sec 6.1 for obtaining the following equation,

$$V_1 = Z_i I_1 + Z_r I_2$$

$$V_2 = Z_f I_1 + Z_0 I_2 \tag{6.9}$$

When $I_2 = 0$, i.e. output port is open-circuited,

$$Z_i = \frac{V_1}{I_1} = \text{(open circuit input impedance)}$$

$$Z_f = \frac{V_2}{I_1} = \text{(open circuit forward transfer impedance)} \tag{6.10}$$

When $I_1 = 0$, i.e. input port is open-circuited,

$$Z_r = \frac{V_1}{I_2} = \text{(open circuit reverse transfer impedance)}$$

$$Z_0 = \frac{V_2}{I_2} = \text{(open circuit output impedance)} \tag{6.11}$$

Thus, the open circuit impedance parameters are defined by Eqs. 6.10 and 6.11.

6.1.3 Admittance Parameters

Equation 6.3 is used to define admittance or Y-parameters. By differentiating Eq. 6.3 in a way similar to that defined in Sec 6.1.1, the following equations are obtained.

$$I_1 = Y_i V_1 + Y_r V_2$$

$$I_2 = Y_f V_1 + Y_0 V_2 \tag{6.12}$$

When $V_2 = 0$, i.e. output port is short-circuited,

$$Y_i = \frac{I_1}{V_1} = \text{(short circuit input admittance)}$$

$$Y_f = \frac{I_2}{V_1} = \text{(short circuit forward transfer admittance)} \tag{6.13}$$

When $V_1 = 0$, i.e. input port is short-circuited,

$$Y_r = \frac{I_1}{V_2} = \text{(short circuit reverse transfer admittance)}$$

$$Y_0 = \frac{I_2}{V_2} = \text{(short circuit output admittance)} \tag{6.14}$$

6.2 TRANSISTOR HYBRID MODEL

Based upon Eq. 6.5, an equivalent circuit for a transistor can be developed. Equation 6.5 is reproduced as

$$V_1 = h_i I_1 + h_r V_2 \tag{6.15}$$

$$I_2 = h_f\ I_1 + h_0 V_2 \tag{6.16}$$

Equation 6.15 represents the input section of the transistor and since each term in this equation has the units of volts, the Kirchoff's voltage law in reverse can be applied to develop a circuit for which this equation can be rewritten. In other words, Eq. 6.15 represents a voltage addition of $h_i\ I_1$ (a voltage across a resistor h_i due to a current I_1) and a voltage generator of voltage $h_r\ V_2$ (a feed-back voltage to the input and proportional to the output voltage V_2). As the voltages are added in series connections, these two voltages constitute the two arms of the circuit connected in series. This equivalent circuit is shown in Fig. 6.2(a).

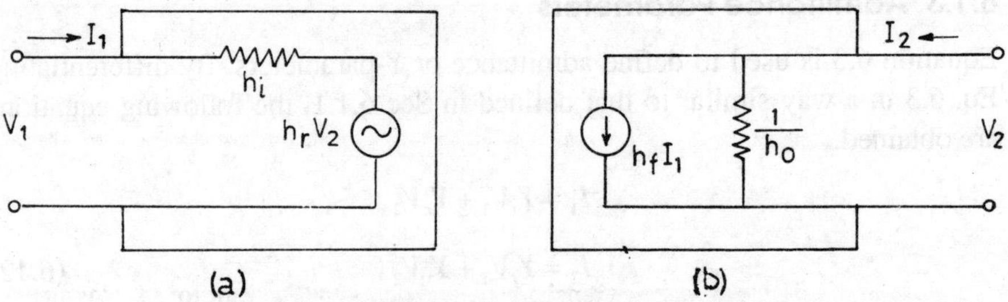

Fig. 6.2. Hybrid equivalent circuit for a transistor
(a) Input section (b) Output section.

Similarly, Eq. 6.16 represents the output section of the transistor and since each term in this equation has the units of current, the Kirchoff's current law in reverse is applied to develop the equivalent circuit for this equation. Equation 6.16 shows an addition of two currents $h_f I_1$ and $h_o\ V_2$. As the currents are added in parallel connections, these two currents constitute the two parallel arms of the circuit, where $h_f\ I_1$ is current source and $h_o\ V_2$ is a current in an impedance $1/h_o$ owing to voltage V_2. This equivalent circuit is shown in Fig. 6.2(b).

As the transistor is used in either of the common emitter, the common base, or the common collector configuration, a second subscript is added to the h-parameter notations to show the mode of transistor connections. For example, the input impedance of a transistor is depicted as h_{ie}, h_{ib}, and h_{ic} in the three configurations respectively. Figure 6.3 shows complete transistor ac equivalent circuits based upon h-parameters for the common emitter and the common base configuration. These circuits are the same for p-n-p or n-p-n transistor.

6.3 IMPEDANCE PARAMETERS CIRCUIT MODEL

Equation 6.9 shows the relationship among the four impedance parameters as

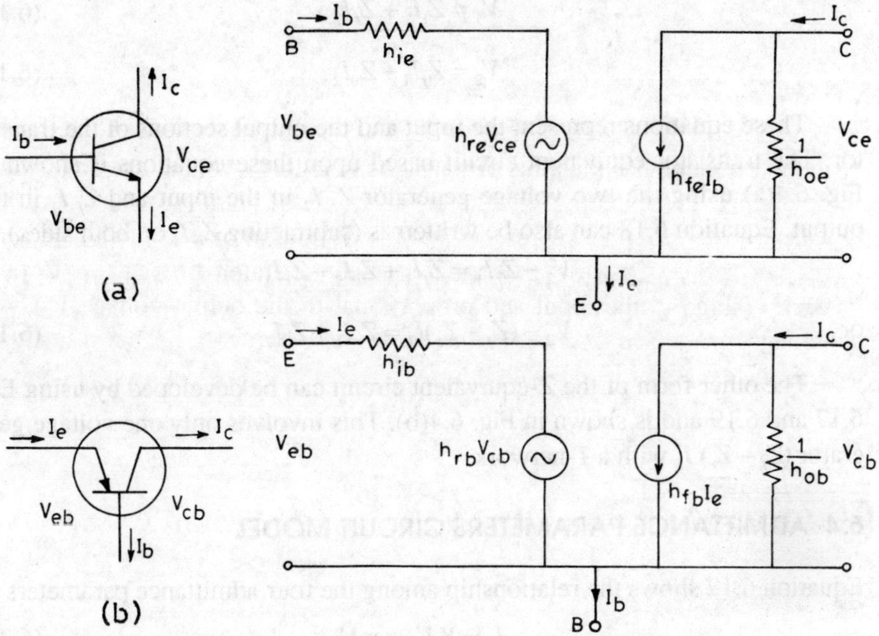

Fig. 6.3. Transistor hybrid equivalent circuit for
(a) Common emitter configuration (b) Common base configuration.

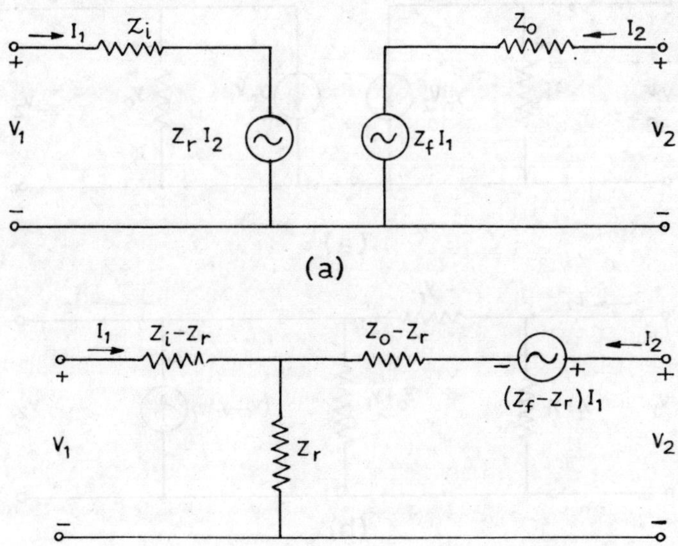

Fig. 6.4. Impedance parameter equivalent circuit for a transistor.

$$V_1 = Z_i I_1 + Z_r I_2 \tag{6.17}$$

$$V_2 = Z_f I_1 + Z_0 I_2 \tag{6.18}$$

These equations represent the input and the output sections of the transistor. The transistor equivalent circuit based upon these equations is shown in Fig. 6.4(a) using the two voltage generator $Z_r I_2$ in the input and $Z_f I_1$ in the output. Equation 6.18 can also be written as (subtracting $Z_r I_1$ on both sides),

$$V_2 - Z_r I_1 = Z_f I_1 + Z_0 I_2 - Z_r I_1$$

or

$$V_2 - (Z_f - Z_r) I_1 = Z_r I_1 + Z_0 I_2 \tag{6.19}$$

The other form of the Z-equivalent circuit can be developed by using Eqs 6.17 and 6.19 and is shown in Fig. 6.4(b). This involves only one voltage generator $(Z_f - Z_r) I_1$ with a *T*-network.

6.4 ADMITTANCE PARAMETERS CIRCUIT MODEL

Equation 6.12 shows the relationship among the four admittance parameters as

$$I_1 = Y_i V_1 + y_r V_2 \tag{6.20}$$

$$I_2 = Y_f V_1 + y_0 V_2 \tag{6.21}$$

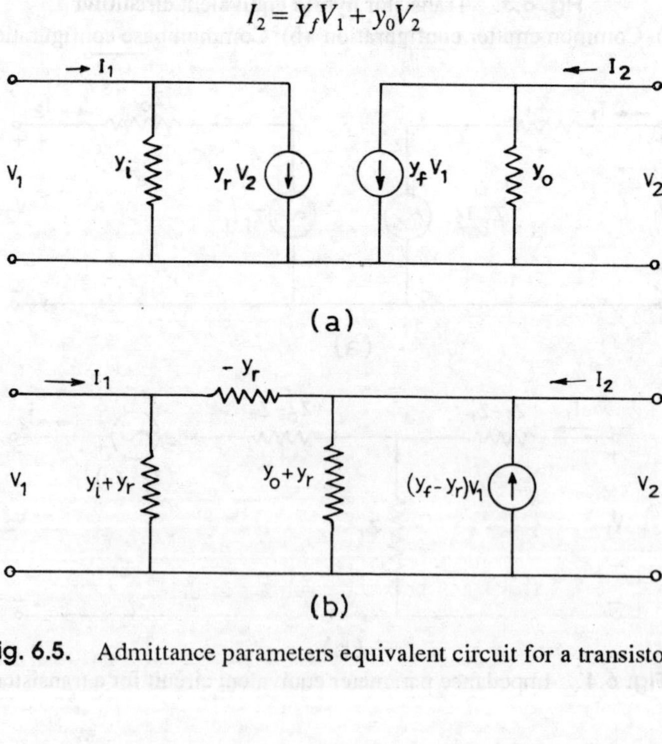

Fig. 6.5. Admittance parameters equivalent circuit for a transistor.

These equations represent the input and the output sections of the transistor. The transistor equivalent circuit based upon these equations is shown in Fig. 6.5(a), using two current generator $Y_r V_2$ and $Y_f V_1$. Eqution 6.21 can also be written as (subtracting $Y_r V_1$ on both the sides),

$$I_2 - Y_r V_1 = Y_f V_1 + Y_0 V_2 - Y_r V_1$$

or

$$I_2 - (Y_f - Y_r) V_1 = Y_r V_1 + Y_0 V_2 \tag{6.22}$$

The other form of Y-equivalent circuit can be developed by using Eqs 6.20 and 6.22 and is shown in Fig. 6.5(b). This involves only one current generator with a π network.

6.5 TRANSCONDUCTANCE CIRCUIT MODEL

The hybrid equivalent circuit for a transistor in common emitter configuration was developed in Fig. 6.3(a). The voltage generator $h_{re} V_{ce}$ is the feedback source, which makes the input section of the circuit dependent on the output. Therefore, this effect is made small in the manufacturing and h_{re} can be taken to be zero. The values of h_{oe} are in the order of 10^{-5} and hence $1/h_{oe} \sim 10^5$ ohms can be dropped from the circuit (high resistance in parallel can be dropped). Thus, the circuit of Fig. 6.3(a) becomes simplified and is shown in Fig. 6.6(a).

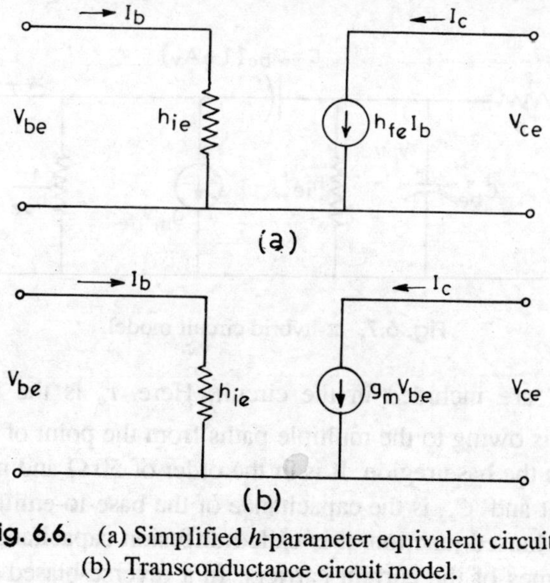

Fig. 6.6. (a) Simplified h-parameter equivalent circuit
(b) Transconductance circuit model.

The transconductance g_m is defined as the ratio of the output current and the input voltage, i.e

$$g_m = \frac{I_C}{V_{be}}$$

or $$I_C = g_m V_{be} \tag{6.23}$$

From Fig. 6.6(a)

$$I_C = h_{fe} I_b \tag{6.24}$$

Hence $$g_m V_{be} = h_{fe} I_b \tag{6.25}$$

Equation 6.25 shows that the current source $h_{fe}I_b$ in the circuit can be replaced by another current source $g_m V_{be}$. This equivalent circuit having $g_m v_{be}$ as the current source is shown in Fig. 6.6(b) and is known as the transconductance model.

6.6 HIGH FREQUENCY π-HYBRID CIRCUIT MODEL

The hybrid circuit model described in Sec. 6.2 is valid only in the low frequency range. For the applications in the high frequency range (above a few hundred kilohertz), the modified hybrid model is known as π-hybrid model and is depicted in Fig. 6.7, h_{re} being assumed to be zero. Additional circuit elements

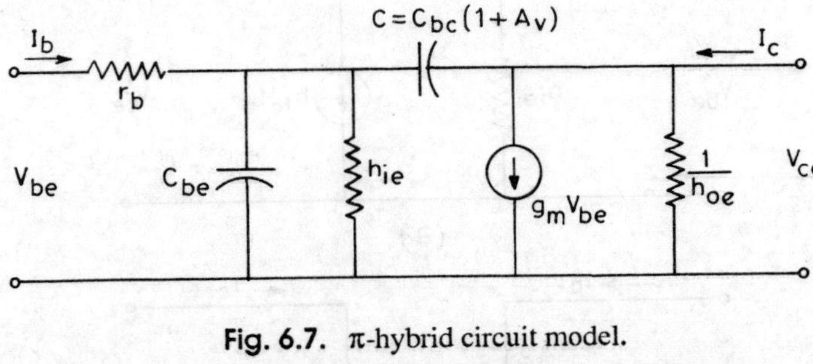

Fig. 6.7. π-hybrid circuit model.

r_b, C_{be}, and C are included in the circuit. Here, r_b is the base spreading resistance and is owing to the multiple paths from the point of base contact to all the point on the base region. It is in the order of 50 Ω and may be dropped from the circuit and C_{be} is the capacitance of the base-to-emitter junction and in the forward-biased junction, it is called diffusion capacitance and is owing to the delay times of the current carriers. In a reverse-biased junction, C_{be} is called the depletion capacitance and is owing to the depletion region. The capacitance C connected between the input and the output of the transistor is equivalent to $C_{bc}(1 + A_v)$ and is known as the Miller capacitance, where C_{bc} is the base-collector capacitance and A_v is the voltage gain of the transistor.

6.7 VOLTAGE AMPLIFIER CIRCUIT MODEL

Sometimes, the voltage source in the output of the Z-parameter equivalent circuit is expressed in the form of the voltage gain of the transistor. For the voltage amplifier circuit model, the equivalent circuit of Fig. 6.4(a) is used, in which the feedback voltage $Z_r I_2$ is ignored, as explained in Sec 6.5. Assuming Z_o to be very small, so that it is also dropped from the circuit (small resistance in series can be ignored). The voltage source can be expressed as,

$$Z_f\, I_1 = \frac{V_2}{I_1}\cdot I_1 = \frac{V_2}{V_1}\cdot V_1 = A_v \cdot V_1 \qquad (6.26)$$

where A_v is the voltage gain of the transistor. In the common emitter configuration, equivalent circuit is shown in Fig. 6.8(a)

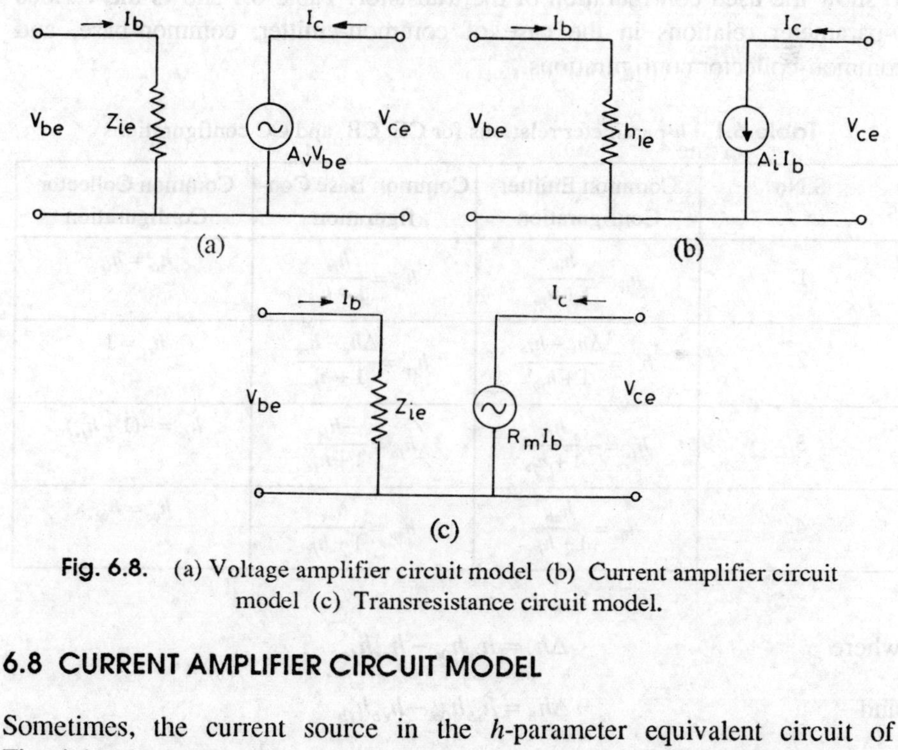

Fig. 6.8. (a) Voltage amplifier circuit model (b) Current amplifier circuit model (c) Transresistance circuit model.

6.8 CURRENT AMPLIFIER CIRCUIT MODEL

Sometimes, the current source in the h-parameter equivalent circuit of Fig. 6.6(a) is expressed in the form of the current gain of the transistor. Therefore, the current source $h_{fe}I_b$ is expressed as $A_i I_b$ since $h_{fe} = A_i$, the current gain. The equivalent circuit is shown in Fig. 6.8(b).

6.9 TRANSRESISTANCE AMPLIFIER CIRCUIT MODEL

If the voltage source in the Z-equivalent circuit is expressed in the form of transresistance $R_m = \frac{V_2}{I_1}$, Eq. 6.26 becomes

$$Z_f I_1 = V_2 = \frac{V_2}{I_1} \cdot I_1 = R_m I_1 \tag{6.27}$$

The transresistance equivalent circuit is shown in Fig. 6.8(c) for a common emitter configuration

6.10 HYBRID PARAMETERS FOR CE, CB, AND CC CONFIGURATIONS

As explained in Sec 6.2, a second subscript is used to the *h*-parameter notation to show the used configuration of the transistor. Table 6.1 shows the various *h*-parameter relations in the case of common-emitter, common-base, and common-collector configurations.

Table 6.1 *h*-parameter relations for CE, CB, and CC configurations

S.No	Common Emitter Configuration	Common Base Configuration	Common Collector Configuration
1.	$h_{ie} = \frac{h_{ib}}{1+h_{fb}}$	$h_{ib} = \frac{h_{ie}}{1+h_{fe}}$	$h_{ic} = h_{ie}$
2.	$h_{re} = \frac{\Delta h_b - h_{rb}}{1+h_{fb}}$	$h_{rb} = \frac{\Delta h_e - h_{re}}{1+h_{fe}}$	$h_{rc} = 1$
3.	$h_{fe} = -\frac{h_{fb}}{1+h_{fb}}$	$h_{fb} = \frac{-h_{fe}}{1+h_{fe}}$	$h_{fc} = -(1+h_{fe})$
4.	$h_{0e} = \frac{h_{ob}}{1+h_{fb}}$	$h_{0b} = \frac{h_{oe}}{1+h_{fe}}$	$h_{0c} = h_{0e}$

where $\Delta h_e = h_{ie} h_{0e} - h_{re} h_{fe}$

and $\Delta h_b = h_{ib} h_{0b} - h_{rb} h_{fb}$

6.11 FET CIRCUIT MODEL

The field effect transistors are voltage-operated devices, whereas the BJT is a current-operated device. The voltage operated device functions by controlling the current with the help of the electrostatic field. Because the FET has a high output impedance, its circuit model has a current generator coupled to an output impedance, as shown in Fig. 6.9(a). The input side of the model is isolated from the output side.

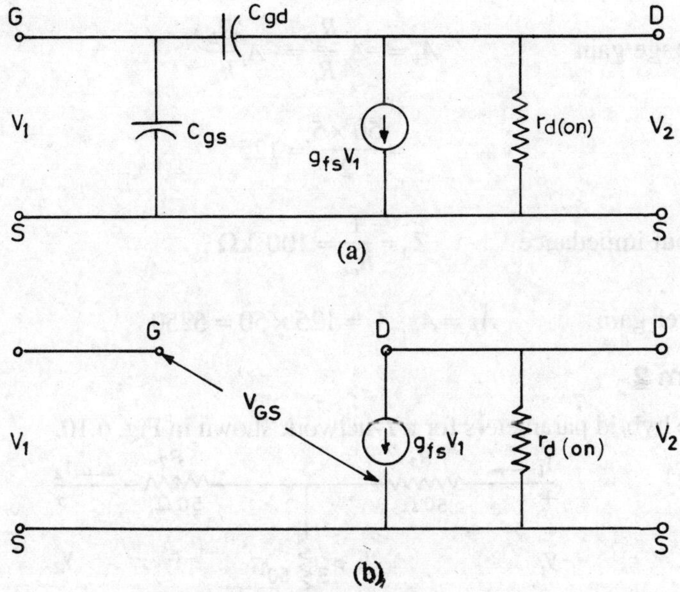

Fig. 6.9. (a) FET equivalent circuit
(b) Simplified form of the circuit.

Because of the high impedance associated with the FET the capacitances between the various elements of the FET become relatively important at the frequencies where they are generally used. The capacitor C_{gs} and C_{gd} represent the gate-to-source and the gate-to-drain capacities respectively. The output resistance from source-to-drain at the point of operation of the transistor is $r_{d(on)}$. At low frequencies, C_{gs} shows high reactance and can be ignored. Similarly, C_{gd} is also dropped at low frequencies, making the gate and the drain isolated. This simplified circuit is shown in Fig. 6.9(b).

6.12 SOLVED PROBLEMS

Problem 1

Calculate the current gain, voltage gain, input and output impedances, and power gain for a transistor connected in CE configuration having $h_{ie} = 2$ kΩ, $h_{re} = 2 \times 10^{-4}$, $h_{fe} = 50$, $h_{oe} = 10^{-5}$ mhos, $R_B = 75$ kΩ, and $R_L = 5$ kΩ.

Solution

According to the definitions of hybrid parameters, as shown in Eqs. 6.7 and 6.8, the current gain

$$A_i = h_{fe} = 50$$

the input impedance $\quad Z_i = h_{ie} = 2\ \text{k}\Omega$

the voltage gain $\quad A_v = -A_i \dfrac{R_L}{R_i} = -A_i \dfrac{R_L}{h_{ie}}$

$$= \frac{50 \times 5}{2} = 125$$

the output impedance $\quad Z_0 = \dfrac{1}{h_{oe}} = 100\ \text{k}\Omega$

the power gain $\quad A_P = A_v \cdot A_i = 125 \times 50 = 6250$

Problem 2

Find the hybrid parameters for a *T*-network shown in Fig. 6.10.

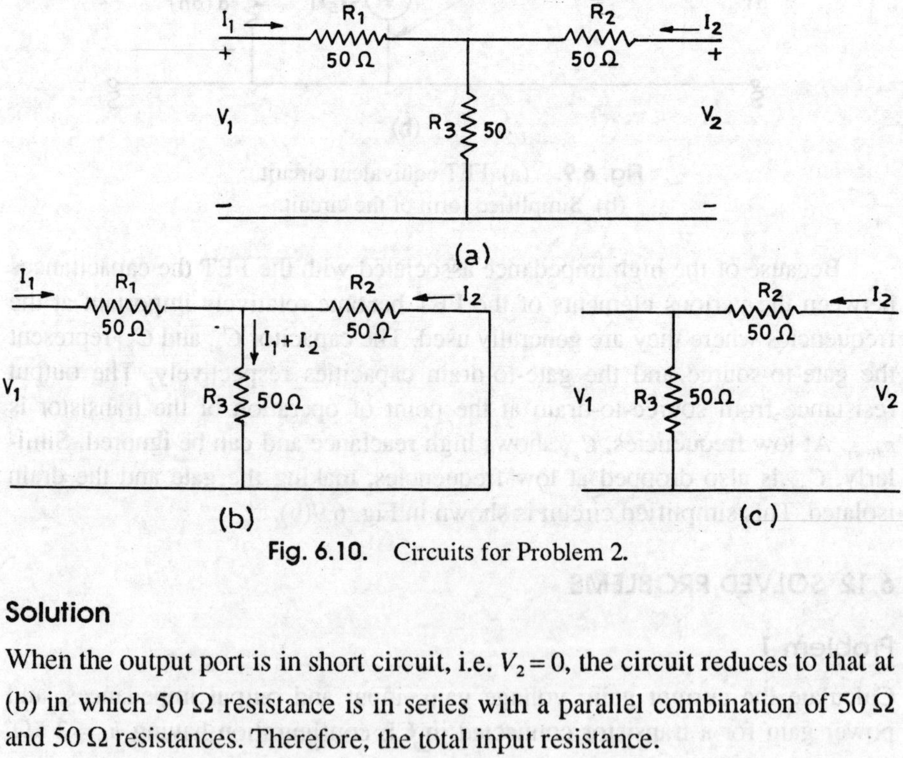

Fig. 6.10. Circuits for Problem 2.

Solution

When the output port is in short circuit, i.e. $V_2 = 0$, the circuit reduces to that at (b) in which 50 Ω resistance is in series with a parallel combination of 50 Ω and 50 Ω resistances. Therefore, the total input resistance.

$$h_i = 50 + \frac{50 \times 50}{50 + 50} = 75\ \Omega$$

As R_2 and R_3 are in parallel combination, the voltage across R_2 = voltage across R_3

or $$-50I_2 = 50(I_1 + I_2)$$

or $$-100I_2 = 50I_1$$

or $$\frac{I_2}{I_1} = h_f = -0.5$$

When the input port is in open circuit, $I_1 = 0$ and the circuit reduces to that at Fig. 6.10 (c) in which R_2 and R_3 are in series. Hence

$$V_2 = 50I_2 + 50I_2 = 100I_2$$

and $$V_1 = 50I_2$$

$\therefore$ $$h_r = \frac{V_1}{V_2} = 0.5$$

and $$h_0 = \frac{I_2}{V_2} = \frac{1}{100} = 0.01 \text{ mhos}$$

Problem 3

Find the h-parameters for a circuit shown in Fig. 6.11.

Solution

When the output port is in short circuit, $V_2 = 0$ and the circuit is reduced to that at (b) in which R_1 and R_2 are in parallel combination and hence

$$h_i = \frac{100 \times 100}{100 + 100} = 50\ \Omega$$

and voltage across R_1 = voltage across R_2

or $$-100I_2 = 100(I_1 + I_2)$$

or $$h = \frac{I_2}{I_1} = -0.5$$

When the input port is in open circuit, $I_1 = 0$ and the circuit reduces to that at (c) in which R_3 is in parallel with $(R_1 + R_2)$. The voltage across $(R_1 + R_2)$ is V_2 and the voltage across R_2 is V_1. Hence

$$V_1 = IR_2 = \frac{V_2}{R_1 + R_2} \cdot R_2$$

$\therefore$ $$h_r = \frac{V_1}{V_2} = \frac{R_2}{R_1 + R_2} = \frac{100}{200} = 0.5$$

From the circuit at Fig. 6.11(c),

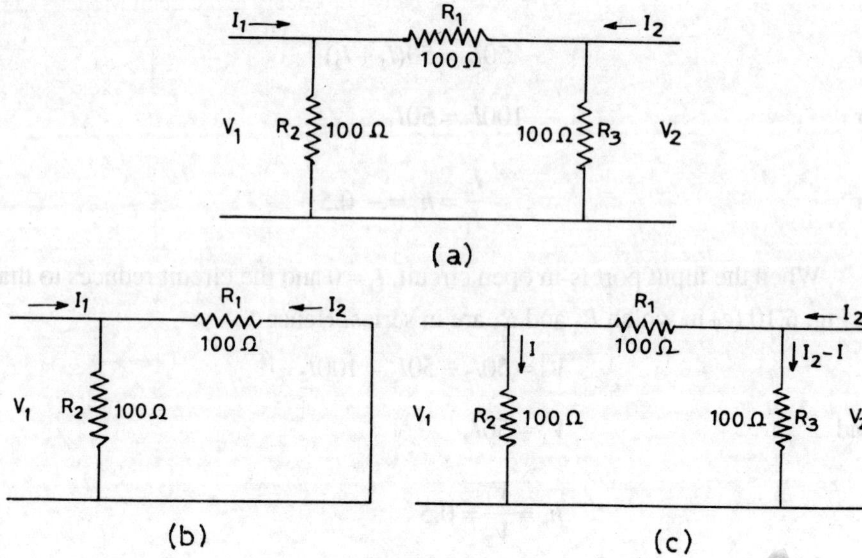

Fig. 6.11. Circuit for Problem 3.

$$V_2 = I(R_1 + R_2) = (I_2 - I)R_3$$

or

$$I(R_1 + R_2 + R_3) = I_2R_3$$

or

$$\frac{V_2}{R_1 + R_2}(R_1 + R_2 + R_3) = I_2R_3$$

$\therefore$

$$h_0 = \frac{I_2}{V_2} = \frac{R_1 + R_2 + R_3}{(R_1 + R_2)R_3}$$

$$= \frac{300}{20000} = 0.015 \text{ mhos}$$

Problem 4

Find the voltage gain, current gain, input and output impedances for the amplifier circuit of Fig. 6.12. The transistor has $h_{ie} = 2$ k and $h_{fe} = 50$.

Solution

The equivalent circuit is shown in Fig. 6.12 (b) in which

$$R_B = \frac{100 \times 1}{100 + 1} \cong 1 \text{ k}\Omega$$

Using current division at the input port,

$$I_b = \frac{1 \times 10^3 \cdot I_1}{(1+2) \times 10^3} = \frac{I_1}{3}$$

Using current division at the output port,

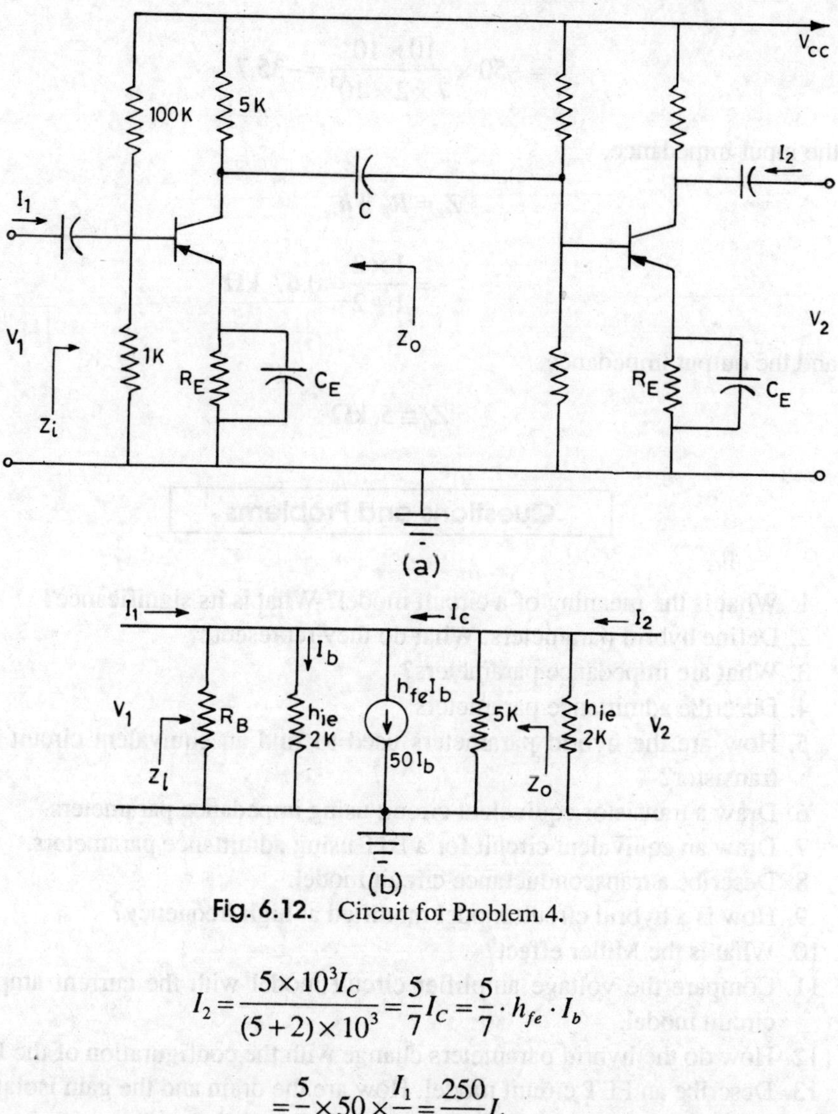

Fig. 6.12. Circuit for Problem 4.

$$I_2 = \frac{5\times 10^3 I_C}{(5+2)\times 10^3} = \frac{5}{7} I_C = \frac{5}{7}\cdot h_{fe}\cdot I_b$$

$$= \frac{5}{7}\times 50\times\frac{I_1}{3} = \frac{250}{21} I_1$$

$\therefore$ Current gain of the amplifier,

$$A_i = \frac{I_2}{I_1} = \frac{250}{21} \cong 12$$

The voltage gain of the amplifier,

$$A_v = -\,h_{fe}\cdot\frac{R_L}{h_{ie}}$$

$$= -50 \times \frac{10 \times 10^3}{7 \times 2 \times 10^3} = -35.7$$

the input impedance,

$$Z_i = R_B \parallel h_{ie}$$

$$= \frac{1 \times 2}{1 + 2} = 0.67 \text{ k}\Omega$$

and the output impedance,

$$Z_0 \cong 5 \text{ k}\Omega$$

Questions and Problems

1. What is the meaning of a circuit model? What is its significance?
2. Define hybrid parameters. What do they represent?
3. What are impedance parameters?
4. Describe admittance parameters.
5. How are the hybrid parameters used to find an equivalent circuit for a transistor?
6. Draw a transistor equivalent circuit using impedance parameters.
7. Draw an equivalent circuit for a BJT using admittance parameters.
8. Describe a transconductance circuit model.
9. How is a hybrid circuit model modified at high frequency?
10. What is the Miller effect?
11. Compare the voltage amplifier circuit model with the current amplifier circuit model.
12. How do the hybrid parameters change with the configuration of the BJT?
13. Describe an FET circuit model. How are the drain and the gain isolated?
14. Can you suggest many circuit models for one transistor?
15. Describe a transresistance amplifier circuit model.
16. What are the units of various h-parameters?
17. Draw a circuit diagram to measure the h-parameters in the laboratory.
18. Find the various characteristic parameters of the CE amplifier if, $h_{ie} = 1.5\text{k}\Omega, h_{re} = 2 \times 10^{-4}, h_{fe} = 75, h_{oe} = 10 \times 10^{-6}$mhos, $R_B = 50$ kΩ, and $R_L = 2$ kΩ.
19. Find the hybrid parameters of a T-network with resistance of 100 Ω in each arm.
20. Find the h-parameters of a π-network with resistance of 200 Ω in each arm.

7

Small Signal Basic Amplifiers

An amplifier is the most popular circuit in electronics. It is a network of circuits working together around a transistor for the purpose of strengthening an electrical signal. Put differently, the function of an amplifier is to accept very small signals at its input and amplify them to be used to perform some given function. When a transistor is used as an amplifier, it has various properties, viz., voltage gain, current gain, input impedance, output impedance, power gain, and phase relationships. In order to describe an amplifier, the above mentioned properties have to be discussed. This chapter deals with the various basic BJT and FET amplifying circuits used for small signal operations in the low frequency range.

7.1 BASIC BJT AMPLIFIER

A basic transistor amplifier is shown in Fig. 7.1(a). As the configuration (CE, CB, or CC) of the transistor is not specified in the circuit, no second subscript is used in the *h*-parameters. The resistance R_s is the total resistance of the signal source and any other resistance in series with the source. The load impedance is shown as Z_L. Figure 7.1(b) shows the *h*-parameter equivalent circuit for the transistor. R_s is dropped from the circuit as it is assumed to be very small.

7.1.1 Current Gain A_i

Writing the Kirchoff's current equation for the output section of the circuit,

$$I_2 = h_f I_1 + h_0 V_2 \tag{7.1}$$

From the circuit,

$$V_2 = -I_2 Z_L \tag{7.2}$$

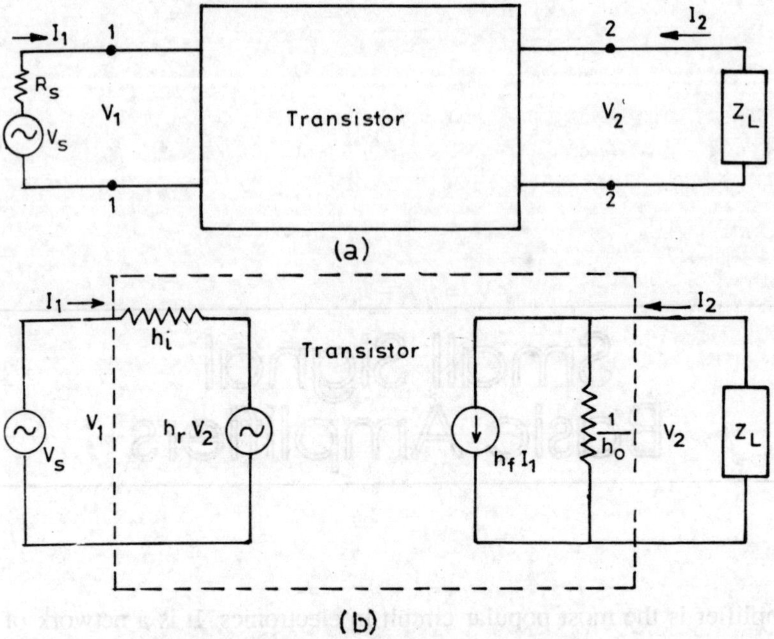

Fig. 7.1. (a) Basic transistor amplifier (b) *h*-Parameter equivalent circuit, replaced for the transistor.

Then, $$I_2 = h_f I_1 - h_o I_2 Z_L$$

or $$I_2(1 + h_O Z_L) = h_f I_1$$

or $$\frac{I_2}{I_1} = A_i = \frac{h_f}{1 + h_O Z_L} \tag{7.3}$$

7.1.2 Voltage Gain A_v

Writing the Kirchoff's voltage equation for the input section of the circuit,

$$V_1 = h_i I_1 + h_r V_2 \tag{7.4}$$

Using the value of I_1 from Eq. 7.3 and of I_2 from Eq. 7.2,

$$V_1 = h_i \frac{(1 + h_O Z_L)}{h_f}\left(-\frac{V_2}{Z_L}\right) + h_r V_2$$

$$= -\frac{h_i(1 + h_O Z_L)V_2}{h_f \cdot Z_L} + h_r V_2$$

$$= V_2\left[\frac{h_i + h_i h_O Z_L - h_r h_f Z_L}{-h_f Z_L}\right]$$

or $$\frac{V_2}{V_1} = A_v = \frac{-h_f Z_L}{h_i + (h_i h_o - h_r h_f) Z_L} \tag{7.5}$$

7.1.3 Input Impedance Z_i

From Eq. 7.4, $$V_1 = h_i I_1 + h_r V_2$$

and from Eq. 7.2, $$V_2 = -I_2 Z_L$$

Hence $$V_1 = h_i I_1 - h_r I_2 Z_L$$

$$= h_i I_1 - h_r Z_L \cdot I_1 \cdot \frac{I_2}{I_1}$$

$$= I_1 \left(h_i - h_r Z_L \cdot \frac{I_2}{I_1} \right)$$

Substituting $\frac{I_2}{I_1} = \frac{h_f}{1 + h_o Z_L}$ from Eq. 7.3,

$$V_1 = I_1 \left(h_i - h_r Z_L \cdot \frac{h_f}{1 + h_o Z_L} \right)$$

or $$\frac{V_1}{I_1} = Z_i = h_i - \frac{h_r h_f Z_L}{1 + h_o Z_L} \tag{7.6}$$

7.1.4 Output Impedance Z_o

When signal source voltage $V_s = 0$,

$$I_1 = \frac{-h_r V_2}{h_i} \tag{7.7}$$

Here, the minus sign indicates that the direction of current I_1 owing to the voltage generator $h_r V_2$ is opposite to the direction shown in the circuit. Substituting the value of I_1 from Eq. 7.7 in Eq. 7.1,

$$I_2 = h_f I_1 + h_r V_2$$

$$= \frac{-h_f h_r V_2}{h_i} + h_o V_2$$

or $$\frac{V_2}{I_2} = Z_o = \frac{1}{h_o - (h_f h_r / h_i)} \tag{7.8}$$

7.1.5 Output Admittance Y_0

From Eq. 7.8,

$$Y_0 = h_0 - \frac{h_f h_r}{h_i} \tag{7.9}$$

7.1.6 Power Gain A_p

The power gain is defined as the product of voltage gain and the current gain, i.e.,

$$A_p = A_v \cdot A_i$$

Using Eqs. 7.3 and 7.5,

$$A_p = \frac{-h_f^2 Z_L}{(1 + h_0 Z_L)\,[h_i + (h_i h_0 - h_r h_f) Z_L]} \tag{7.10}$$

7.2 COMMON EMITTER BJT AMPLIFIER

The circuit for common emitter BJT amplifier is shown in Fig. 7.2(a). The various parameters of common emitter amplifier are expressed using Eqs 7.3, 7.5, 7.6, 7.8, and 7.10 as

$$A_{ie} = \frac{h_{fe}}{1 + h_{oe} Z_L}$$

$$A_{ve} = \frac{-h_{fe} Z_L}{h_{ie} + (h_{ie} h_{oe} - h_{re} h_{fe}) Z_L}$$

$$Z_{ie} = h_{ie} - \frac{h_{fe} h_{re} Z_L}{1 + h_{oe} Z_L}$$

$$Z_{oe} = \frac{1}{h_{oe} - \frac{h_{fe} h_{re}}{h_{ie}}}$$

$$A_{pe} = \frac{-h_{fe}^2 Z_L}{(1 + h_{oe} Z_L)\,[h_{ie} + (h_{ie} h_{oe} - h_{fe} h_{re}) Z_L)]}$$

Assuming h_{re} and h_{0e} to be zero, as discussed in Sec. 6.5 of Chapter 6, the above equations become,

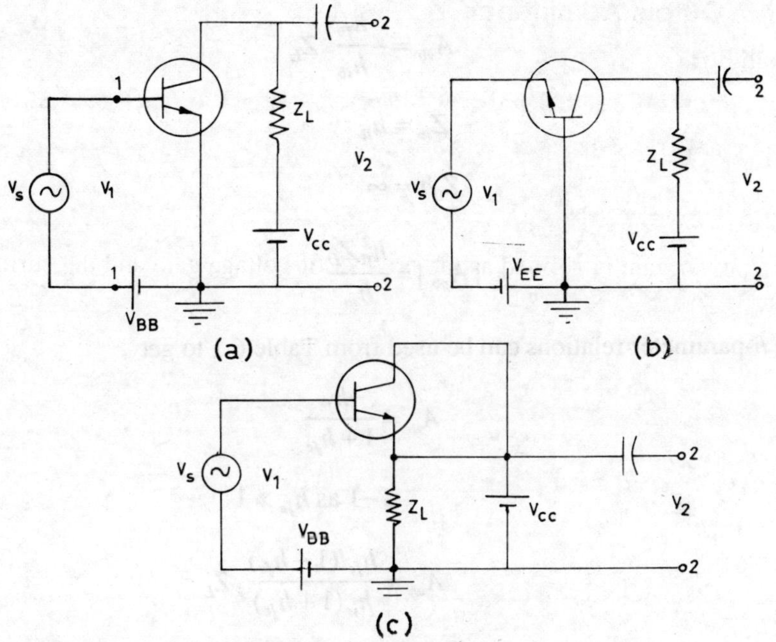

Fig. 7.2. (a) CE amplifier (b) CB amplifier (c) CC amplifier.

$$A_{ie} = h_{fe} \tag{7.11}$$

$$A_{ve} = -\frac{h_{fe}}{h_{ie}} Z_L = -g_m Z_L \tag{7.12}$$

$$Z_{ie} = h_{ie} \tag{7.13}$$

$$Z_0 e = \infty \tag{7.14}$$

$$A_{pe} = -\frac{h_{fe}^2 Z_L}{h_{ie}}$$

$$= \frac{h_{fe}^2 Z_L}{h_{ie}} \text{ in magnitude} \tag{7.15}$$

7.3 COMMON BASE BJT AMPLIFIER

The circuit for common base amplifier is shown in Fig. 7.2(b). Using Eqs. 7.11, 7.12, 7.13, 7.14 and 7.15, the various parameters of the amplifier are,

$$A_{ib} = h_{fb}$$

$$A_{vb} = -\frac{h_{fb}}{h_{ib}} \cdot Z_L$$

$$Z_{ib} = h_{ib}$$

$$Z_0 b = \infty$$

$$|A_{pb}| = \frac{h_{fb}^2 Z_L}{h_{ib}}$$

The h-parameter relations can be used from Table 6.1 to get

$$A_{ib} = \frac{-h_{fe}}{1+h_{fe}}$$

$$\cong -1 \text{ as } h_{fe} \gg 1 \tag{7.16}$$

$$A_{vb} = \frac{h_{fe}/(1+h_{fe})}{h_{ie}(1+h_{fe})} \cdot Z_L$$

$$= \frac{h_{fe}}{h_{ie}} \cdot Z_L$$

$$= g_m Z_L \tag{7.17}$$

$$Z_{ib} = h_{ib} = \frac{h_{ie}}{1+h_{fe}}$$

$$= \frac{h_{ie}}{h_{fe}} \text{ as } h_{fe} \gg 1$$

$$= \frac{1}{g_m} \tag{7.18}$$

$$Z_{ob} = \infty \tag{7.19}$$

$$A_{pb} = A_{ib} \cdot A_{vb}$$

$$= g_m' Z_L \tag{7.20}$$

7.4 COMMON COLLECTOR BJT AMPLIFIER

The circuit for common collector transistor amplifier is shown in Fig. 7.2(c). The Kirchoff's voltage equation for the input section is written as

$$V_1 = V_{be} + V_2$$

As V_{be} is an extremely small value,

$$V_1 \cong V_2 \tag{7.21}$$

This equation shows that the output voltage is almost equal to the input voltage. The output voltage follows the input voltage and therefore, it is also called the emitter follower circuit.

The open circuit reverse transfer voltage gain is defined using Eq.7.21, as

$$h_{rc} = \frac{V_1}{V_2} \cong 1 \tag{7.22}$$

This shows that in the case of common collector amplifier h_r is not zero but equals 1. Therefore, applying Eqs.7.3, 7.5, 7.6, 7.8 and 7.10, the various parameters can be written as

$$A_{ic} = \frac{h_{fc}}{1 + h_{oc}Z_L}$$

$$A_{vc} = \frac{-h_{fc}Z_L}{h_{ic} + (h_{ic}h_{oc} - h_{rc}h_{fc})Z_L}$$

$$Z_{ic} = h_{ic} - \frac{h_{fc}h_{rc}Z_L}{1 + h_{oc}Z_L}$$

$$Z_{oc} = \frac{1}{h_{oc} - \frac{h_{fc}h_{rc}}{h_{ic}}}$$

$$A_{pc} = \frac{-h_{fc}^2 Z_L}{(1 + h_{oc}Z_L)[h_{ic} + (h_{ic}h_{oc} - h_{rc}h_{fc})Z_L]}$$

The h-parameter relations from Table 6.1 are used and also $h_{oc} \cong 0$ and $h_{rc} \cong 1$ are assumed to get,

$$A_{ic} = h_{fc} \cong -(1 + h_{fe})$$

$$\cong -h_{fe} \text{ as } h_{fe} \gg 1 \tag{7.23}$$

$$A_{vc} = \frac{-h_{fc}Z_L}{h_{ic} - h_{fc}Z_L}$$

$$= \frac{(1 + h_{fe})Z_L}{h_{ie} + (1 + h_{fe})Z_L}$$

$$\cong \frac{h_{fe}Z_L}{h_{ie} + h_{fe}Z_L} \text{ as } h_{fe} \gg 1$$

$$\cong \frac{h_{fe}/h_{ie} \cdot Z_L}{1 + (h_{fe}/h_{ie}) \cdot Z_L}$$

$$= \frac{g_m Z_L}{1 + g_m Z_L} \tag{7.24}$$

$$Z_{ic} = h_{ic} - \frac{h_{fc}Z_L}{1}$$

$$\cong h_{ie} + (1 + h_{fe})Z_L$$

$$\cong h_{ie} + h_{fe}Z_L \text{ as } h_{fe} \gg 1$$

$$= h_{ie}(1 + g_m Z_L) \tag{7.25}$$

$$Z_{oe} = \frac{1}{-h_{fc}/h_{ic}} = -\frac{h_{ic}}{h_{fc}}$$

$$= \frac{h_{ie}}{1 + h_{fe}} \tag{7.26}$$

$$A_{pc} = \frac{-h_{fc}^2 Z_L}{h_{ic} - h_{fc} \cdot Z_L}$$

$$= \frac{-(1 + h_{fe})^2 Z_L}{h_{ie} + (1 + h_{fe})Z_L}$$

$$= \frac{-h_{fe}^2 Z_L}{h_{ie}\left(1 + \frac{h_{fe}}{h_{ie}} Z_L\right)} \quad \text{as} \quad h_{fe} \gg 1$$

$$= \frac{-h_{fe} g_m Z_L}{1 + g_m Z_L} \quad \text{as} \quad g_m = \frac{h_{fe}}{h_{ie}}$$

or

$$|A_{pc}| = \frac{h_{fe} g_m Z_L}{1 + g_m Z_L} \tag{7.27}$$

7.5 COMPARISON OF CE, CB, AND CC AMPLIFIERS

The various parameters obtained in the above sections for the common emitter, common base, and common collector BJT amplifiers are summarized in Table 7.1 to present a comparison of the performance of the BJT in the three configurations. The approximations followed are h_{oe}, h_{ob}, $h_{oc} \cong 0$, h_{re}, $h_{rb} = 0$, and $h_{rc} = 1$, as discussed in the text above

Table. 7.1 Summary of basic BJT amplifiers

Transistor Parameter	Common-emitter Configuration	Common-base Configuration	Common-collector Configuration
Current Gain A_i	h_{fe}	-1	$-h_{fe}$
Voltage Gain A_v	$-g_m Z_L$	$g_m Z_L$	$\frac{g_m Z_L}{1+g_m Z_L} \cong 1$
Input Impedance Z_i	h_{ie}	$\frac{1}{g_m}$	$h_{ie}(1+g_m Z_L)$
Output Impedance Z_o	High	High	$\frac{h_{ie}}{1+h_{fe}}$ (low)
Power Gain A_p	$\frac{h_{fe}^2 Z_L}{h_{ie}}$	$g_m Z_L$	$\frac{h_{fe} g_m Z_L}{1+g_m Z_L}$

7.6 APPLICATIONS OF CE AMPLIFIER

The CE configuration is generally used in transistor amplifiers because of its ability to provide high voltage, current, and power gains. Another important feature is that its input and output impedances are normally of the same order and therefore, more than one transistor can be used in cascade without using impedance matching transformers.

7.7 APPLICATIONS OF CB AMPLIFIERS

The CB amplifier shows low input impedance and the current gain is also in the range of one. Therefore, the CB amplifier is highly suitable for microphone amplifier where the input currents are very low. The frequency response of the CB amplifier is much larger than the CE amplifier; therefore, it is more suitable for the radiofrequency amplifiers.

7.8 APPLICATIONS OF CC AMPLIFIERS

The CC amplifier has unit voltage gain and the current gain is more than unity, hence it has power gain greater than one. As it has high input impedance and low output impedance, it is used as an impedance matching device with power gain, and no matching transformer is needed. The frequency response of CC amplifier is also extremely wide, so it can be used for higher frequencies.

7.9 BASIC FET AMPLIFIERS

A field effect transistor can be used in a circuit in three possible configurations, viz. common source, common drain, and common gate. The most popular circuits are the common source and the common drain configurations. The FET is biased to operate in the linear region.

7.9.1 Common Source Amplifier

The circuit for common source amplifier and its equivalent circuit are shown in Fig. 7.3. The FET biasing for the linear region of operation is provided by R_{G1} and R_{G2}. The capacitor C_s is used to bypass the source resistor R_s. The current generator is $g_{fs}.V_{gs}$ and the drain resistance is $r_{d(on)}$. The input resistance of the amplifier is the parallel combination of R_{G1} and R_{G2}, i.e.

$$R_i = \frac{R_{G1} \cdot R_{G2}}{R_{G1} + R_{G2}} \tag{7.28}$$

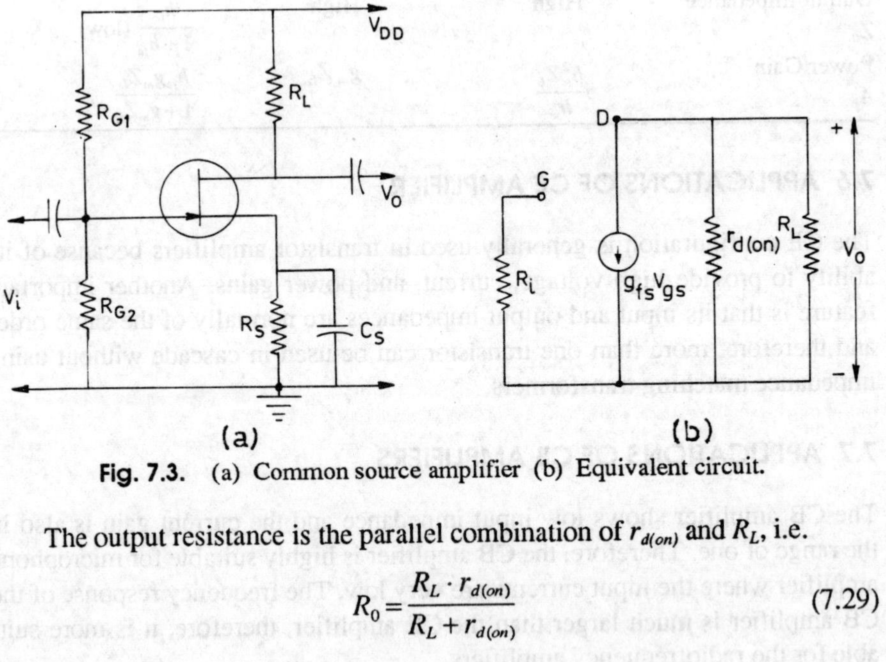

Fig. 7.3. (a) Common source amplifier (b) Equivalent circuit.

The output resistance is the parallel combination of $r_{d(on)}$ and R_L, i.e.

$$R_0 = \frac{R_L \cdot r_{d(on)}}{R_L + r_{d(on)}} \tag{7.29}$$

The output voltage is given by,

$$V_0 = -g_{fs} V_{gs} R_0$$

The minus sign indicates that the voltage V_0 is opposite in sign to the voltage produced by the current $g_{fs} V_{gs}$.

or
$$V_0 = -\frac{g_{fs} \cdot V_{gs} R_L r_{d(on)}}{R_L + r_{d(on)}} \tag{7.30}$$

The input voltage $V_i = V_{gs}$ (7.31)

Hence, the voltage gain

$$A_v = \frac{V_0}{V_i} = -\frac{g_{fs} \cdot R_L r_{d(on)}}{R_L + r_{d(on)}}$$

If the drain resistance is much greater than the load resistance,

$$A_v = -\, g_{fs} R_L \quad (7.32)$$

The minus sign shows that the input and the output voltages are 180° out of phase.

7.9.2 Common Drain Amplifier (Source Follower Circuit)

Figure 7.4 shows the common drain amplifier and its equivalent circuit. The output voltage is taken across the source resistor R_s. Since R_s and $r_{d(on)}$ are parallel, the output resistance is

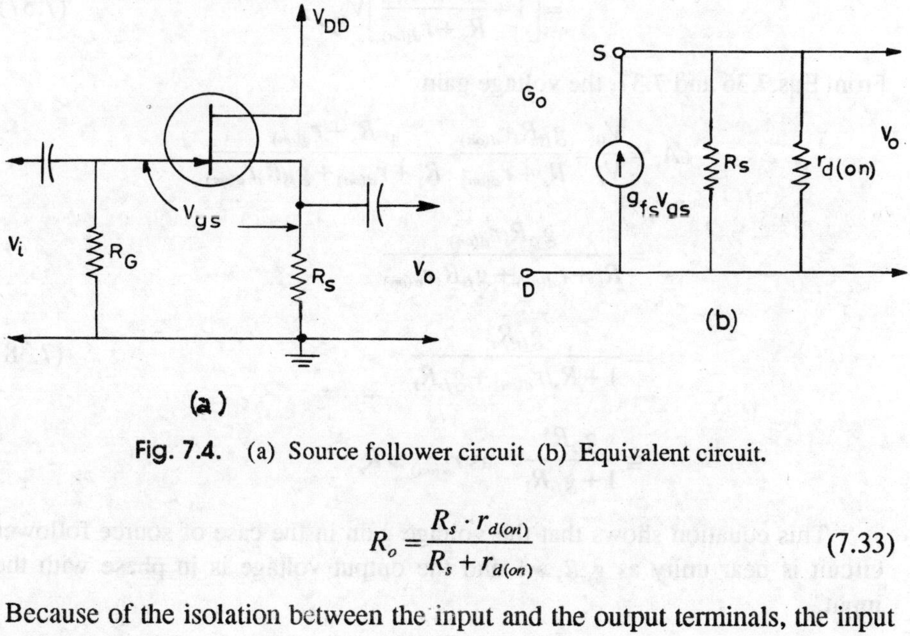

Fig. 7.4. (a) Source follower circuit (b) Equivalent circuit.

$$R_o = \frac{R_s \cdot r_{d(on)}}{R_s + r_{d(on)}} \quad (7.33)$$

Because of the isolation between the input and the output terminals, the input resistance is

$$R_{in} = R_G \quad (7.34)$$

From the input section of the circuit, the Kirchoff's voltage equation is

$$V_i = V_{gs} + V_0 \tag{7.35}$$

or
$$V_0 = V_i - V_{gs}$$

or
$$V_0 \cong V_i \quad \text{if } V_{gs} \text{ is small}$$

This shows that the output or the source voltage follows the input voltage and that is why the circuit is called the source follower circuit. From the circuit,

$$V_0 = g_{fs} \cdot V_{gs} \cdot R_0 = g_{fs} V_{gs} \cdot \frac{R_s r_{d(on)}}{R_s + r_{d(on)}} \tag{7.36}$$

Substituting this value of V_o in Eq.7.35,

$$V_i = V_{gs} + g_{fs} V_{gs} \cdot \frac{R_s r_{d(on)}}{R_s + r_{d(on)}}$$

$$= \left[1 + \frac{g_{fs} R_s r_{d(on)}}{R_s + r_{d(on)}}\right] V_{gs} \tag{7.37}$$

From Eqs.7.36 and 7.37, the voltage gain

$$A_v = \frac{V_0}{V_i} = \frac{g_{fs} R_s r_{d(on)}}{R_s + r_{d(on)}} \cdot \frac{R_s + r_{d(on)}}{R_s + r_{d(on)} + g_{fs} R_s r_{d(on)}}$$

$$= \frac{g_{fs} R_s r_{d(on)}}{R_s + r_{d(on)} + g_{fs} R_s r_{d(on)}}$$

$$= \frac{g_{fs} R_s}{1 + R_s / r_{d(on)} + g_{fs} R_s} \tag{7.38}$$

$$\cong \frac{g_{fs} R_s}{1 + g_{fs} R_s} \quad \text{as } r_{d(on)} \gg R_s$$

This equation shows that the voltage gain in the case of source follower circuit is near unity as $g_{fs} R_s \gg 1$ and the output voltage is in phase with the input.

7.10 SOLVED PROBLEMS

Problem 1

Find the values of the current gain, voltage gain, power gain, and input and output impedance for a BJT connected in CE configuration if $h_{fe} = 60$, $h_{ie} = 1\ \text{k}\Omega$, $h_{re} = 2 \times 10^{-4}$, $h_{oe} = 20 \times 10^{-6}$ mhos, and $Z_L = 2\ \text{k}\Omega$.

Solution

Using Eqs 7.11 to 7.15, we get current gain

$$A_{ie} = h_{fe} = 60$$

voltage gain $$A_{ve} = -\frac{h_{fe}}{h_{ie}} \cdot Z_L$$

$$= -\frac{60}{1} \times 2 = -120$$

input impedance $$Z_{ie} = h_{ie} = 1 \text{ k}\Omega$$

output impedance $$Z_{oe} = Z_L \parallel \frac{1}{h_{oe}} \cong Z_L = 2 \text{ k}\Omega$$

power gain $$P_{oe} = \frac{h_{fe}^2}{h_{ie}} \cdot Z_L$$

$$= \frac{60 \times 60}{1} \times 2 = 7.2 \times 10^3$$

Problem 2

Find the values of current gain, voltage gain, power gain , and input and output impedances for a BJT as in problem 1 connected as an emitter follower.

Solution

The emitter follower circuit is also called *common-collector amplifier*. Therefore, using Eqs.7.23–7.27, we get current gain

$$A_{ic} = -h_{fe} = -\ 60$$

voltage gain $$A_{vc} = \frac{(h_{fe}/h_{ie})Z_L}{1 + \frac{h_{fe}}{h_{ie}} \cdot Z_L}$$

$$= \frac{\frac{60 \times 2}{1}}{1 + \frac{60 \times 2}{1}} = \frac{120}{121} \cong 1$$

input impedance

$$Z_{ic} = h_{ie} + h_{fe} Z_L$$

$$= (1 + 60 \times 2) = 121 \text{ k}\Omega$$

output impedance $$Z_{oc} = \frac{h_{ie}}{1 + h_{fe}} = \frac{1}{1 + 60} = 16.4\ \Omega$$

power gain $$A_{pc} = \frac{-h_{fe}^2 Z_L}{h_{ie}\left(1+\frac{h_{fe}}{h_{ie}}Z_L\right)} = \frac{-60\times 60\times 2}{1+\left[1+\frac{60\times 2}{1}\right]}$$

$$= 59.5$$

Problem 3

Find the input impedance for the circuit of Fig.7.5(a). $h_{fe} = 75$

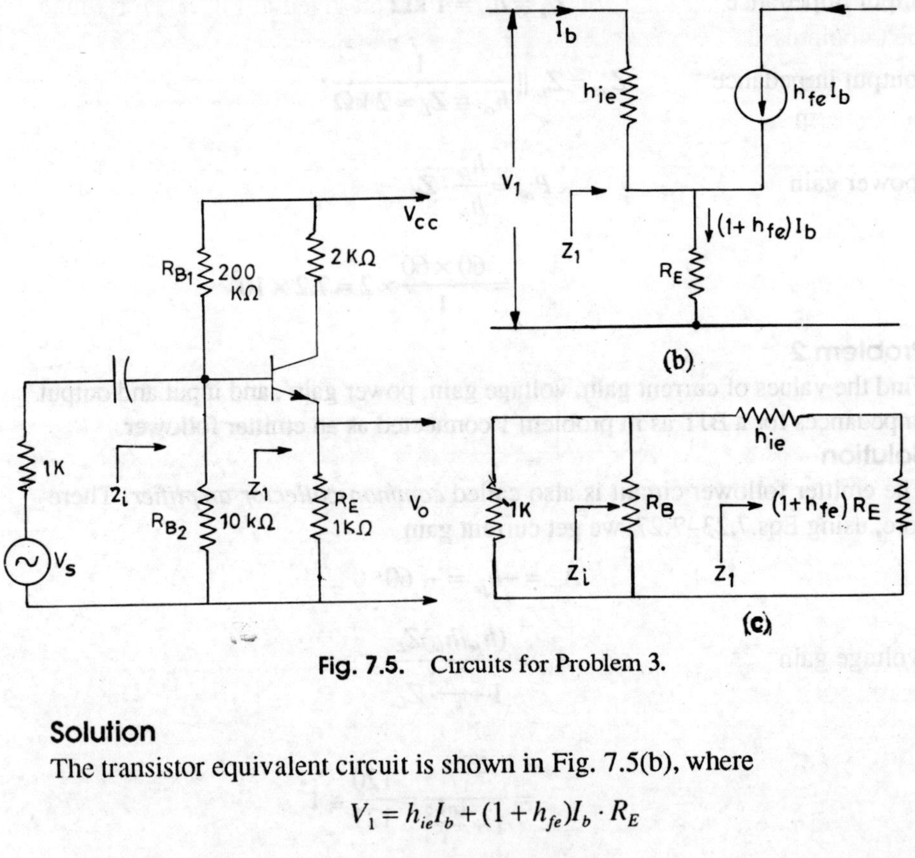

Fig. 7.5. Circuits for Problem 3.

Solution

The transistor equivalent circuit is shown in Fig. 7.5(b), where

$$V_1 = h_{ie}I_b + (1+h_{fe})I_b \cdot R_E$$

$$= I_b[h_{ie} + (1+h_{fe})R_E]$$

$$\therefore \quad \frac{V_1}{I_b} = Z_1 = h_{ie} + (1+h_{fe})R_E$$

$$= 1 + (1+75)\times 1 = 77 \text{ k}\Omega$$

From Fig. 7.5(c), the input impedance

$$Z_i = R_B \parallel Z_1 \text{ where } R_B = R_{B_1} \parallel R_{B_2}$$

$$= \frac{200 \times 10}{210}$$

$$\cong 10 \text{ k}\Omega$$

$$\therefore \qquad Z_i = \frac{10 \times 77}{10 + 77} = 8.85 \text{ k}\Omega$$

Problem 4

Find the voltage gain for a source follower circuit given in Fig. 7.6 for which the Q-point is at

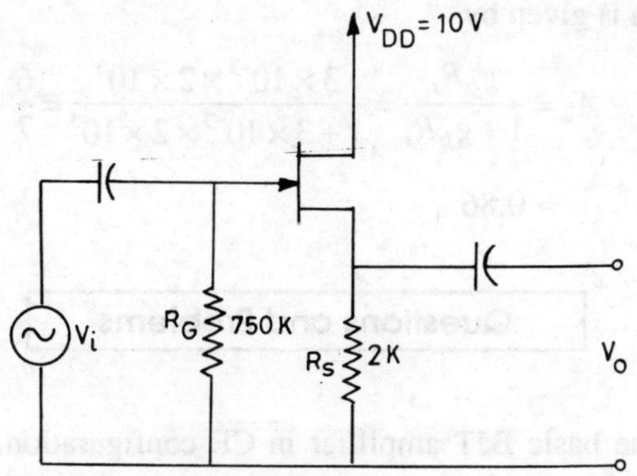

Fig. 7.6. Circuit for Problem 4

$$V_{GSQ} = -2.5 \text{ V}, \quad I_{DSQ} = 1.0 \text{ mA}$$

$$I_{DSS} = 15 \text{ mA}, \ V_p = -5 \text{ V}$$

Solution

Using the Schockley's equation above the pinch-off for the JFET, we get

$$I_D = I_{DSS}\left(1 - \frac{V_{Gs}}{V_p}\right)^2 \tag{7.39}$$

where I_{DSS} and V_p are the pinch off current and voltage respectively. The trans-conductance for JFET is defined as

$$g_{fs} = \left.\frac{\partial I_D}{\partial V_{GS}}\right|_{(V_{DS} \text{ is a constant})}$$

$\therefore$ Differentiating Eq. 7.39, we get

$$g_{fs} = -\frac{2I_{DSS}}{V_p}\left(1 - \frac{V_{GS}}{V_p}\right)$$

and at the Q-point,

$$g_{fs} = -\frac{2I_{DSS}}{V_p}\left(1 - \frac{V_{GSQ}}{V_p}\right)$$

$$= -\frac{2 \times 15 \times 10^{-3}}{-5}\left[1 - \frac{(-2.5)}{(-5)}\right]$$

$$= 6 \times 10^{-3}(1 - 0.5)$$

$$= 3 \times 10^{-3} \text{ mhos}$$

The voltage gain is given by

$$A_v = \frac{g_{fs}R_s}{1 + g_{fs}R_s} = \frac{3 \times 10^{-3} \times 2 \times 10^3}{1 + 3 \times 10^{-3} \times 2 \times 10^3} = \frac{6}{7}$$

$$= 0.86$$

Questions and Problems

1. Describe the basic BJT amplifier in CE configuration. Find the expressions for voltage gain, current gain, power gain, and input and output impedances.
2. Find A_i, A_v, Z_i, Z_O, and A_p for a common base BJT amplifier.
3. What is an emitter follower circuit? Describe it.
4. Make a comparative study of the CE, CB, and CC amplifier circuits.
5. Explain a common source FET amplifier.
6. Why is a common drain amplifier also called a source follower circuit? Explain its configuration and working.
7. Find the various circuit characteristics for a CE amplifier if $h_{ie} = 1$ kΩ, $h_{fe} = 75$, $h_{re} = 10 \times 10^{-4}$, $h_{Oe} = 2 \times 10^{-6}$, and $Z_L = 2.5$ kΩ.
8. In Problem 7, calculate the various circuit parameters for an emitter follower.
9. Repeat the solved Problem 3 with the following changes:
   $h_{ie} = 1.5$ k and $h_{fe} = 50$
10. Find the voltage gain and transconductance of the JFET used in the circuit of Fig.7.6 if, $I_{DSS} = 20$ mA and $R_s = 2.5$ kΩ.

8

Multistage Amplifiers

A single-stage amplifier, as discussed in Chapter 7, is limited in its gain. In several practical applications, more gain is required than can be derived from a single transistor stage. This gain can be power, voltage, current, or transconductance. In order to increase the gain, the multistage amplifiers are used, in which several transistors are connected together. The first transistor in the amplifier circuit increases the input signal to a specific value which in turn, is applied to the input section of the second transistor. Thus, the signal is increased further. In this way, two or more transistors can be coupled to increase the gain of the amplifier. There are various techniques commonly used to couple two transistor stages.

8.1 RESISTANCE-CAPACITANCE COUPLING

The most common method of coupling the two transistor stages of an amplifier is *RC* coupling, as shown in Fig. 8.1. The output of Q_1 is fed to the input of Q_2 through a dc blocking capacitor *C*. The advantage of *RC* coupling is that only the *ac* signals from the first stage are fed to the second stage, and the changes in the dc operating points of the different stages are not disturbed. As the capacitor *C* prevents the dc voltage components of the output of the first stage from appearing at the input of the second stage, the reactance of *C* must be small compared to the input resistance of the second transistor stage, otherwise a large signal voltage drop will take place across *C*. The disadvantage of *RC* coupling is its low efficiency (ratio of *ac* power output and dc power input), because the dissipation of dc power takes place in the load resistors.

The dc biasing for the first stage is established by R_{B_1}, R_{E_1} and R_{L_1}. Similarly, the dc biasing of the second stage is established by R_{B_2}, R_{E_2} and R_{L_2}. The voltage gain of the first stage, when isolated from the second stage, is approximately equal to R_{L_1}/R_{E_1}. Ignoring R_L, the gain of the second stage is $\sim R_{L_2}/R_{E_2}$.

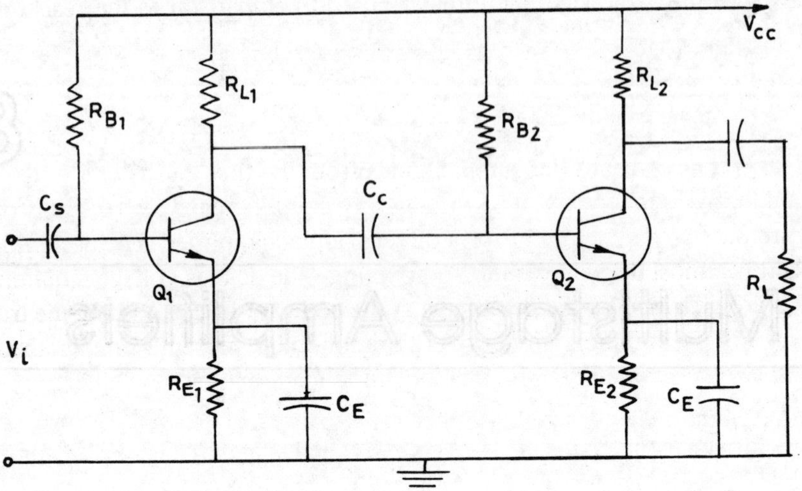

Fig. 8.1. *R-C* coupling in amplifier.

8.2 TRANSFORMER COUPLING

Instage coupling of two transistor stages by means of a transformer is shown in Fig. 8.2. The primary of the transformer acts as a load impedance of the first stage, while the secondary of the transformer feeds the ac signals to the second transistor. As there is no load resistance to dissipate power, the power efficiency of the transformer coupled amplifiers is the maximum. The use of the transformer is important because it provides the required impedance matching also between the two stages to facilitate the maximum power transfer from the first stage to the second. The disadvantages of the transformer coupling are its

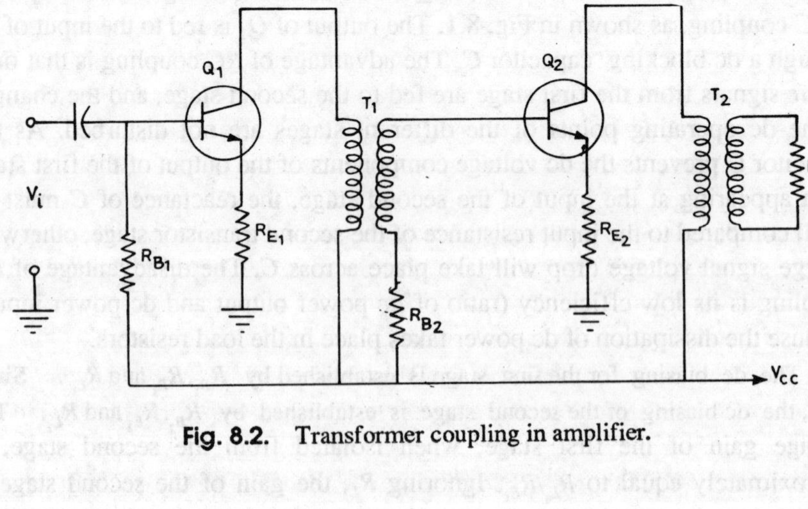

Fig. 8.2. Transformer coupling in amplifier.

heavy weight, large size, and greater cost in comparison to the resistors and capacitors.

8.3 DIRECT COUPLING

Direct coupling means the output signals from the first transistor are fed to the second transistor directly, i.e., without any coupling capacitor or transformer. Figure 8.3 shows the circuit for a direct coupled amplifier. The design of this circuit is almost the same as in *R-C* coupling. The dc bias voltage requirements of Q_2 are provided by the dc voltage at the collector of Q_1. As both the transistors are in the common emitter configuration in the present example, the input and the output voltages are in phase. The ac load seen by the collector of Q_1 is the parallel combination of R_{L_1} and the input resistance of the Q_2. The overall gain is the product of the gains of the two individual stages. The main advantage of the direct coupling is that it does not involve reactive coupling elements and it provides a very flat frequency vs gain response for the range of frequencies for which it is designed. The disadvantage of direct coupling is that each succeeding stage requires a higher source voltage than the preceding one. Also, the direct coupling suffers from poor efficiency owing to high power losses in the load resistors. These losses become even more if the number of stages is increased. The impedance matching is also a problem in this coupling.

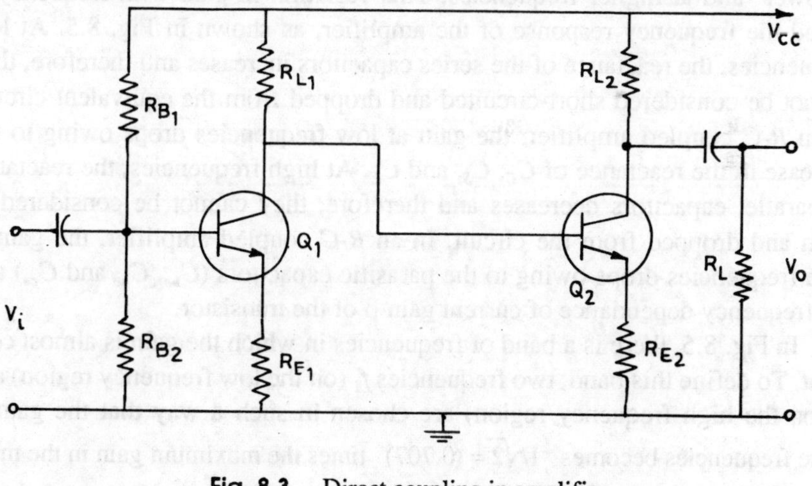

Fig. 8.3. Direct coupling in amplifier.

8.3.1 Darlington Pair

The most common example of the direct coupling is the Darlington pair shown in Fig. 8.4. In all respects, the two transistors behave as a single transistor with total β as the product of β of the two transistors. The voltage gain is approximately equal to one as both the transistors act as the emitter follower.

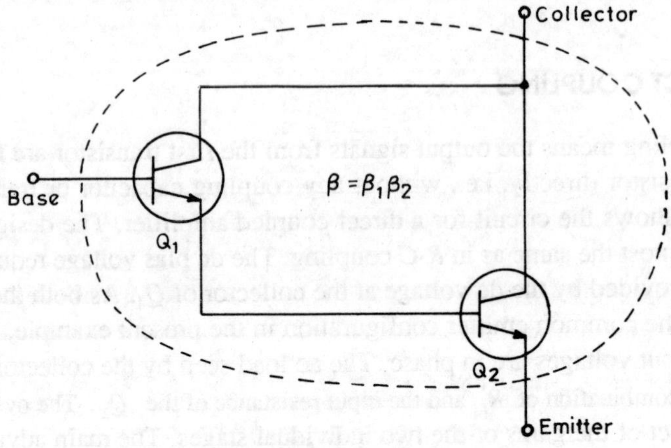

Fig. 8.4. Darlington pair.

8.4 FREQUENCY RESPONSE OF *R-C* COUPLED AMPLIFIER

If the frequency of the applied ac signal varies, the gain of the amplifier changes in both single as well as multistage amplifiers. In an *R-C* coupled amplifier, the gain remains constant at the mid-frequency range, while it drops at lower and at higher frequencies. This variation in gain with frequency is called the frequency response of the amplifier, as shown in Fig. 8.5. At low frequencies, the reactance of the series capacitors increases and therefore, they cannot be considered short-circuited and dropped from the equivalent circuit. In an *R-C* coupled amplifier, the gain at low frequencies drops owing to the increase in the reactance of C_C, C_S, and C_E. At high frequencies, the reactance of parallel capacitors decreases and therefore, they cannot be considered as open and dropped from the circuit. In an *R-C* coupled amplifier, the gain at high frequencies drops owing to the parasitic capacitors (C_{be}, C_{bc} and C_{ce}) and the frequency dependance of current gain β of the transistor.

In Fig. 8.5, there is a band of frequencies in which the gain is almost constant. To define this band, two frequencies f_1 (on the low frequency region) and f_2 (on the high frequency region) are chosen in such a way that the gain at these frequencies becomes $1/\sqrt{2} = (0.707)$ times the maximum gain in the mid-frequency range. These frequencies are generally called the corner, band, break, cut-off, or half-power frequencies. At these frequencies, the output power also becomes half the power at the mid-frequency. This can be understood as follows:

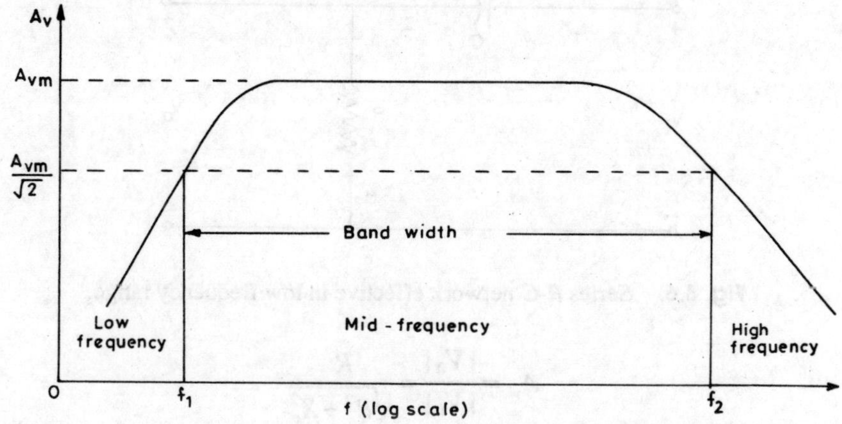

Fig. 8.5. Frequency response for *R-C* coupled amplifier

The output power in the mid range

$$P_{om} = \frac{V_0^2}{R_0} = \frac{|A_{vm} \cdot V_i|^2}{R_o}$$

and at the half-power frequencies

$$P_{ol} \text{ or } P_{oh} = \frac{\left|\frac{1}{\sqrt{2}} A_{vm} \cdot V_i\right|^2}{R_0}$$

$$= \frac{P_{om}}{2}$$

The bandwidth or pass band of the amplifier is defined as

$$B \cdot W = f_2 - f_1 \tag{8.1}$$

8.5 LOW FREQUENCY RESPONSE

The low frequency response of an *R-C* coupled amplifier is owing to the series capacitors C_S, C_C or C_E. Figure 8.6 shows a general *RC* circuit involving a capacitor *C*, whose reactance X_C is very small at mid- and high frequencies and therefore, approximated to be a short circuit and ineffective in these frequency ranges. But at low frequencies, the reactance of *C* becomes large enough to cause a voltage drop across *C*. This drop in voltage V_o results in decreasing gain ($A_v = V_0/V_i$). From the circuit of Fig. 8.6, the voltage gain at the low frequency.

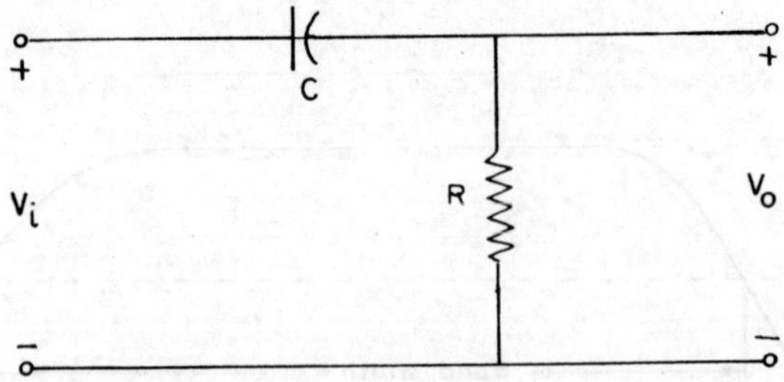

Fig. 8.6. Series *R-C* network effective in low frequency range.

$$A_{vl} = \frac{|V_0|}{|V_i|} = \frac{R}{\sqrt{R^2 + X_c^2}}$$

As at mid-frequency $X_C = 0$, the gain becomes

$$A_{vm} = 1$$

Hence, the expression for the gain at low frequency can be written as

$$\frac{A_{vl}}{A_{vm}} = \frac{R}{\sqrt{R^2 + X_c^2}}$$

or $$\frac{1}{\sqrt{2}} = \frac{R}{\sqrt{R^2 + X_c^2}} \quad \text{as} \quad \frac{A_{vl}}{A_{vm}} = \frac{1}{\sqrt{2}}$$

or $$|X_c| = R \tag{8.2}$$

or $$\frac{1}{2\pi f_1 C} = R$$

or $$f_1 = \frac{1}{2\pi RC} \tag{8.3}$$

This defines the lower half-power frequency and Eq. 8.2 gives the condition for the half-power frequencies. The voltage gain in low frequency range can be rewritten as

$$\frac{A_{vl}}{A_{vm}} = \frac{R}{\sqrt{R^2 + X_c^2}} = \frac{R}{R - jX_c}$$

where $X_C = \dfrac{1}{\omega C}$ and $j_2 = -1$

Hence $$\frac{A_{vl}}{A_{vm}} = \frac{1}{1 - jX_c / R} = \frac{1}{1 - j\frac{1}{\omega CR}}$$

$$= \frac{1}{1 - j\frac{1}{2\pi fCR}} = \frac{1}{1 - j(f_1/f)} \tag{8.4}$$

$$= \frac{1}{\sqrt{1 + (f_1/f)^2}} \angle \tan^{-1}(f_1/f) \tag{8.5}$$

where $\frac{1}{\sqrt{1 + (f_1/f)^2}}$ is the magnitude of A_{vl} and $\tan^{-1}(f_1/f)$ is the phase by which V_0 leads V_i.

8.5.1 Role of Capacitor C_S

From Fig. 8.1, the equivalent circuit around C_S can be drawn as shown in Fig. 8.7(a). The condition for defining lower half-power frequency as in Eq. 8.2 is

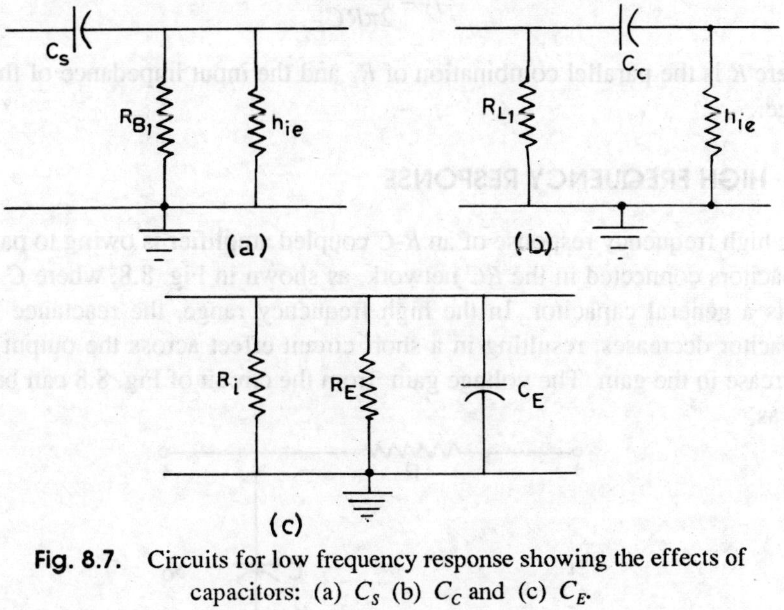

Fig. 8.7. Circuits for low frequency response showing the effects of capacitors: (a) C_S (b) C_C and (c) C_E.

$$R_i = X_{CS} = \frac{1}{2\pi f_1 C_S}$$

or $$f_1 = \frac{1}{2\pi R_i C_S} \tag{8.6}$$

where $R_i = R_{B_1} \parallel h_{ie}$ of the first transistor.

8.5.2 Role of Capacitor C_C

The equivalent circuit showing the effect of C_C is shown in Fig. 8.7(b). The lower half-power frequency is defined as in Eq.8.2

$$R_{L_1} + h_{ie} = X_{C_c} = \frac{1}{2\pi f_1 C_C}$$

where h_{ie} is the input impedance of the second stage and $R_{B_2} \gg h_{ie}$. Hence,

$$f_1 = \frac{1}{2\pi(R_{L_1} + h_{ie})C_c} \tag{8.7}$$

8.5.3 Role of Capacitor C_E

The equivalent circuit showing the effect of C_E is shown in Fig.8.7(c). The lower half-power frequency is defined as in Eq. 8.3

$$f_1 = \frac{1}{2\pi R C_E} \tag{8.8}$$

where R is the parallel combination of R_E and the input impedance of the first stage.

8.6 HIGH FREQUENCY RESPONSE

The high frequency response of an *R-C* coupled amplifier is owing to parasitic capacitors connected in the *RC* network, as shown in Fig. 8.8, where *C* represents a general capacitor. In the high frequency range, the reactance of the capacitor decreases, resulting in a short circuit effect across the output and a decrease in the gain. The voltage gain from the circuit of Fig. 8.8 can be written as,

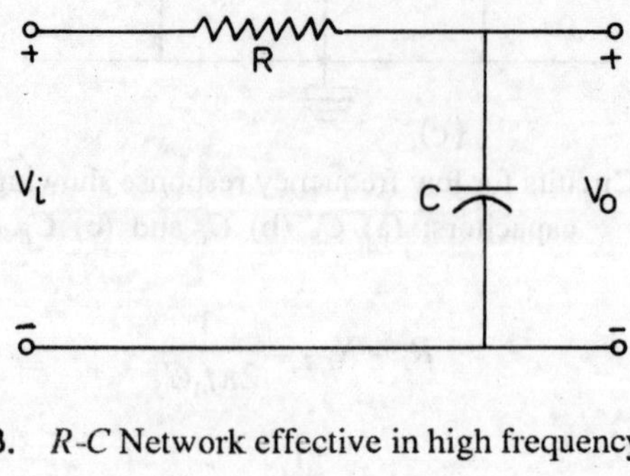

Fig. 8.8. *R-C* Network effective in high frequency range.

$$A_{vh} = \frac{|V_0|}{|V_i|} = \frac{X_c}{\sqrt{R^2 + X_c^2}}$$

In the mid-frequency range, X_C will be very large and acts as open, so, $R << X_C$ gives

$$A_{vm} = 1$$

Hence, the expression for A_{vh} can also be written as

$$\frac{A_{vh}}{A_{vm}} = \frac{X_C}{\sqrt{R^2 + X_C^2}}$$

At the upper half-power frequencies

$$\frac{A_{vh}}{A_{vm}} = \frac{1}{\sqrt{2}}$$

or

$$\frac{X_c}{\sqrt{R^2 + X_c^2}} = \frac{1}{\sqrt{2}}$$

or

$$|X_c| = R$$

or

$$\frac{1}{2\pi f_2 C} = R$$

or

$$f_2 = \frac{1}{2\pi RC} \tag{8.9}$$

This defines the upper half-power frequency. The voltage gain in the high frequency range can also be written as

$$\frac{A_{vh}}{A_{vm}} = \frac{-jX_C}{R - jX_C} = \frac{1}{-R/(jX_C) + 1}$$

$$= \frac{1}{1 + j\omega CR} \quad \text{as } X_C = \frac{1}{\omega C}$$

$$= \frac{1}{1 + j2\pi f \cdot CR} \quad \text{as } \omega = 2\pi f$$

$$= \frac{1}{1 + j(f/f_2)} \tag{8.10}$$

or

$$\frac{A_{vh}}{A_{vm}} = \frac{1}{\sqrt{1 + (f/f_2)^2}} \angle \tan^{-1}(f/f_2) \tag{8.11}$$

where $\dfrac{1}{\sqrt{1+(f/f_2)^2}}$ is the magnitude of A_{vh} and $\tan^{-1}(f/f_2)$ is the phase by which V_o leads V_i.

Figure 8.9 shows various capacitors which become effective at high frequencies. The capacitors C_S, C_E, and C_C are not shown in the equivalent circuit as they become ineffective at high frequencies. The upper half power frequency in an *R-C* coupled amplifier as given by Eq. 8.9 can be written for the input section as

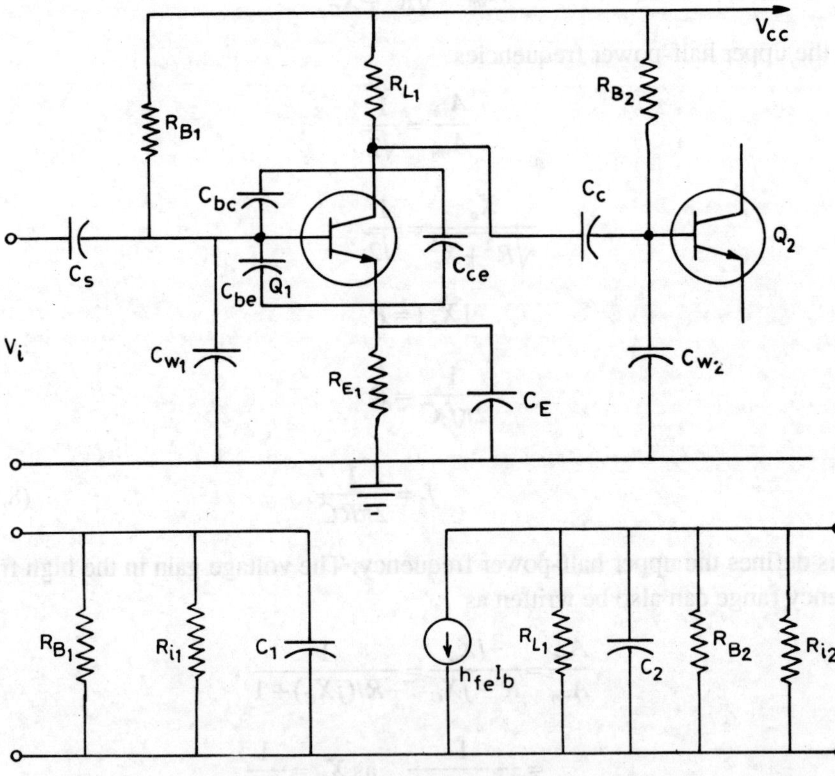

Fig. 8.9. *R-C* Coupled amplifier circuit showing the presence of various capacitors effective at high frequency.

$$F_{2(i)} = \frac{1}{2\pi R_1 C_1} \tag{8.12}$$

with $$R_1 = R_{B_1} \parallel R_{i_1} \quad \text{and}$$

$$C_1 = C_{W_1} + C_{be} + C_M$$

where R_{i_1} is the input resistance of the first stage and C_{W_1} is the wiring capacitance as shown in the figure. C_M is a capacitor known as the Miller capacitor, which is equal to $(1 + A_v C_{bc})$ and is effective at high frequencies. Similarly, the upper half-power frequency for the output section of the amplifier can be written as

$$f_{2(0)} = \frac{1}{2\pi R_2 C_2} \tag{8.13}$$

with
$$R_2 = R_{L_1} \parallel R_{B_2} \parallel R_{i_2}$$

and
$$C_2 = C_{W_2} + C_{ce}$$

Here R_{i_2} is the input resistance of the second stage and C_{W_2} is the wiring capacitance as shown in the figure.

8.7 DECIBEL AS UNIT OF GAIN

The term bel was taken from the name of Alexander Graham Bell and the decibel (*dB*) is defined as 10 dB = 1 bel. The gain of an amplifier is most expressed in decibels as

$$A\,(\text{dB}) = 10 \log_{10} \frac{P_{\text{out}}}{P_{\text{in}}} \tag{8.14}$$

where P_{out} and P_{in} are the output and the input powers. To express the decibel gain in terms of voltages, the input and the output powers are replaced by

$$P_{\text{out}} = \frac{V_0^2}{R_0} \quad \text{and} \quad P_{\text{in}} = \frac{V_i^2}{R_i}$$

where V_O and V_i are the output and the input voltages, and R_O and R_i are the output and the input resistances. Hence, Eq.8.14 becomes

$$A\,(\text{dB}) = 10 \log_{10} \frac{V_0^2}{R_0} \cdot \frac{R_i}{V_i^2}$$

$$= 10 \log_{10} \left(\frac{V_0}{V_i}\right)^2 \left(\frac{R_i}{R_0}\right)$$

$$= 20\log_{10}\left(\frac{V_o}{V_i}\right) + 10\log_{10}\left(\frac{R_i}{R_o}\right)$$

If the input and the output resistances are made equal, the decibel voltage gain becomes

$$A_v\,(\text{dB}) = 20\log_{10}\frac{V_0}{V_i} \tag{8.15}$$

If the input power P_{in} is accepted as 1 mW as reference, the decibel gain is expressed as

$$A\,(\text{dBm}) = 10\log_{10}P_{\text{out}} \tag{8.16}$$

where P_{out} is measured in mW.

The most important advantage of expressing gain in decibels is in the multistage amplifiers. For example, the voltage gain of a multistage amplifier is given by

$$A_v = A_{v_1}A_{v_2}A_{v_3}\ldots\ldots$$

where $A_{v_1}, A_{v_2}, A_{v_3}$.... are the voltage gains of the first, second, third.... stages respectively. Taking the log on both sides,

$$20\log_{10}A_v = 20\log_{10}A_{v_1} + 20\log_{10}A_{v_2} + 20\log_{10}A_{v_3} + \ldots..$$

Expressing in decibels

$$A_v(\text{dB}) = A_{v_1}(\text{dB}) + A_{v_2}(\text{dB}) + A_{v_3}(\text{dB}) + \ldots. \tag{8.17}$$

This equation shows that the decibel gain of a multistage amplifier is simply the sum of the decibel gains of each stage.

8.8 BODE PLOT FOR *R-C* COUPLED AMPLIFIER

The magnitude of the voltage gain of an *R-C* coupled amplifier in the low frequency range is expressed as (Eq. 8.5),

$$|A_{vl}| = \frac{1}{\sqrt{1+(f_1/f)^2}}$$

which can be expressed in decibels as

$$|A_{vl}|_{\text{dB}} = 20\log_{10}\frac{1}{\sqrt{1+(f_1/f)^2}}$$

$$= 20\log_{10}[1+(f_1/f)^2]^{-1/2}$$

$$= -10\log_{10}[1 + (f_1/f)^2] \qquad (8.18)$$

For frequencies much lower than f_1, i.e., $f << f_1$, one can be ignored in the above expression and the gain becomes

$$|A_{vl}|_{dB} = -20\log_{10}(f_1/f) \qquad (8.19)$$

A plot of this equation $|A_{vl}|_{dB}$ *vs* $\log_{10}(f/f_1)$ is shown in Fig. 8.10(a), which is a straight line *AB* with a slope of 20 dB per decade. Equation 8.18 also shows that $|A_{vl}|_{dB} = 0\text{dB}$ for $f \gg f_1$ and this 0dB line is expressed in Fig. 8.10(a) as *BC*. These two lines *AB* and *BC* meet at *B* where $f = f_1$ also known as corner frequency, but the gain at $f = f_1$ must become $\frac{1}{\sqrt{2}} = (-3\text{dB})$ expressed as point *D*, according to the definition of lower half-power frequency f_1. Hence, incorporating this point *D*, the frequency response in the low-frequency region becomes as *ADC* shown in the figure. This piecewise linear plot of the asymptotes and associated break point is called the Bode plot.

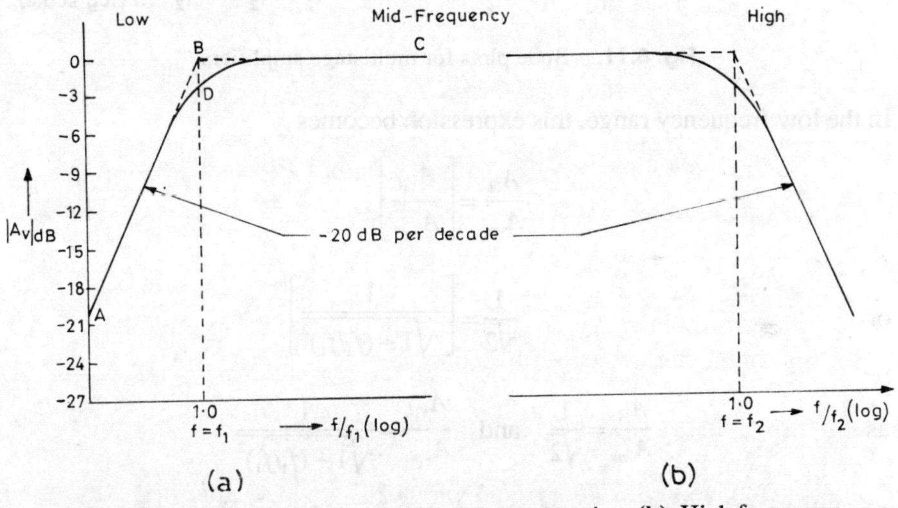

Fig. 8.10. Bode plots for: (a) Low frequency region (b) High frequency region.

Similarly, the Bode plot can be drawn as shown in Fig. 8.10(b) for high frequency region using Eq. 8.11.

8.9 BODE PLOT FOR MULTISTAGE AMPLIFIER

The voltage gain for a *n*-stage amplifier can be expressed as

$$A_v = A_{v_1} A_{v_2} A_{v_3} \ldots A_{v_n}$$

If all the stages are assumed to be identical,

$$A_v = (A_{v_1})^n \tag{8.20}$$

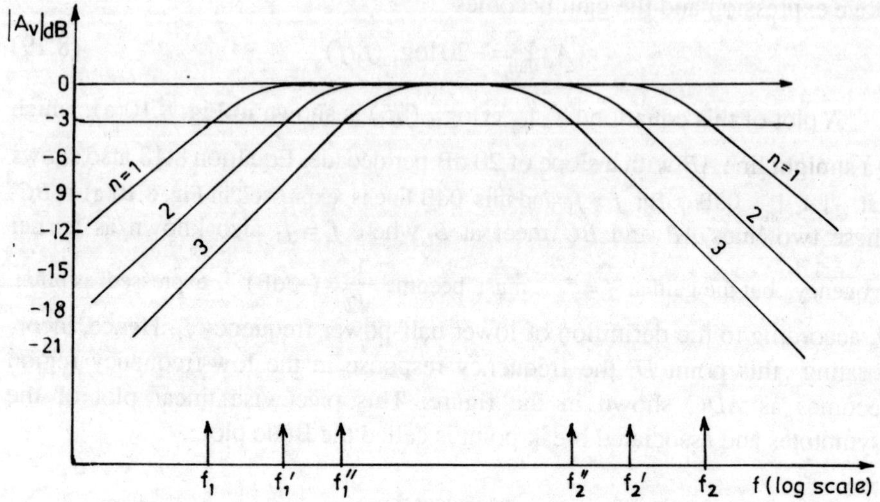

Fig. 8.11. Bode plots for multistage amplifiers.

In the low frequency range, this expression becomes

$$\frac{A_{vl}}{A_{vm}} = \left[\frac{A_{v_1 l}}{A_{v_1 m}}\right]^n$$

or

$$\frac{1}{\sqrt{2}} = \left[\frac{1}{\sqrt{1+(f_1/f_1')^2}}\right]^n$$

as

$$\frac{A_{vl}}{A_{vm}} = \frac{1}{\sqrt{2}} \quad \text{and} \quad \frac{A_{v_1 l}}{A_{v_1 m}} = \frac{1}{\sqrt{1+(f_1/f_1')^2}}$$

from Eq. 8.5, with f_1' as the lower cut-off frequency for an n-stage amplifier. Hence

$$\left[1+(f_1/f_1')^2\right]^{n/2} = 2^{1/2}$$

or

$$1+(f_1/f_1')^2 = 2^{1/n}$$

or

$$f_1' = \frac{f_1}{(2^{1/n}-1)} \tag{8.21}$$

Similarly, the higher cut-off frequency for an n-stage amplifier becomes

$$f_2' = f_2\sqrt{2^{1/n} - 1} \tag{8.21}$$

For $n = 2$ (say), $\sqrt{2^{1/n} - 1} = 0.64$. This shows that the lower and upper cut-off frequencies become 1/0.64 = 1.55 and 0.64 times respectively the frequencies for a single stage. Hence, the bandwidth decreases with the addition of more stages in the amplifier but gain becomes more and more. Figure 8.11 shows Bode plots for multistage amplifiers.

The bandwidth of an amplifier can be increased by adding more transistor stages, provided the gain is kept constant for single as well as multistage conditions. In other words, if a single-stage amplifier has a gain of 100 with a bandwidth of 10 kHz, the same gain can be obtained by a two-stage amplifier having a gain of 10 for each stage (as 10 × 10 = 100). But the fixed gain-band width product principle (100 × 10 kHz = 10^5) dictates that the bandwidth for each stage increases by a factor of 10 as the gain is reduced by 10.

8.10 SOLVED PROBLEMS

Problem 1

Find the emitter current, input impedance, and output impedance for the circuit of Fig. 8.12.

Solution

The emitter resistance is

$$= h_{fe}(2\,k + 1\,k)$$

$$= 50 \times 3 = 150\ \text{k}\Omega$$

The base potential

$$V_B = \frac{(10\,k \mid 150\,k)V_{cc}}{10\,k \mid 150\,k + 100\,k}$$

$$= \frac{10 \times 25}{110} = 2.27\ \text{V}$$

∴ the emitter potential

$$V_E = V_B - V_{BE} = 2.27 - 0.6$$

$$= 1.67\ \text{V}$$

Hence, the emitter current,

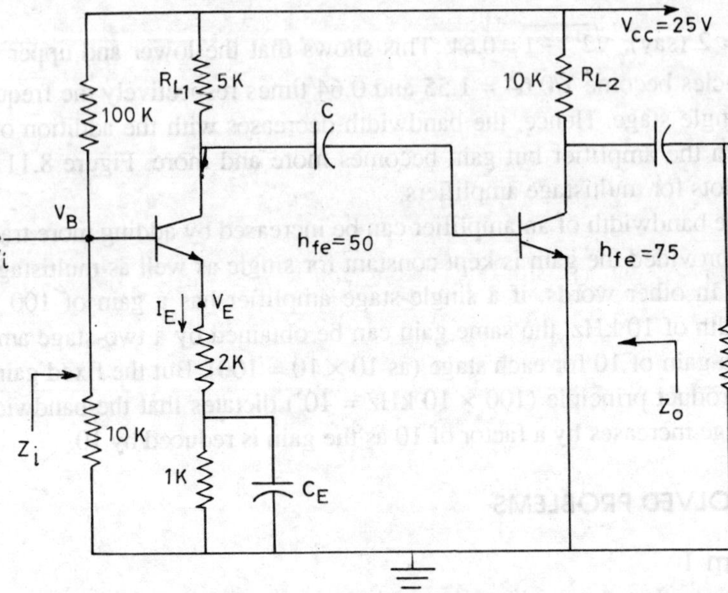

Fig. 8.12. Circuit for problem 1.

$$I_E = \frac{1.67}{(2+1)\times 10^{-3}} = 0.55 \text{ mA}$$

the input impedance

$Z_i = 100\ k \parallel 10\ k \parallel (h_{fe} \times 2\ k)$ as $1k$ resistance is bypassed by capacitor C_E.

$$\therefore \qquad Z_i = 100\ k \parallel 10\ k \parallel 100\ k$$

$$= 8.34\ k\Omega$$

the output impedance (when $V_i = 0$)

$$Z_0 = R_{L_2} = 10\ \text{k}\Omega$$

Problem 2

Find the current gain for a Darlington pair of identical transistors if $h_{fe} = 50$, $h_{0e} = 10 \times 10^{-6}$, and $R_E = 2\ \text{k}\Omega$.

Solution

In a Darlington pair configuration, the input impedance Z_{i_2} is in the order of

$1/h_{0e_1}$ and therefore, h_{0e} is not neglected. Hence, the current gain for a common-emitter amplifier as in Eq. 7.3 is given by

$$A_i = \frac{h_{fe}}{1 + h_{oe} Z_L}$$

for a circuit of Darlington pair

$$Z_L = Z_{i_2} = h_{fe} R_E$$

Hence, for the first transistor, the current gain

$$A_{i_1} = -\frac{h_{fe}}{1 + h_{oe} \cdot h_{fe} \cdot R_E}$$

$$= \frac{50}{1 + (10 \times 10^{-6} \times 50 \times 2 \times 10^3)}$$

$$= \frac{50}{2} = 25$$

The current gain for the second transistor is

$$A_{i_2} \cong h_{fe} = 50$$

Hence total current gain

$$A_i = A_{i_1} \cdot A_{i_2} = 25 \times 50 = 1250$$

Problem 3

If the decibel gain is 10, find the power gain of the amplifier.

Solution

The gain of the amplifier is defined as

$$A(\text{dB}) = 10 \log \frac{P_2}{P_1}$$

or

$$\log \frac{P_2}{P_1} = \frac{A(\text{dB})}{10} = \frac{10}{10} = 1$$

or power gain, $\dfrac{P_2}{P_1} = 10.$

Problem 4

The input and output powers of an amplifier are 10 W and 1000 W respectively, when an input voltage applied is 1 kV and the output impedance is 50 Ω. Find the power gain and voltage gain in decibel.

Solution

The decibel gain is defined as

$$A_p(\text{dB}) = 10\log_{10}\frac{P_2}{P_1}.$$

$$= 10\log_{10}(100)$$

$$= 20 \text{ dB}$$

and the voltage gain in decibel is defined as

$$A_v(\text{dB}) = 20\log_{10}\frac{V_2}{V_1}$$

$$= 20\log_{10}\frac{\sqrt{\text{Power output } \times \text{output impedance}}}{V_1}$$

$$= 20\log_{10}\frac{\sqrt{1000\times 50}}{1000}$$

$$= 20\log_{10}(0.223)$$

$$= -13 \text{ dB}$$

Problem 5

Determine the lower half-power frequencies f_1 (C_S), f_1 (C_C), and f_1 (C_E) for the circuit of Fig.8.13.

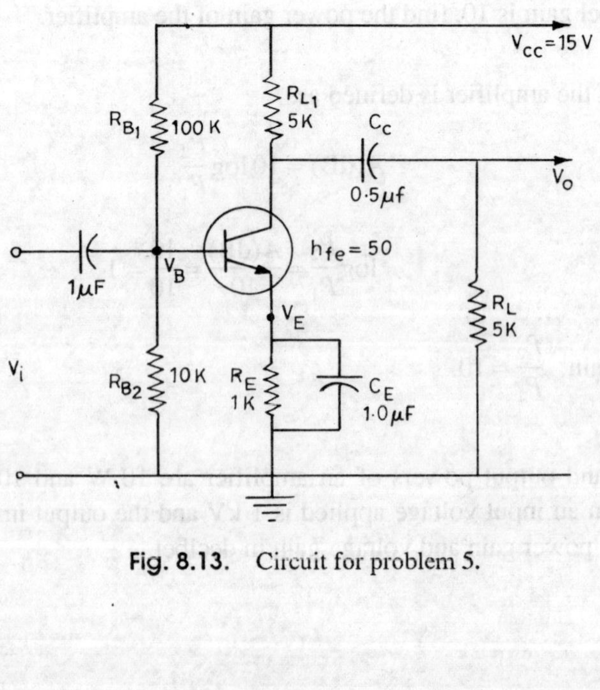

Fig. 8.13. Circuit for problem 5.

Solution

The base potential

$$V_B = \frac{R_{B_2} \cdot V_{CC}}{R_{B_1} + R_{B_2}} = \frac{10 \times 15}{110} = 13.6 \text{ V}$$

The emitter current

$$I_E = \frac{V_E}{R_E} = \frac{V_B - V_{BE}}{R_E}$$

$$= \frac{1.36 - 0.7}{1 \times 10^3} = 0.66 \text{ mA}$$

The dynamic resistance between base and emitter is

$$r_d = \frac{26 \text{ mV}}{I_E} = \frac{26}{0.66} = 39.4 \ \Omega$$

The input impedance

$$Z_i = R_{B_1} \parallel R_{B_2} \parallel h_{fe} \cdot r_e$$

$$= 100 \parallel 10 \parallel 1.97$$

$$= 5.17 \text{ k}\Omega$$

the lower half-power frequency owing to capacitor C_s is

$$f_1(C_s) = \frac{1}{2\pi Z_i C_s}$$

$$= \frac{7}{2 \times 22 \times 5.17 \times 10^3 \times 10^{-6}}$$

$$= 31 \text{ Hz}$$

The lower half-power frequency owing to coupling capacitor C_c is

$$f_1(C_c) = \frac{1}{2\pi(R_{L_1} + R_L)C_c}$$

$$= \frac{7}{2 \times 22 \times 10 \times 10^3 \times 0.5 \times 10^{-6}}$$

$$= 32 \text{ Hz}$$

The lower half-power frequency due to C_E is

$$f_1(C_E) = \frac{1}{2\pi R C_E}$$

where
$$R = R_E \parallel r_d$$

$$= \frac{1 \times 10^3 \times 39.4}{1039.4} = 37.9\ \Omega$$

$$\therefore \quad f_1(C_E) = \frac{7}{2 \times 22 \times 37.9 \times 10 \times 10^{-6}}$$

$$= 420 \text{ Hz}$$

Problem 6

Calculate the upper half-power frequency for a four-stage amplifier with the upper half-power frequency of one stage as 3 MHz.

Solution

The higher cut-off frequency or the upper half-power frequency for a multistage amplifier is given by

$$f_2' = f_2\sqrt{2^{1/n} - 1} \text{ where } n = 4$$

$$= 3 \times \sqrt{2^{1/4} - 1}$$

$$= 3 \times 0.44 = 1.32 \text{ MHz}$$

Problem 7

Find the lower cut-off frequency for a four-stage amplifier with individual f_1 = 30 Hz.

Solution

The lower cut-off frequency for a multistage amplifier is given by

$$f_1' = \frac{f_1}{\sqrt{2^{1/n} - 1}} \quad \text{where} \quad n = 4$$

$$= \frac{30}{0.44} = 68 \text{ Hz}$$

Problem 8

A two-stage amplifier has a lower 3 dB frequency for individual stage as 30 Hz. Calculate f_1 for the full amplifier.

Solution

The lower cut-off frequency, also called lower 3 dB frequency, for a multistage amplifier is given by

$$f_1' = \frac{f_1}{\sqrt{2^{1/n} - 1}} \text{ where } n = 2$$

$$= \frac{20}{0.64} = 47 \text{ Hz}$$

Questions and Problems

1. How is the gain increased in an amplifier?
2. What do you mean by interstage coupling in amplifier circuits?
3. Describe the various types of coupling techniques used in amplifiers to increase gain.
4. What is the meaning of a Darlington pair ? where is it used?
5. Describe the frequency response of a *R-C* coupled amplifier.
6. What are corner frequencies?
7. Why does the gain decrease in the low frequency range in an *R-C* amplifier?
8. Which factors are responsible for the loss of gain in high frequency range in a *R-C* coupled amplifier?
9. What do you mean by the Miller effect?
10. Find the expression for the gain of *R-C* amplifier at the lower half-power frequency.
11. Find the expression for the gain of an *R-C* coupled amplifier at the upper half-power frequency.
12. Why are the lower or upper half-power frequencies also called 3 dB frequencies?
13. Define decibel unit gain. Why is it preferred?
14. Draw Bode plots for an *R-C* amplifier.
15. What is the modification in Bode plot for a multistage amplifier over a single-stage amplifier?
16. Find the emitter current and input and output impedances for the circuit of Fig.8.12 if the two transistors are interchanged.
17. If the current gain obtained from a Darlington pair of identical transistors is 2500 and the emitter resistance is 1 kΩ, find the value of h_{oe}.
18. If the power gain of an amplifier is 100, find the dB gain.
19. Find the voltage gain in dB in the above Problem 18.
20. Following the circuit of Fig. 8.13, find f_1 (C_S), f_1 (C_C), and f_1 (C_E) if h_{fe} = 75 and V_{cc} = 12 V.
21. Find the lower and upper half-power frequencies for a five-stage amplifier with one stage f_1 = 20 Hz and f_2 = 2.5 MHz.

9

Feedback Amplifiers

In feedback amplifiers, a sample of the output voltage or current is fed back to the input of the amplifier. If this sample output is applied in phase to the input, the feedback is called regenerative or the positive feedback. On the other hand, if the sample output is applied out of phase to the input, the feedback is called the degenerative or the negative feedback. The positive feedback is used in oscillators, while the negative feedback is used in amplifiers to stabilize their gain, increase their frequency response, increase or decrease their input and output resistances, and reduce their distortion. The disadvantage of the negative feedback is that it reduces the overall gain of the amplifier.

9.1 TYPES OF FEEDBACK

The sample of the output fed back to the input can be a function of the output voltage or output current. In the former case, the feedback is known as voltage feedback and in the latter case, it is called current feedback. If both the feedbacks are used in an amplifier, it is called compound feedback.

9.1.1 Voltage Feedback

Voltage feedback is further divided into two types:

(a) Voltage Series Feedback

If the sample voltage as a function of output voltage is fed back to the input in series, it is called voltage series feedback. The block diagram for this type of feedback is shown in Fig. 9.1(a), where A is the voltage gain of the amplifier and β is the fraction of the output voltage fedback to the input. Taking the amplifier and feedback network as two-port networks, there are two input ports and two output ports in this voltage amplifier, and from the figure, it is clear that the input ports are connected in series, while the output ports in parallel. So, this amplifier is also referred to as a series-parallel feedback configuration. In this case, the feedback voltage is a function of output voltage. That is why, it is also called voltage amplifier.

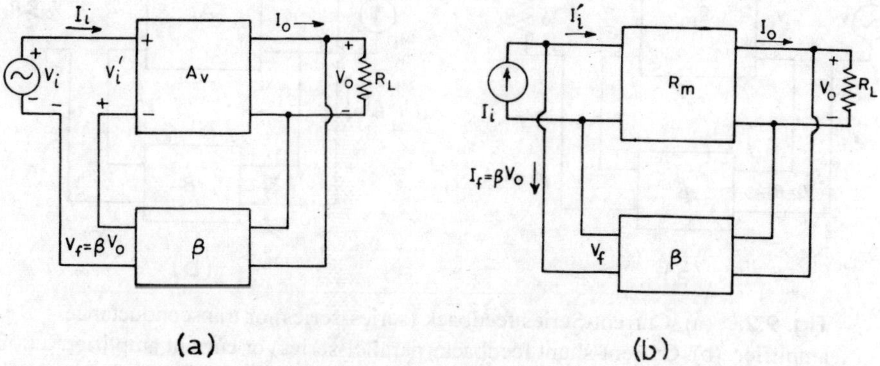

Fig. 9.1. (a) Voltage series feedback (series-parallel) or voltage amplifier (b) Voltage shunt feedback (parallel-parallel) or transresistance amplifier.

(b) Voltage Shunt Feedback

If the sample current as a function of output voltage is applied to the input in parallel, it is known as voltage shunt feedback. The block diagram for this feedback is shown in Fig. 9.1(b). The input as well as the output ports are connected in parallel and therefore, the circuit is called a parallel–parallel feedback configuration. As in this case, the feedback current is a function of output voltage, the circuit is known as transresistance amplifier.

9.1.2 Current Feedback

It is further divided into two catagories:

(a) Current Series Feedback

If the sample voltage as a function of output current is applied to the input in series, it is known as current series feedback. The block diagram for this feedback is shown in Fig. 9.2(a). Both the input and the output ports are connected in series and therefore, this circuit is also known as a series-series feedback configuration. As the feedback voltage is a function of the output current, the circuit model is known as the transconductance amplifier.

(b) Current Shunt Feedback

If the sample current as a function of output current is fed back to the input in parallel, it is called a current shunt feedback. The block diagram for this feedback is shown in Fig. 9.2(b). In this case, the input and the output ports are connected in parallel and in series respectively, therefore, the circuit is called a parallel series feedback configuration. As the feedback current is a function of output current, the circuit model is known as the current amplifier.

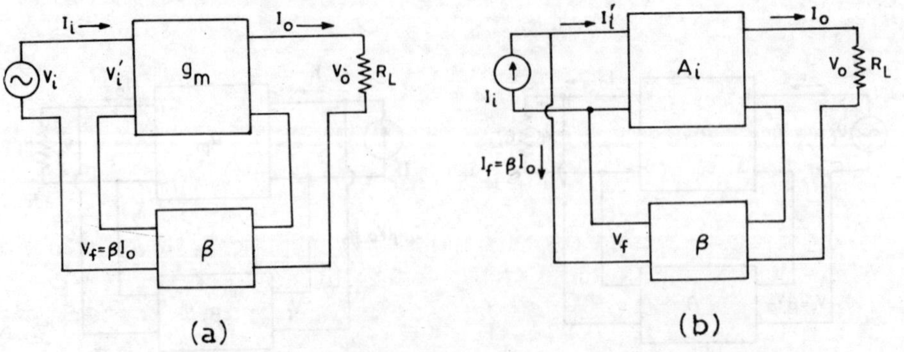

Fig. 9.2. (a) Current-Series feedback (series-series) or transconductance amplifier (b) Current-shunt feedback (parallel-series) or current amplifier.

9.2 VOLTAGE AMPLIFIER WITH NEGATIVE FEEDBACK

The block diagram of a voltage amplifier with negative feedback is shown in Fig. 9.3, depicting three blocks of basic amplifier, feedback network, and mixer network. The input signal V_i and feedback signal V_f are mixed to form a signal V_i' which is applied to the input of the amplifier. The output voltage is V_0. The feedback voltage βV_o is fed to the mixer in a polarity opposite to that of the input signal V_i for negative feedback. The gain of the amplifier without feedback is accepted as A_v and β is the feedback factor representing the fraction of the output voltage fed back to the input.

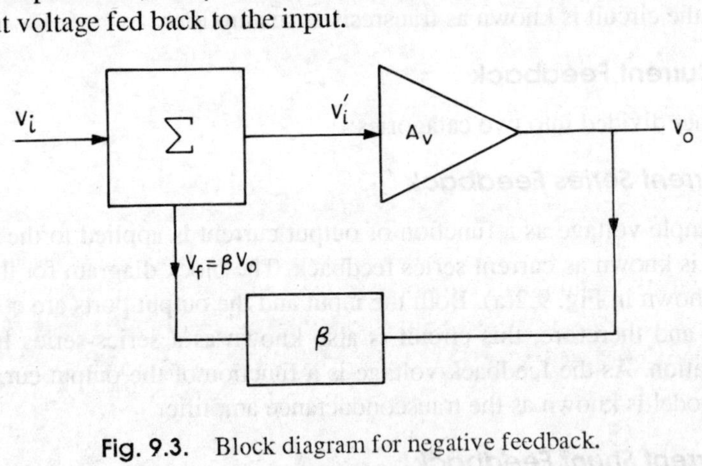

Fig. 9.3. Block diagram for negative feedback.

9.2.1 Voltage Gain with Feedback

The open loop gain of basic amplifier is given by

$$A_v = \frac{V_0}{V_i} \tag{9.1}$$

The overall gain of the amplifier after feedback is

$$A_{vf} = \frac{V_O}{V_i} \tag{9.2}$$

The feedback voltage V_f is given by

$$V_f = \beta V_O \tag{9.3}$$

From the figure, the input voltage to the basic amplifier is the difference of input voltage and the feedback voltage for negative feedback, i.e.,

$$V_i' = V_i - V_f \tag{9.4}$$

Substituting the values of V_i', V_i and V_f from Eqs 9.1, 9.2 and 9.3,

$$\frac{V_O}{A_v} = \frac{V_O}{A_{vf}} - \beta V_O$$

or $$\frac{1}{A_v} = \frac{1 - A_{vf}\beta}{A_{vf}}$$

or $$A_{vf} = A_v - A_v A_{vf}\beta$$

or $$A_{vf}(1 + A_v\beta) = A_v$$

or $$A_{vf} = \frac{A_v}{1 + A_v\beta} \tag{9.5}$$

This equation represents that the gain of an amplifier is reduced when negative feedback is applied.

9.2.2 Input Resistance with Feedback

In order to study the effect of negative feedback on the input resistance of an amplifier, the equivalent circuit is shown in Fig. 9.4. It is assumed that the feedback network does not load the amplifier. The input resistance with and without feedback are given by

$$R_{if} = \frac{V_i}{I_i} \quad \text{and} \quad R_i = \frac{V_i'}{I_i}$$

These equations along with the figure show that the voltage V_i' appears across the resistance R_i, while the voltage V_i appears across a resistance R_{if}, which would replace R_i if the feedback is considered to exist. The current in both the cases is I_i. Hence,

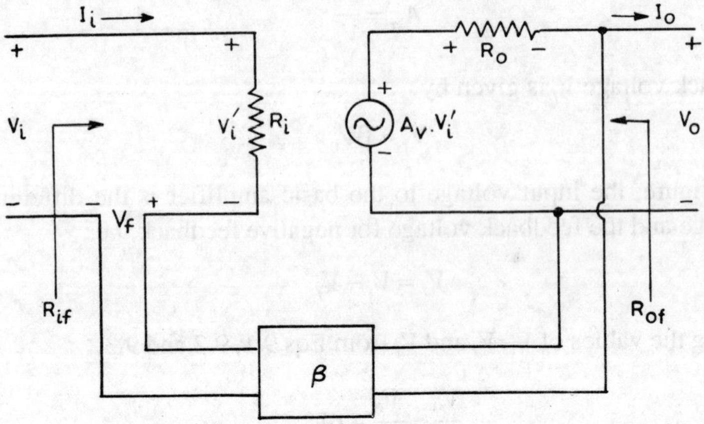

Fig. 9.4. Equivalent circuit for voltage amplifier with negative feedback.

$$R_{if} = \frac{V_i}{I_i}$$

$$= \frac{V_i' + V_f}{I_i} \text{ (as } V_i' = V_i - V_f)$$

$$= \frac{V_i' + \beta V_0}{I_i} \text{ (as } V_f = \beta V_0)$$

From the figure,

$$V_0 = A_v \cdot V_i'$$

Thus

$$R_{if} = \frac{V_i'}{I_i} + \frac{\beta A_v V_i'}{I_i}$$

$$= \frac{V_i'}{I_i}(1 + A_v\beta)$$

or

$$R_{if} = R_i(1 + A_v\beta) \left(\text{as } R_i = \frac{V_i'}{I_i} \right) \tag{9.6}$$

This equation shows that the input resistance for a voltage amplifier with negative feedback is increased by a factor of $(1 + A_v\beta)$.

9.2.3 Output Resistance with Feedback

The output resistance with feedback R_{0f} using Fig. 9.4 when input V_i is short-circuited ($V_i = 0$) is

$$R_{of} = -\frac{V_0}{I_0} \tag{9.7}$$

Here, the minus sign shows that the potential drop $I_o R_{of}$ owing to I_o is opposite to the polarity of V_O shown in the figure. The voltage equation for the output section can be written as

$$A_v V_i' - I_0 R_0 = V_0$$

From this equation, the output resistance without feedback can be defined as

$$R_0 = \frac{A_v V_i' - V_0}{I_0}$$

The value of V_i' is given by

$$V_i' = V_i - V_f$$

$$= -V_f \text{ (as } V_i \text{ is short-circuited)}$$

$$= -\ \beta V_0 \text{ (as } V_f = \beta V_0\text{)}$$

Hence,

$$R_0 = \frac{-\ A_v \beta V_0 - V_0}{I_0}$$

$$= -\frac{V_0}{I_0}(1 + A_v \beta)$$

$$= R_{of}(1 + A_v \beta) \text{ (from Eq. (9.7))}$$

or

$$R_{of} = \frac{R_0}{1 + A_v \beta} \tag{9.8}$$

This equation shows that the output resistance of a voltage amplifier with negative feedback is decreased by a factor of $(1 + A_v\beta)$.

9.2.4 BJT Circuit for Voltage Amplifier with Negative Feedback

Figure 9.5 shows a circuit configuration for a multistage BJT voltage amplifier with negative feedback. The gains of the two stages are A_{v_1} and A_{v_2} respectively. The two identical stages are connected in a common emitter configuration. Therefore, the input V_i and the output V_O are in phase. The voltage that the first stage amplifies is the difference between V_i and V_f; thus the amplifier uses negative feedback. From the circuit, the feedback voltage V_f can be written as

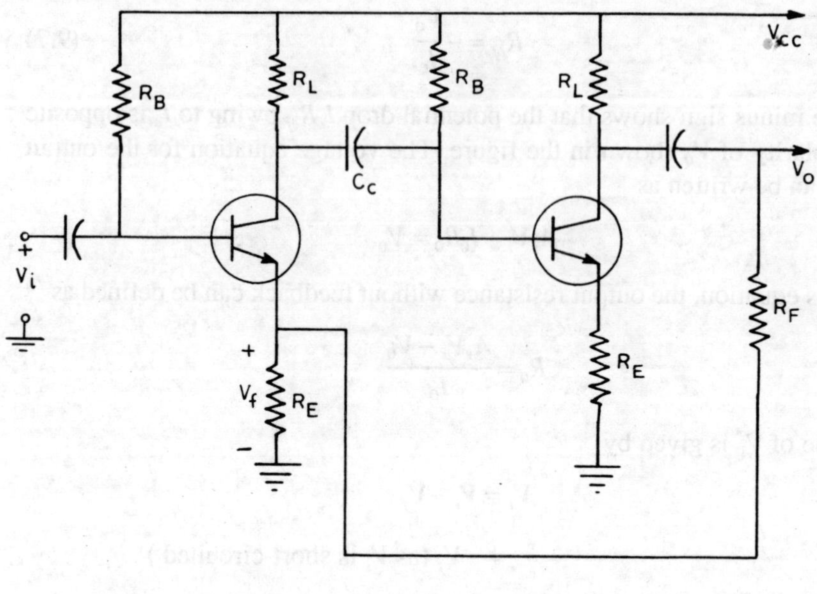

Fig. 9.5. BJT Circuit for voltage amplifier with negative feedback.

$$V_f = \frac{R_E}{R_F + R_E} \cdot V_0$$

Hence
$$\beta = \frac{R_E}{R_F + R_E} \tag{9.9}$$

The gain of the common emitter amplifier as given in Eq. 7.12 of Chapter 7 is $-g_m R_L$. Therefore,

$$A_v = A_{v_1} \cdot A_{v_2}$$

$$= (-g_m R_L)(-g_m R_L)$$

$$= g_m^2 R_L^2 \tag{9.10}$$

as the two stages are taken identical. From Eq. 9.5, the gain with feedback can be given as

$$A_{vf} = \frac{A_v}{1 + A_v\beta}$$

As $A_v\beta$ is much greater than unity,

$$A_{vf} = \frac{A_v}{A_v\beta} = \frac{1}{\beta}$$

or $$A_{vf} = 1 + \frac{R_F}{R_E} \text{ (using Eq. (9.9))} \tag{9.11}$$

The input and the output resistances with feedback are given by Eqs 9.6 and 9.8 respectively as,

$$R_{if} = R_i g_m^2 R_L^2 \cdot \frac{R_E}{R_F + R_E} \tag{9.12}$$

and $$R_{of} = \frac{R_0(R_F + R_E)}{g_m^2 R_L^2 \cdot R_E} \tag{9.13}$$

9.3 TRANSRESISTANCE AMPLIFIER WITH NEGATIVE FEEDBACK

The block diagram for a transresistance amplifier with negative feedback is shown in Fig. 9.1 (b). The feedback current I_f is a function of the output voltage and is subtracted from the input current I_i. The rest of the current flows into the amplifier.

9.3.1 Transresistance Gain with Feedback

The transresistance gain with and without feedback are given by

$$R_{mf} = \frac{V_0}{I_i} \tag{9.14}$$

and $$R_m = \frac{V_0}{I_i'} \tag{9.15}$$

The currents in the input section are governed by

$$I_i - I_f = I_i'$$

or $$I_i - \beta V_0 = I_i' \tag{9.16}$$

Substituting the value of V_0 from Eq. 9.15,

$$I_i - \beta R_m I_i' = I_i'$$

or $$I_i = I_i'(1 + \beta R_m)$$

Substituting this value of I_i in Eq. 9.14,

$$R_{mf} = \frac{V_0}{I_i'(1 + \beta R_m)}$$

or $$R_{mf} = \frac{R_m}{1+\beta R_m} \tag{9.17}$$

This equation shows that the transresistance gain is reduced by a factor of $(1 + \beta R_m)$ when negative feedback is applied.

9.3.2 Input Resistance with Feedback

The equivalent circuit for a transresistance amplifier is shown in Fig. 9.6. The input resistance with and without feedback are given from the figure as,

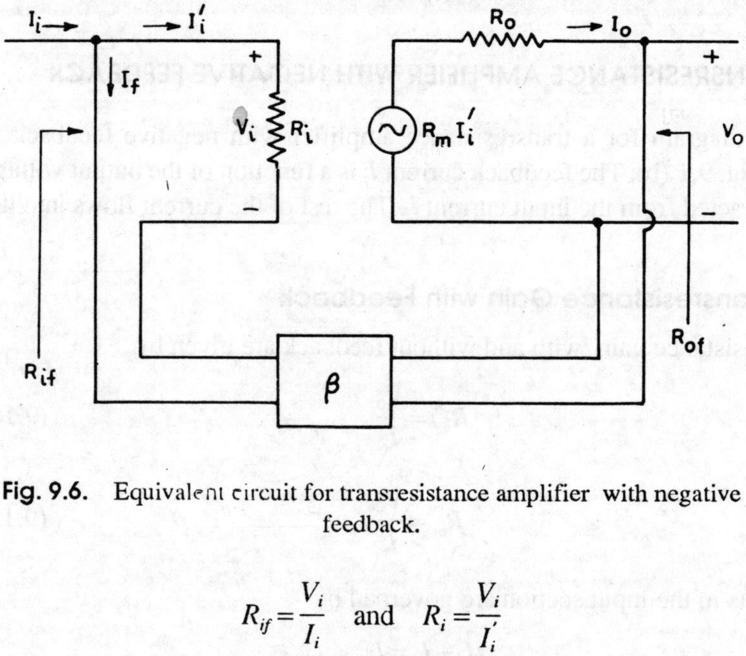

Fig. 9.6. Equivalent circuit for transresistance amplifier with negative feedback.

$$R_{if} = \frac{V_i}{I_i} \quad \text{and} \quad R_i = \frac{V_i}{I_i'}$$

Substituting the value of I_i from Eq. 9.16,

$$R_{if} = \frac{V_i}{I_i' + \beta V_0}$$

From Eq. 9.15, $V_O = R_m I_i'$.

Therefore $$R_{if} = \frac{V_i}{I_i' + \beta R_m I_i'}$$

$$= \frac{V_i}{I_i'} \cdot \frac{1}{1 + \beta R_m}$$

or
$$R_{if} = \frac{R_i}{1 + \beta R_m} \tag{9.18}$$

This shows that the input resistance of a transresistance amplifier is reduced by a factor of $(1 + \beta R_m)$ when negative feedback is applied.

9.3.3 Output Resistance with Feedback

From Fig. 9.6, the output resistance with feedback when input current is zero is defined as,

$$R_{of} = -\frac{V_o}{I_o} \tag{9.19}$$

The voltage equation for the output section can be written as

$$R_m I_i' - I_0 R_0 = V_0$$

or
$$R_0 = \frac{R_m I_i' - V_0}{I_0}$$

$$= \frac{R_m(I_i - \beta V_0) - V_0}{I_0} \qquad \text{(from Eq. 9.16)}$$

$$= \frac{-R_m \cdot \beta V_0 - V_0}{I_0} \qquad (\text{as } I_i = 0)$$

$$= \frac{-V_0}{I_0}(1 + \beta R_m)$$

$$= R_{of}(1 + \beta R_m) \qquad \text{(from Eq. 9.19)}$$

or
$$R_{of} = \frac{R_0}{1 + \beta R_m} \tag{9.20}$$

This equation shows that the output resistance of a transresistance amplifier is decreased by a factor of $(1 + \beta R_m)$ when negative feedback is applied.

9.3.4 BJT Circuit for Transresistance Amplifier with Negative Feedback

The circuit for the transresistance amplifier using BJT is shown in Fig. 9.7. The resistor R_f provides the required feedback. As BJT in common-emitter configuration introduces a phase difference of 180° between the input and the output, the feedback in this circuit is negative. The open loop transresistance gain as defined in Eq. 9.15 is

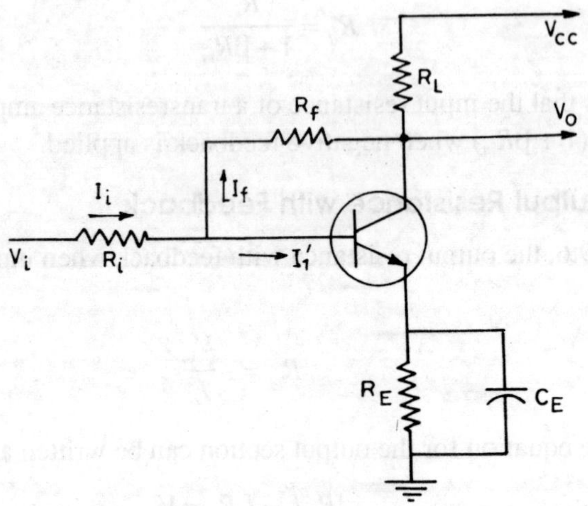

Fig. 9.7. BJT circuit for transresistance amplifier with negative feedback.

$$R_m = \frac{V_0}{I_i'} = \frac{I_c R_L}{I_b} \text{ (as } I_i' = I_b)$$

but $$I_c = h_{fe} I_b,$$

therefore, $$R_m = \frac{h_{fe} I_b R_L}{I_b}$$

or $$R_m = h_{fe} R_L \tag{9.21}$$

The feedback current I_f is defined as

$$I_f = \beta V_0$$

From the circuit, if the input resistance is considered low, the feedback current is

$$I_f = \frac{V_0}{R_f}$$

Hence $$\beta V_0 = \frac{V_0}{R_f}$$

or $$\beta = \frac{1}{R_f} \tag{9.22}$$

Using Eqs 9.18, 9.21 and 9.22, the input resistance with negative feedback is

$$R_{if} = \frac{R_i}{1+\beta R_m}$$

$$= \frac{h_{ie}}{1+\frac{1}{R_f}\cdot h_{fe}R_L} \quad (\text{as } R_i = h_{ie})$$

or
$$R_{if} = \frac{h_{ie}}{1+\frac{h_{fe}R_L}{R_f}} \tag{9.23}$$

Similarly, the output resistance with feedback can be written using Eqs 9.20, 9.21 and 9.22 as,

$$R_{of} = \frac{R_0}{1+\frac{h_{fe}R_L}{R_f}} \tag{9.24}$$

9.4 TRANSCONDUCTANCE AMPLIFIER WITH NEGATIVE FEEDBACK

The block diagram for a transconductance amplifier with negative feedback is shown in Fig. 9.2(a). In this amplifier, the feedback voltage V_f is a function of the output current I_0.

9.4.1 Transconductance Gain with Feedback

The transconductance gains with and without feedback are:

$$g_{mf} = \frac{I_0}{V_i} \tag{9.25}$$

and
$$g_m = \frac{I_0}{V_i'} \tag{9.26}$$

The feedback voltage V_f is given by

$$V_f = \beta I_0$$

The voltage equation for the input section of the current can be written as

$$V_i' = V_i - V_f \tag{9.27}$$

Hence, from Eqs. 9.25 and 9.27,

$$g_{mf} = \frac{I_o}{V_i' + V_f}$$

$$= \frac{I_o}{V_i' + \beta I_o}$$

$$= \frac{I_o}{V_i'}\left(\frac{1}{1 + \beta \cdot \frac{I_o}{V_i'}}\right)$$

or
$$g_{mf} = \frac{g_m}{1 + \beta g_m} \quad \left(\text{as } g_m = \frac{I_o}{V_i'}\right) \tag{9.28}$$

This equation shows that the transconductance gain is decreased by a factor $(1 + \beta g_m)$ when negative feedback is applied.

9.4.2 Input Resistance with Feedback

The equivalent circuit for a transconductance amplifier is shown in Fig. 9.8. The input resistance with and without feedback are

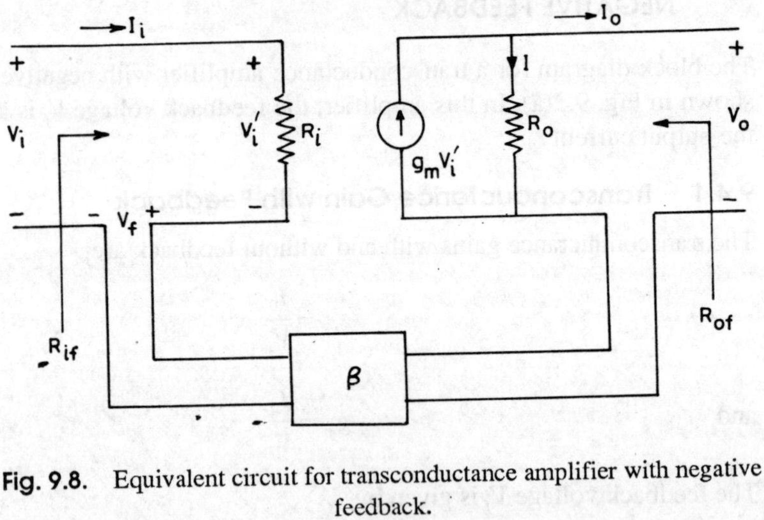

Fig. 9.8. Equivalent circuit for transconductance amplifier with negative feedback.

$$R_{if} = \frac{V_i}{I_i} \text{ and } R_i = \frac{V_i'}{I_i}$$

From Eq. 9.27,

$$V_i = V_i' + V_f$$

$$= V_i' + \beta I_0 \text{Thus}$$

$$R_{if} = \frac{V_i' + \beta I_0}{I_i}$$

$$= \frac{V_i}{I_i} + \frac{\beta g_m V_i}{I_i} \quad \text{(from Eq. 9.26)}$$

$$= \frac{V_i'}{I_i}(1 + \beta g_m)$$

or

$$R_{if} = R_i(1 + \beta g_m) \tag{9.29}$$

Hence, the input resistance of a transconductance amplifier is increased by $(1 + \beta g_m)$ when negative feedback is applied.

9.4.3 Output Resistance with Feedback

From Fig. 9.8, the output resistance with feedback when input is short circuit, i.e $V_i = 0$, is defined as

$$R_{of} = -\frac{V_0}{I_0} \tag{9.30}$$

The minus sign shows that the polarity of V_0 is taken as opposite to that owing to I_0. In the output section of the circuit,

$$V_0 = (g_m V_i' - I_0)R_0$$

$$= [g_m(V_i - V_f) - I_0]R_0 \quad \text{(from Eq. 9.27)}$$

$$= [-g_m V_f - I_0]R_0 \quad (\text{as } V_i = 0)$$

$$= -\,(g_m \beta I_0 + I_0)R_0$$

$$= -\,I_0(1 + \beta g_m)R_0$$

or

$$-\frac{V_0}{I_0} = (1 + \beta g_m)R_0$$

From Eq. 9.30,

$$R_{of} = R_0(1 + \beta g_m) \tag{9.31}$$

This shows that the output resistance of a transconductance amplifier is increased by $(1 + \beta g_m)$ with negative feedback.

9.4.4 BJT Circuit for Transconductance Amplifier with Negative Feedback

The transconductance amplifier using a BJT is shown in Fig. 9.9. The feedback voltage V_f is across R_E. As the feedback voltage V_f in this amplifier is a function of the output current I_c, it is possible only when the current flowing through R_E, I_e must be approximately equal to I_c. Therefore,

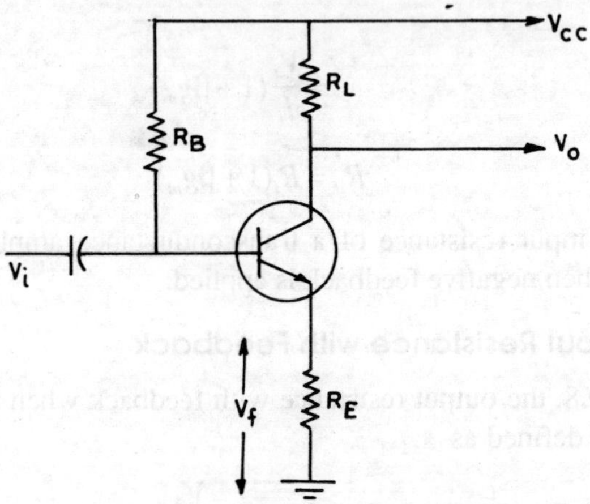

Fig. 9.9. BJT circuit for transconductance amplifier with negative feedback.

$$V_f = R_E \cdot I_e = R_E \cdot I_c$$

Also, $$V_f = \beta I_0 = \beta I_c$$

From these equations

$$\beta = R_E \tag{9.32}$$

The transconductance open loop gain is defined as

$$g_m = \frac{I_0}{V_i}$$

$$= \frac{I_c}{h_{ie} I_b} \quad (\text{as } V_i' = h_{ie} I_b = V_{be})$$

or $$g_m = \frac{h_{fe}}{h_{ie}} \tag{9.33}$$

From Eqs. 9.29, 9.32 and 9.33, the input resistance with feedback is given by

$$R_{if} = R_i(1 + \beta g_m)$$

$$= h_{ie}\left(1 + R_E \cdot \frac{h_{fe}}{h_{ie}}\right) \quad (\text{as } R_i = h_{ie})$$

or

$$R_{if} = h_{ie} + R_E \cdot h_{fe} \tag{9.34}$$

From Eqs. 9.31, 9.32 and 9.33, the output resistance with feedback is given by,

$$R_{of} = R_0(1 + \beta g_m)$$

$$= \frac{1}{h_{0e}}\left(1 + R_E \cdot \frac{h_{fe}}{h_{ie}}\right) \left(\text{as } R_0 = \frac{1}{h_{0e}}\right) \tag{9.35}$$

9.5 CURRENT AMPLIFIER WITH NEGATIVE FEEDBACK

The block diagram for a current amplifier with negative feedback is shown in Fig. 9.2(b). The feedback current is a function of output current in this amplifier.

9.5.1 Current Gain with Feedback

The current gains with and without feedback are given by,

$$A_{if} = \frac{I_0}{I_i} \quad \text{and} \quad A_i = \frac{I_0}{I'_i}$$

The currents in the input section are governed by the relation

$$I_i = I_i' + \beta I_0 \tag{9.36}$$

Therefore,

$$A_{if} = \frac{I_0}{I_i' + \beta I_0}$$

$$= \frac{I_0}{I_i'}\left(\frac{1}{1 + \beta \cdot \frac{I_0}{I_i'}}\right)$$

or

$$A_{if} = \frac{A_i}{1 + \beta A_i} \tag{9.37}$$

Thus, the current gain in a current amplifier is reduced by $(1 + \beta A_i)$ when the negative feedback is applied.

9.5.2 Input Resistance with Feedback

The equivalent circuit for a current amplifier is shown in Fig. 9.10. The input resistances with and without feedback are given by

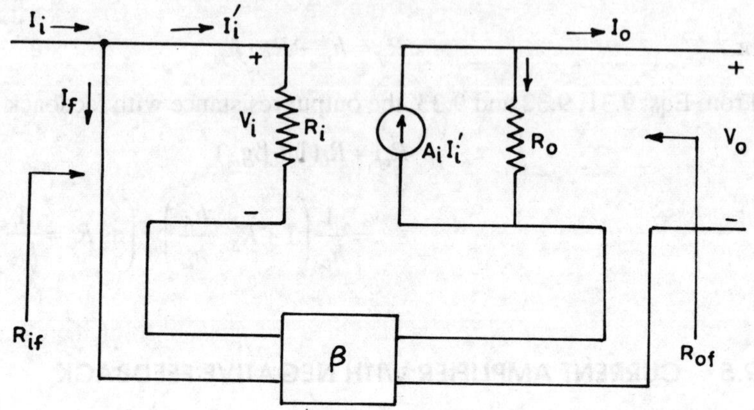

Fig. 9.10. Equivalent circuit for current amplifier with negative feedback.

$$R_{if} = \frac{V_i}{I_i} \text{ and } R_i = \frac{V_i}{I_i'}$$

Using Eq. 9.36,

$$R_{if} = \frac{V_i}{I_i' + \beta I_0}$$

$$= \frac{V_i}{I_i'\left(1 + \beta \frac{I_0}{I_i'}\right)}$$

or
$$R_{if} = \frac{R_i}{1 + \beta A_i} \tag{9.38}$$

where $A_i = \frac{I_0}{I_i'}$. Hence, the input resistance in a current amplifier is reduced by $(1 + \beta A_i)$ when the negative feedback is applied.

9.5.3 Output Resistance with Feedback

From Fig. 9.10, the output resistance with feedback when input current is zero, is given by

$$R_{of} = -\frac{V_0}{I_0} \tag{9.39}$$

In the output section of the circuit,

$$V_0 = (A_i I_i' - I_0)R_0$$

$$= (-A_i \beta I_e - I_0)R_0 \quad \text{(from Eq. 9.36 with } I_i = 0)$$

$$= -I_0(1 + \beta A_i)R_0$$

or

$$-\frac{V_0}{I_0} = (1 + \beta A_i)R_0$$

From Eq. 9.39,

$$R_{0f} = R_0(1 + \beta A_i) \tag{9.40}$$

This indicates that the output resistance of a current amplifier increases with negative feedback by a factor of $(1 + \beta A_i)$.

9.5.4 BJT Circuit for Current Amplifier with Negative Feedback

The circuit for a multistage current amplifier using BJT is shown in Fig. 9.11. The sample of output current I_f is taken from the emitter resistor of the second stage and fed back to the input current I_i in parallel through a feedback resistor R_f. Hence, it is assumed that the output current I_0 is always equal to the emitter current which is divided into I_e and I_f as R_E and R_f are in parallel configuration. Applying current division,

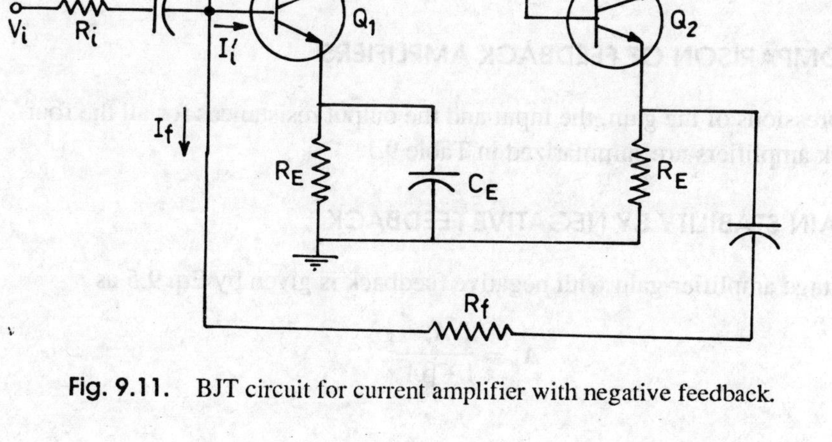

Fig. 9.11. BJT circuit for current amplifier with negative feedback.

$$I_f = \frac{R_E}{R_E + R_f} I_0$$

or
$$\frac{I_f}{I_0} = \frac{R_E}{R_E + R_f}$$

where I_f / I_0 is the feedback parameter β. Hence

$$\beta = \frac{R_E}{R_E + R_f} \tag{9.41}$$

From Eq. 9.37, the current gain with feedback is given by

$$A_{if} = \frac{A_i}{1 + A_i \beta} \tag{9.42}$$

$$\cong \frac{1}{\beta} \quad (\text{as } A_i \gg 1)$$

or
$$A_{if} = \frac{R_E + R_f}{R_E} \tag{9.43}$$

Also, the open loop current gain

$$A_i = A_{i_1} \cdot A_{i_2}$$

$$= h_{f_{e_1}} \cdot h_{f_{e_2}}$$

Hence, using Eqs 9.41 and 9.42

$$A_{if} = \frac{h_{f_{e_1}} \cdot h_{f_{e_2}}}{1 + h_{f_{e_1}} h_{f_{e_2}} \cdot \frac{R_E}{R_E + R_f}} \tag{9.44}$$

9.6 COMPARISON OF FEEDBACK AMPLIFIERS

The expressions of the gain, the input and the output resistances for all the four feedback amplifiers are summarized in Table 9.1.

9.7 GAIN STABILITY BY NEGATIVE FEEDBACK

The voltage amplifier gain with negative feedback is given by Eq. 9.5 as

$$A_{vf} = \frac{A_v}{1 + \beta A_v}$$

Table. 9.1 Comparison of four negative feedback amplifiers

Sr. No.	Types of feedback	Feedback voltage/current	Gain	Input resistance R_{if}	Output resistance R_{of}
1.	Voltage amplifier (voltage-series feedback)	$V_f = \beta V_o$	$A_{vf} = \dfrac{A_v}{1+\beta A_v}$	$R_i(1+\beta A_v)$	$\dfrac{R_0}{1+\beta A_v}$
2.	Transresistance amplifier (voltage-shunt feedback)	$I_f = \beta V_0$	$R_{mf} = \dfrac{R_m}{1+\beta R_m}$	$\dfrac{R_i}{1+\beta R_m}$	$\dfrac{R_0}{1+\beta R_m}$
3.	Trans conductance amplifier (current-series feedback)	$V_f = \beta I_0$	$g_{mf} = \dfrac{g_m}{1+\beta g_m}$	$R_i(1+\beta g_m)$	$R_0(1+\beta g_m)$
4.	Current amplifier (current shunt feedback)	$I_f = \beta I_0$	$A_{if} = \dfrac{A_i}{1+\beta A_i}$	$\dfrac{R_i}{1+\beta A_i}$	$R_0(1+\beta A_i)$

If βA_v is set much greater than one, i.e. if $\beta A_v >> 1$,

$$A_{vf} \cong \frac{A_v}{\beta A_v} = \frac{1}{\beta} \tag{9.45}$$

This shows that as long as $\beta A_v \gg 1$, the overall gain of the amplifier is approximately $1/\beta$ which is independent of open loop gain A_v. If the open loop gain A_v changes owing to replacement of BJT (with same number), ageing components, change in temperature, change in supply voltage, and so on, the overall gain after negative feedback remains stabilized as it is independent of A_v. Thus, gain stability is achieved by negative feedback.

In amplifiers, the open loop gain drops off at lower as well as higher frequencies as described in Chapter 8. This behaviour indicates that the condition $\beta A_v \gg 1$ does not hold good for the lower and higher frequencies and as a result, the negative feedback does not provide gain stabilization in these frequency ranges.

9.8 REDUCTION IN NOISE AND NONLINEAR DISTORTION BY NEGATIVE FEEDBACK

The amount of noise signal (such as power supply hum) and nonlinear distortion (owing to nonlinearity of the amplifier characteristics) are reduced by a factor of $(1 + \beta A_v)$ when negative feedback is used. This is because the feedback network feeds back not only a portion of the output voltage, but also a portion of the distortion in the opposite phase, which appears in the output after amplification and reduces the amount of the distortion present in the output. The distortion after negative feedback is expressed as

$$D_f = \frac{D}{1+\beta A_v} \tag{9.46}$$

where D is the distortion without feedback.

9.9 EFFECT OF NEGATIVE FEEDBACK ON THE FREQUENCY RESPONSE

When negative feedback is applied, the frequency response of an amplifier is increased. The lower half power frequency f_1 is decreased by a factor of $(1 + \beta A_v)$, while the upper half-power frequency f_2 is increased by a factor of $(1 + \beta A_v)$. Hence, these two frequencies with feedback are given by

$$f_{1f} = \frac{f_1}{1+\beta A_v} \tag{9.47}$$

and

$$f_{2f} = f_2(1+\beta A_v) \tag{9.48}$$

Hence, the bandwidth of the amplifier $f_{2f} - f_{1f}$ increases by negative feedback as shown in Fig. 9.12.

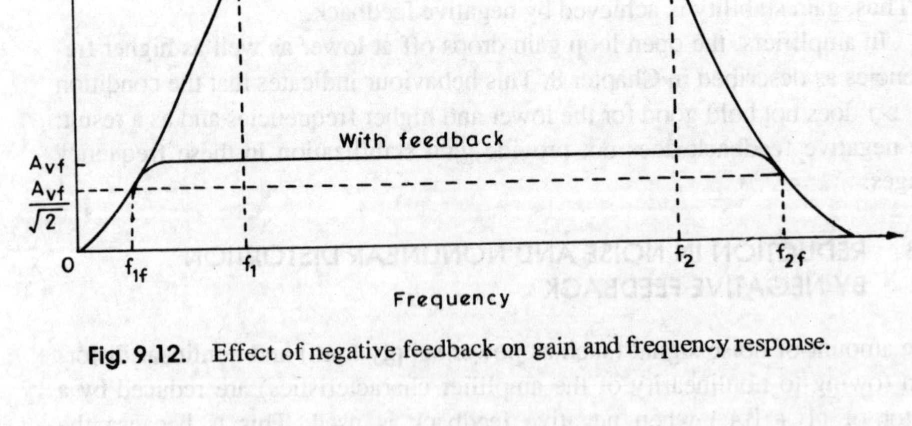

Fig. 9.12. Effect of negative feedback on gain and frequency response.

9.10 SOLVED PROBLEMS

Problem 1

Find the gain of a negative feedback amplifier if gain without feedback A_v = 500 and feedback factor $\beta = 0.1$.

Solution

The gain of an amplifier after feedback (negative) is given by

$$A_{vf} = \frac{A_v}{1 + A_v \cdot \beta} = \frac{500}{1 + (500 \times 0.1)}$$

$$= 9.8$$

Problem 2

Calculate the change in the gain of negative feedback amplifier with $A_v = 500$ and $\beta = 0.1$ if the gain A_v changes by 25%.

Solution

The gain of a negative feedback amplifier is given by

$$A_{vf} = \frac{A_v}{1 + A_v \cdot \beta}$$

Differentiating the above equation

$$\frac{dA_{vf}}{A_f} = \frac{1}{A_v \cdot \beta} \frac{dA_v}{A_v} \quad \text{for } A_v \cdot \beta \gg 1$$

$$= \frac{1}{500 \times 0.1} \times 25\% = 0.5\%$$

Hence, the gain after negative feedback changes by 0.5% only even if the intrinsic gain of the amplifier changes by 25%. This shows the gain stabilization owing to negative feedback.

Problem 3

Find the gain, input impedance, and output impedance of a voltage series feedback amplifier if $A_v = 750$, $\beta = 0.1$, $R_i = 1\ \text{k}\Omega$, and $R_o = 10\ \text{k}\Omega$.

Solution

The gain of a voltage series feedback amplifier is given by

$$A_{vf} = \frac{A_v}{1 + A_v \cdot \beta} = \frac{750}{1 + 750 \times 0.1} = 9.87$$

The input impedance after feedback is

$$R_{if} = R_i(1 + A_v \cdot \beta)$$

$$= 1(1 + 750 \times 0.1) = 76\ \text{k}\Omega$$

and the output impedance after feedback is

$$R_{of} = \frac{R_0}{1 + A_v \cdot \beta} = \frac{10\ \text{k}\Omega}{1 + 750 \times 0.1}$$

$$= 131.6\ \Omega$$

Problem 4

Calculate the voltage gain and input and output impedances with and without negative feedback in an emitter follower circuit with the following parameters: $R_E = 1$ kΩ, $h_{ie} = 1$ kΩ, $h_{fe} = 50$, and $h_{oe} = 2 \times 10^{-5}$ mhos.

Solution

In an emitter follower circuit

$$Z_{if} = h_{ie}(1 + g_m R_E)$$

$$= h_{ie} + h_{fe} R_E$$

$$= 1 + 50 \times 1 = 51\ \text{k}\Omega$$

$$Z_{of} = \frac{h_{ie}}{1 + h_{fe}} = \frac{1000}{1 + 50} = 19.6\ \Omega$$

and

$$A_{vf} = \frac{g_m R_E}{1 + g_m R_E} = \frac{\frac{50}{1} \times 1}{1 + \frac{50}{1} \times 1} = 0.98$$

If feedback is not applied, then only the BJT parameters will be applicable, i.e.,

$$Z_{ie} = h_{ie} = 1\ \text{k}\Omega$$

$$Z_{0e} = \frac{1}{h_{0e}} = \frac{1}{2 \times 10^{-5}} = 0.5\ \text{k}\Omega$$

and

$$A_{ve} = g_m R_E = \frac{50}{1} \times 1 = 50$$

Problem 5

Find the gain of voltage amplifier with negative feedback following the circuit of Fig. 9.5, where

$$R_F = 10\ \text{k}\Omega \text{ and } R_E = 1\ \text{k}\Omega$$

Solution

The amplifier gain with negative feedback can approximately be given by

$$A_{vf} \approx \frac{1}{\beta}$$

where
$$\beta = \frac{R_E}{R_F + R_E} = \frac{1}{11}$$

$\therefore$
$$A_{vf} = 11$$

Problem 6

Calculate the voltage gain and input and output impedances for a transconductance amplifier with the following parameters:
$h_{ie} = 1$ kΩ, $h_{fe} = 50$, $h_{oe} = 2 \times 10^{-5}$ mhos, and $R_E = 1$ kΩ.

Solution

In a transconductance amplifier

$$\beta_m = R_E = 1 \text{ k}\Omega$$

and the transconductance

$$g_m = \frac{h_{fe}}{h_{ie}} = \frac{50}{1 \times 10^{+3}} = 50 \times 10^{-3} \text{ mhos}$$

Hence, the input impedance

$$R_{if} = h_{ie} + h_{fe} \cdot R_E$$

$$= 1 + 50 \times 1 = 51 \text{ k}\Omega$$

and the output impedance

$$R_{0f} = \frac{1}{h_{0e}}(1 + g_m R_E)$$

$$= \frac{1}{2 \times 10^{-5}}(1 + 50 \times 10^{-3} \times 1 \times 10^{3})$$

$$= 2.55 \text{ M}\Omega$$

Problem 7

Find the voltage gain and input and output impedances in a transresistance amplifier with

$h_{ie} = 1$ kΩ, $h_{oe} = 2 \times 10^{-5}$ mhos, $R_L = 2$ kΩ, $R_E = 1$ kΩ, $R_i = 5$ kΩ, and $R_f =$ 50 kΩ.

Solution

Following Eq. 9.23, the input impedance of a transresistance amplifier

$$R_{if} = \frac{h_{ie}}{1+\frac{h_{fe}R_L}{R_f}} = \frac{1}{1+\frac{50\times 2}{50}} \text{ k}\Omega$$

$$= 334 \ \Omega$$

Equation 9.24 defines the output impedance as

$$R_{of} = \frac{R_0}{1+\frac{h_{fe}R_L}{R_f}} \quad \text{where } R_0 = \frac{1}{h_{0e}}$$

$$= \frac{\frac{1}{2\times 10^{-5}}}{1+\frac{50\times 2}{50}} = \frac{10^5}{6} \ \Omega$$

$$= 16.7 \text{ k}\Omega$$

The transresistance after feedback is

$$R_{mf} = \frac{R_m}{1+\beta \cdot R_m} \cong \frac{1}{\beta} \quad (\text{as } \beta R_m \gg 1)$$

By definition

$$R_{mf} = \frac{\text{Output voltage}}{\text{Input current}} = \frac{V_0}{I_b}$$

$$\therefore \qquad \frac{V_0}{I_b} = \frac{1}{\beta}$$

$$= R_f \quad (\text{by Eq. 9.22})$$

The input voltage

$$V_i = I_b \cdot R_i$$

$\therefore$ voltage gain

$$A_v = \frac{V_0}{V_i} = \frac{I_b R_f}{I_b R_i} = \frac{R_f}{R_i}$$

$$= \frac{50}{5} = 10$$

Problem 8

Find the voltage gain for a current BJT amplifier of the circuit of Fig. 9.11 if $R_i = 5$ kΩ, $R_f = 4.7$ kΩ, $R_E = 100\ \Omega$, and $R_L = 10$ kΩ.

Solution

The voltage gain in the above mentioned circuit (Fig. 9.11) is given by

$$A_{vf} = \frac{V_0}{V_i} \quad \text{where } V_0 = I_c R_L \text{ and } V_i = I_b R_i$$

$$\therefore \quad A_{vf} = \frac{I_c R_L}{I_b R_i} = A_i \frac{R_L}{R_i}$$

where
$$A_i = \frac{1}{\beta} = \frac{R_E + R_f}{R_E} \quad \text{(from Eq. 9.41)}$$

$$\therefore \quad A_{vf} = \frac{R_E + R_f}{R_E} \cdot \frac{R_L}{R_i}$$

$$= \frac{100 + 4700}{100} \times \frac{10 \times 10^3}{5 \times 10^3}$$

$$= 96$$

Problem 9

Show the effect of negative feedback on the lower and upper half-power frequencies in an amplifier if feedback factor $\beta = 0.1$ and gain $A_v = 100$.

Solution

The lower half-power frequency

$$f_{1f} = \frac{f_1}{1 + A_v \cdot \beta} = \frac{f_1}{1 + 10} = \frac{f_1}{11}$$

and the upper half-power frequency

$$f_{2f} = f_2(1 + A_v \beta) = 11 f_2$$

Typically, if $f_1 = 110$ Hz and $f_2 = 0.5$ MHz, then $f_{1f} = 10$ Hz and $f_{2f} = 5.5$ MHz. Hence, the bandwidth increases by approximately 5 MHz.

Questions and Problems

1. What do you mean by feedback? What are regenerative and degenerative feedbacks?
2. Describe voltage and current feedbacks.
3. How many types of feedback do you know? Explain them.
4. What is the effect of negative feedback on the gain of an amplifier?

5. How do the input and output impedances change with negative feedback?
6. Distinguish between open loop and closed loop gains.
7. Describe a *BJT* circuit for voltage amplifier with negative feedback.
8. Which amplifier circuit will you use if the feedback voltage is a function of the output voltage? Describe it.
9. Which amplifier circuit is preferred if the feedback current is a function of output voltage? Describe it.
10. Describe a current amplifier with negative feedback.
11. What is the effect of negative feedback on the bandwidth of an amplifier.
12. How is the distortion reduced by negative feedback?
13. Discuss the role of negative feedback in gain stabilization.
14. Find the intrinsic gain of amplifiers if gain after negative feedback is 10 and $\beta = 0.09$ and 0.1.
15. Show the gain stabilization of a negative feedback amplifier by taking any value of A_v and β.
16. Calculate the voltage gain, and input and output impedances for a voltage series feedback amplifier if $A_v = 500$, $\beta = 0.1$, $R_i = 1.5$ kΩ, and $R_O =$ 7.5 kΩ.
17. Find the voltage gain and input and output impedances with and without negative feedback in an emitter follower circuit if, $R_E = 1.5$ kΩ, $h_{ie} = 1$ kΩ, $h_{fe} = 75$, $h_{Oe} = 2.5 \times 10^{-5}$ mhos.
18. If $\beta = 0.1$, find the gain of a voltage amplifier with negative feedback.
19. Find the voltage gain and input and output impedances for a transconductance amplifier with the following circuit parameters: $h_{ie} = 1.5$ kΩ, $h_{fe} = 75$, $h_{oe} = 2.5 \times 10^{-5}$ mhos, and $R_E = 1$ kΩ.
20. Repeat problem 19 for a transresistance amplifier if $R_L = 2.5$ kΩ, $R_i = 5$ kΩ, and $R_f = 50$ kΩ
21. Find the increase in bandwidth of an amplifier by the application of negative feedback if $A_v = 500$, $\beta = 0.09$, $f_1 = 20$ Hz and $f_2 = 1$ MHz.
22. Calculate the voltage and current gains for a current amplifier if $R_i = 2$ kΩ, $R_f = 2.5$ kΩ, $R_E = 150$ Ω, and $R_L = 5$ kΩ.

10

Power (Large Signal Low Frequency) Amplifier

In the earlier chapters, small signal amplifiers have been described, where the input signals are generally small, i.e the signal voltage and current are small. Therefore, the amount of power handling capacity and power efficiency have little importance. The only requirements of small signal amplifiers are linearity and gain. When large input signals are to be amplified for the operation of output devices such as speakers and motors, the amplifier used must be capable of handling large amount of power and its efficiency of converting input dc power to output ac power must be high. Such an amplifier is known as a power amplifier. Hence, the important requirements of power amplifiers are the power efficiency of the circuit, the maximum amount of power which the circuit can handle, and impedance matching between amplifier output and the load, so that the maximum power is transferred to the output device.

10.1 CLASSIFICATION OF POWER AMPLIFIERS

The power amplifiers are classified on the basis of their operating point as class A, B, C, AB, D, E, F, G, or H. In other words, this classification is based upon the duration of the collector current flow with respect to the input signal, which is assumed to be a full 360°.

In class-A amplifiers, there is always collector current regardless of the time in the cycle of the applied signal. The operating point Q is selected at the middle of the linear region of the transistor as shown in Fig. 10.1(a). This class of amplifiers is considered to be linear because output signal is an exact replica of the input signal.

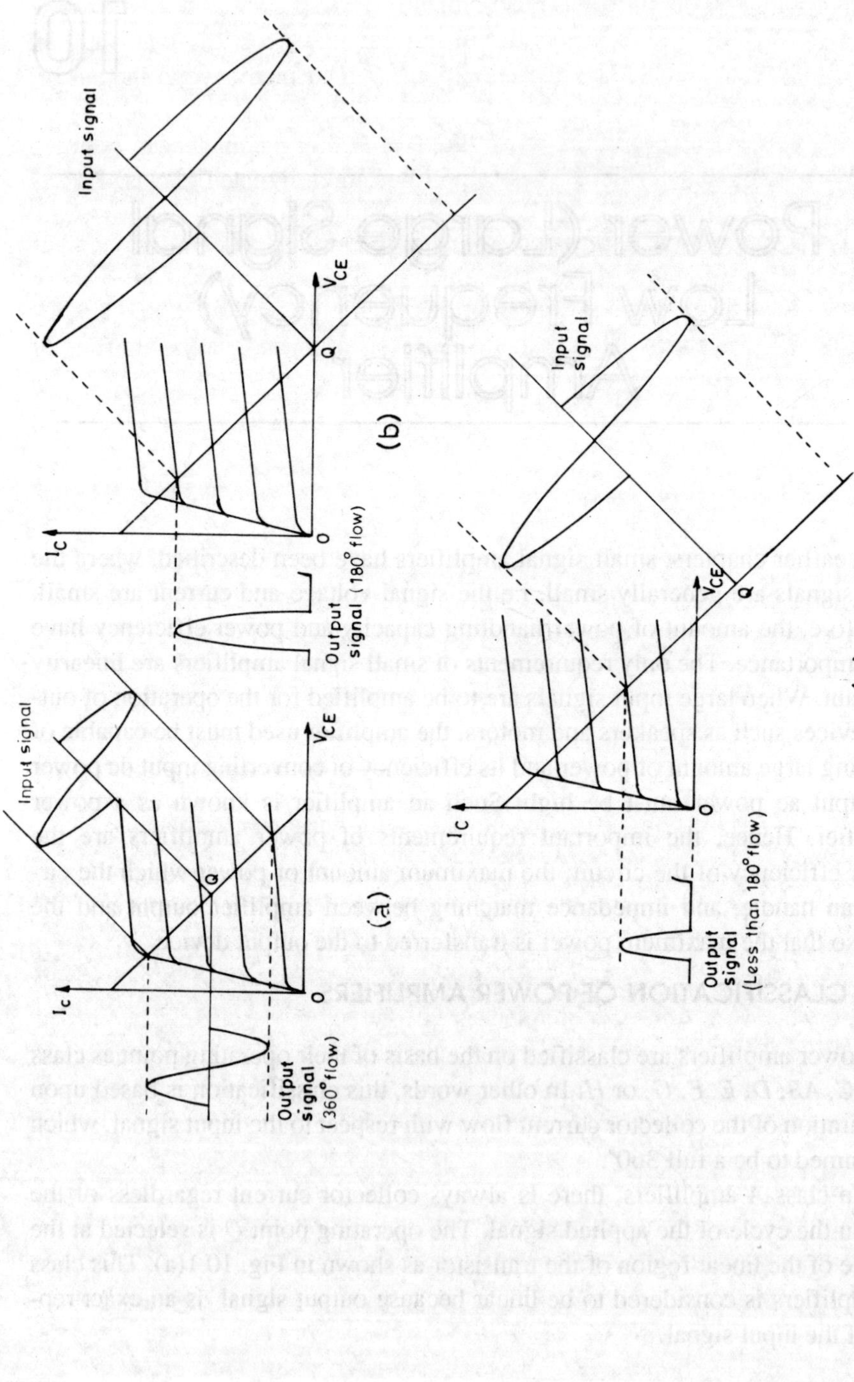

Fig 10.1. Operating conditions for power amplifiers (a) Class-A, (b) Class-B, (c) Class-C.

In class-*B* amplifiers, the collector current flows only for one half of the input signal. The operating point *Q* is located near the cut-off, as shown in Fig. 10.1(b). In order to get output current for full cycle, two class-*B* amplifiers are used in a combination known as push-pull.

In class-*C* amplifiers, the collector current flows for less than the half cycle of the input signal. The operating point *Q* is located beyond the cut-off point, as shown in Fig.10.1(c).

There are some amplifiers which have operating parameters in between class-*A* and class-*B* amplifiers. These amplifiers are designated as class-*AB*.

10.2 EFFICIENCY OF A POWER AMPLIFIER

Figure 10.2 shows a typical power amplifier circuit. The power P_S supplied by the battery V_{CC} is distributed in both the load resistor R_L and the transistor as P_R and P_D respectively, i.e.,

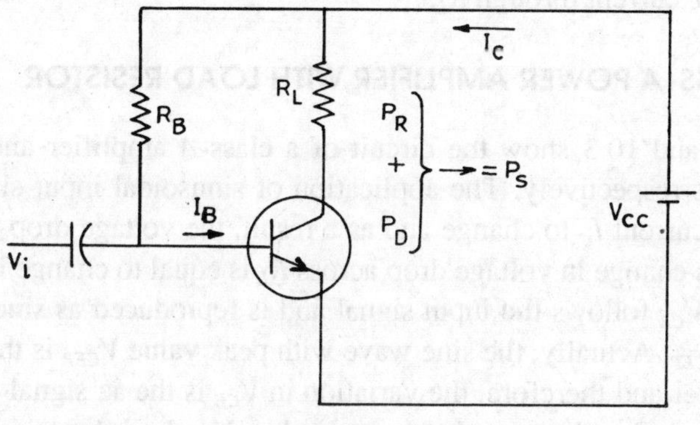

Fig. 10.2. Class-*A* power amplifier circuit showing power distribution.

$$P_S = P_R + P_D \tag{10.1}$$

In an amplifier circuit with input signal, the current drawn from the battery V_{CC} has a wave shape having an ac signal component riding on a dc bias component. That is, the *dc* voltage source in an amplifier circuit furnishes a dc plus an ac current, which is not the case with circuits having only passive elements such as resistors, capacitors, and inductors. Hence the power across R_L is the sum of the powers owing to ac and dc currents, i.e.,

$$P_R = P_{RD} + P_{RA}$$

The power owing to ac current P_{RA} is the only useful component in an amplifier because it represents the ac output power P_0, i.e.,

$$P_{RA} = P_0$$

Hence $$P_S = P_0 + P_{RD} + P_D \quad (10.2)$$

The terms P_{RD} and P_D represent the losses in the circuit. In fact, the operation of the power amplifier is to convert as much of the dc power P_S drawn from the battery into the ac power P_0 across R_L. Hence, the efficiency of the circuit is defined as,

$$\% \text{ Efficiency} = \eta = \frac{P_{0(ac)}}{P_{S(dc)}} \times 100 \quad (10.3)$$

The powers P_S and P_o can be expressed as

$$P_S = V_{CC} \cdot I_C \quad (10.4)$$

and $$P_0 = I^2_{C(rms)} \cdot R_L \quad (10.5)$$

where I_C is the average current drawn from the battery and $I_{C(rms)}$ is the r.m.s value of the *ac* current through R_L.

10.3 CLASS-A POWER AMPLIFIER WITH LOAD RESISTOR

Figures 10.2 and 10.3 show the circuit of a class-*A* amplifier and its output characteristics respectively. The application of sinusoidal input signal causes the collector current I_C to change and as a result, the voltage drop across R_L is changed. This change in voltage drop across R_L is equal to change in V_{CE}. Thus the potential V_{CE} follows the input signal and is reproduced as sine wave with peak value V_{CEP}. Actually, the sine wave with peak value V_{CEP} is the end result of the amplifier and therefore, the variation in V_{CE} is the ac signal output voltage V_o whose peak value may be designated as V_{op}. In order to get the maximum possible symmetrical peak swing for V_o, the operating point must be at the centre of the load line or,

$$V_{OP} = \frac{V_{CC}}{2}$$

The corresponding values of V_{CE} and I_C will be

$$I_C = \frac{V_{CC}}{2R_L} \text{ and } V_{CE} = \frac{V_{CC}}{2}$$

The maximum r.m.s. value of the ac output power is equal to

$$P_{0(ac)} = \frac{V_{CE}}{\sqrt{2}} \cdot \frac{I_C}{\sqrt{2}}$$

$$= \frac{V_{CC}}{2\sqrt{2}} \cdot \frac{V_{CC}}{2\sqrt{2}R_L}$$

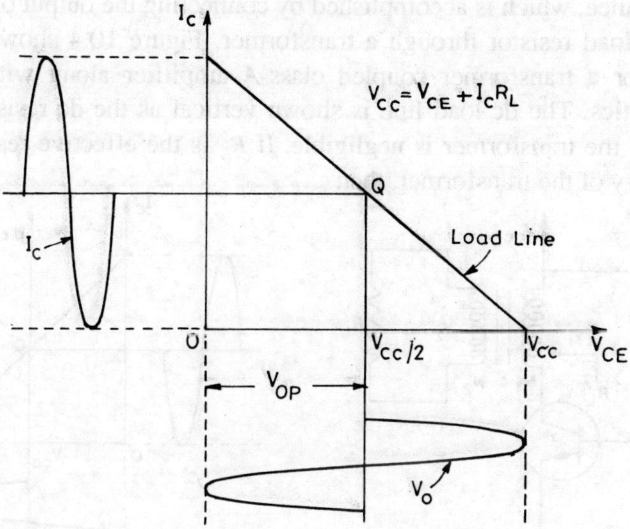

Fig. 10.3. Output characteristics of a class-*A* power amplifier.

or
$$P_{0(ac)} = \frac{V_{CC}^2}{8R_L} \tag{10.6}$$

The dc input power will be

$$P_{S(dc)} = V_{CC} \cdot I_C$$

$$= V_{CC} \cdot \frac{V_{CC}}{2R_L}$$

or
$$P_{S(dc)} = \frac{V_{CC}^2}{2R_L} \tag{10.7}$$

Using Eqs. 10.6 and 10.7, the maximum efficiency of the class-*A* power amplifier is

$$\eta = \frac{P_{0(ac)}}{P_{S(dc)}} \times 100\%$$

$$= 25\% \tag{10.8}$$

Hence, this type of amplifier converts only 25 percent of the dc input power into ac output power.

10.4 CLASS-A POWER AMPLIFIER WITH A TRANSFORMER

The efficiency of a class-*A* amplifier can be increased by eliminating the *dc* load resistance, which is accomplished by connecting the output of the transistor to the load resistor through a transformer. Figure 10.4 shows the circuit diagram for a transformer coupled class-*A* amplifier along with its output characteristics. The dc load line is shown vertical as the dc resistance of the primary of the transformer is negligible. If R_L' is the effective resistance seen into primary of the transformer, then

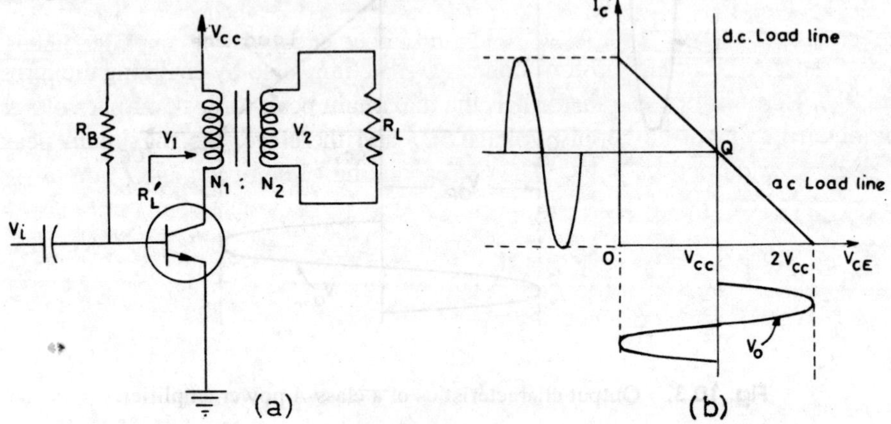

Fig. 10.4. Transformer coupled class-*A* amplifier (a) Circuit (b) Output characteristics.

$$\frac{R_L'}{R_L} = \frac{V_1/I_1}{V_2/I_2}$$

$$= \frac{V_1}{I_1} \cdot \frac{I_2}{V_2}$$

$$= \left(\frac{V_1}{V_2}\right) \cdot \left(\frac{I_2}{I_1}\right)$$

$$= \frac{N_1}{N_2} \cdot \frac{N_1}{N_2}$$

or $$\frac{R_L'}{R_L} = \left(\frac{N_1}{N_2}\right)^2 \tag{10.9}$$

where N_1 and N_2 are the number of turns in the primary and the secondary of the transformer. Equation 10.9 shows that the ratio of the transformer input and output resistance varies directly as the square of the transformer turns ratio,

$$a = \frac{N_1}{N_2}$$

$$\therefore \qquad \frac{R_L'}{R_L} = a^2$$

or
$$R_L' = a^2 R_L \qquad (10.10)$$

Hence, the load R_L appears as R_L' in the output circuit of the amplifier. In this way, a transformer is used for impedance matching also by changing the turns ratio *a*. Because of the transformer, the maximum peak-to-peak output voltage is equal to twice the supply voltage V_{CC} and therefore, the maximum peak value of the output $V_{OP} = V_{CC}$. The corresponding value of V_{CE} and I_C are

$$I_C = \frac{V_{CC}}{R_L'} \text{ and } V_{CE} = V_{CC}$$

The maximum r.m.s. of the ac output power,

$$P_{0(ac)} = \frac{V_{CE}}{\sqrt{2}} \cdot \frac{I_C}{\sqrt{2}}$$

$$= \frac{V_{CC}}{\sqrt{2}} \cdot \frac{V_{CC}}{\sqrt{2} R_L'}$$

$$= \frac{V_{CC}^2}{2R_L'} \qquad (10.11)$$

The dc input power,

$$P_{S(dc)} = V_{CC} \cdot I_C$$

$$= V_{CC} \cdot \frac{V_{CC}}{R_L'}$$

$$= \frac{V_{CC}^2}{R_L'} \qquad (10.12)$$

Using Eqs.10.11 and 10.12, the maximum efficiency of class-*A* power amplifier with transfomer coupling is given by

$$\eta = \frac{P_{0(ac)}}{P_{S(dc)}} \times 100\%$$

$$= 50\% \qquad (10.13)$$

Hence, the efficiency becomes double in comparison to the direct coupled class- *A* amplifier.

10.5 CLASS-*B* PUSH-PULL AMPLIFIER WITH A TRANSFORMER

As described in Sec.10.1, in the class-*B* power amplifier, the operating point is selected at the cut-off point, and the output current flows only for half of the input signal. Consequently, the extreme distortion is introduced owing to one half cycle of the non-operation period. Though the class-*B* operation is more efficient than class-*A*, it is not practical owing to high distortion. But a low-distortion, high-efficiency operation can be achieved using a second transistor operating in the class-*B* condition and during the period of nonoperation of the first transistor. Such a circuit connection is known as a push-pull class-*B* amplifier as shown in Fig.10.5. This circuit requires an input transformer with centre-tapping in the secondary to produce opposite polarity signals to be fed to the two transistor inputs, and an output transformer to couple the load resistor. Thus, the voltage at the base of Q_1 is 180° out of phase with the voltage at the base of Q_2. In this condition, each transistor conducts only during the portion of the cycle when its base is positive with respect to the emitter. The currents flowing through the collectors are shown as I_{C_1} and I_{C_2}, which combine to give the output current in the primary of the output transformer for a full cycle of the input signal. As the operating point for each transistor is $\left(V_{CC}, \frac{I_{C\max}}{2}\right)$, the ac output current can be written as

$$I_C = \frac{I_{C\max}}{2}$$

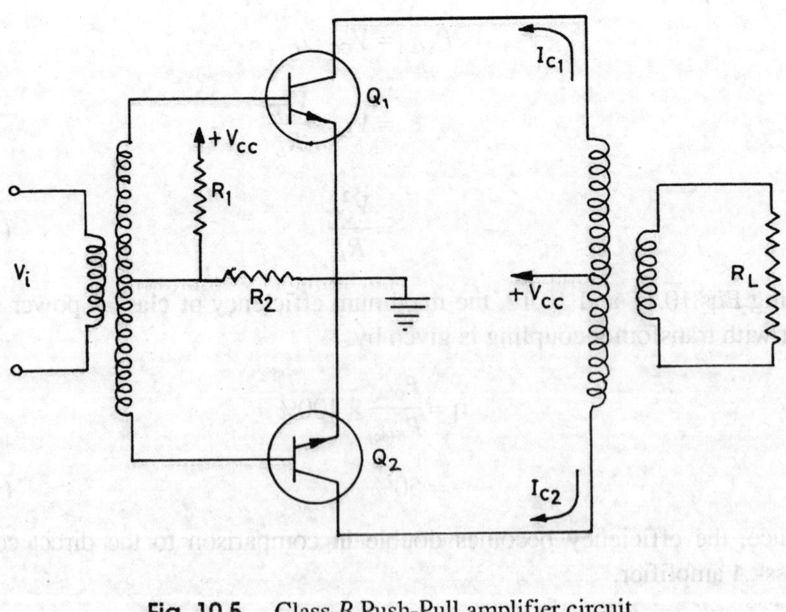

Fig. 10.5. Class-*B* Push-Pull amplifier circuit.

Figure 10.6 shows the collector characteristics of the two transistors, combined in such a way that the output current flows during the full cycle of the input signal. As the output current in the primary of the output transformer is like a full-wave rectified signal, its average value will be

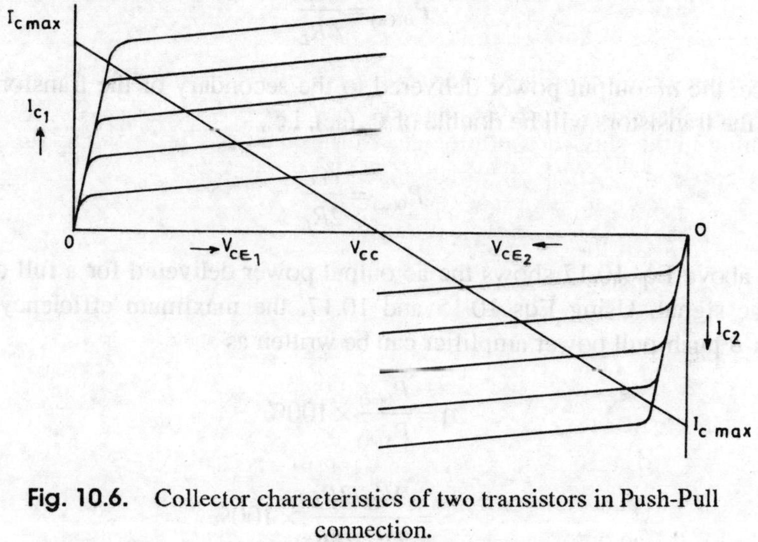

Fig. 10.6. Collector characteristics of two transistors in Push-Pull connection.

$$I_{dc} = \frac{2}{\pi} \cdot I_{C\max}$$

with $I_{C\max} = \dfrac{V_{CC}}{R_L''}$, where R_L'' is the load reflected from the secondary to the half primary of the output transformer. Hence

$$I_{dc} = \frac{2V_{CC}}{\pi R_L''} \tag{10.14}$$

The amount of the input dc power for the full cycle of the ac signal is given by

$$P_{S(dc)} = V_{CC} \cdot I_{dc}$$

$$= \frac{2V_{CC}^2}{\pi R_L} \tag{10.15}$$

The r.m.s. ac output power delivered to the secondary of the output transfomer by one transistor (i.e. for one-half of the input signal) can be written as

$$P_{01(ac)} = \left(\frac{V_{CC}}{\sqrt{2}}\right)\left(\frac{I_C}{\sqrt{2}}\right)$$

$$= \left(\frac{V_{CC}}{\sqrt{2}}\right)\left(\frac{I_{C\max}}{2\sqrt{2}}\right)\left(\text{as } I_c = \frac{I_{C\max}}{2}\right)$$

$$= \left(\frac{V_{CC}}{\sqrt{2}}\right)\left(\frac{V_{CC}}{2\sqrt{2}R_L''}\right)\left(\text{as } I_{C\max} = \frac{V_{CC}}{R_L''}\right)$$

or
$$P_{01(ac)} = \frac{V_{CC}^2}{4R_L''} \tag{10.16}$$

Hence, the *ac* output power delivered to the secondary of the transformer by both the transistors will be double of P_{01}(ac), i.e.,

$$P_{0(ac)} = \frac{V_{CC}^2}{2R_L''} \tag{10.17}$$

This above Eq. 10.17 shows the ac output power delivered for a full cycle of the ac signal. Using Eqs 10.15 and 10.17, the maximum efficiency of the class-*B* push-pull power amplifier can be written as

$$\eta = \frac{P_{0(ac)}}{P_{S(dc)}} \times 100\%$$

$$= \frac{V_{CC}^2/2R_L''}{2V_{CC}^2/\pi R_L''} \times 100\%$$

$$= \frac{\pi}{4} \times 100\%$$

$$= 78.5\% \tag{10.18}$$

when the input signal is not able to give maximum output current swing $I_{C\max}$, the efficiency becomes less than 78.5%.

10.6 CLASS-*B* TRANSFORMERLESS PUSH-PULL AMPLIFIER

In the class-*B* push-pull amplifier described in Sec. 10.5 two transfomers are involved, one in the input stage for phase reversal and the other in the output stage. As transformers are costly and physically large, several other circuit arrangements to replace transformers have been developed.

10.6.1 Phase Splitter to Replace Input Transfomer

In order to operate the two stage push-pull circuit in the class-*B* operation, two opposite phase input signals are required. In Sec.10.5, a centre tapped transformer is used to obtain these opposite phase signals. Figure 10.7 shows a circuit known as paraphase circuit, in which the outputs at the collector and the emitter are taken, as the two outputs are out of phase by 180°. As the gain for the signal from the emitter is one for the emitter follower circuit, the gain for the collector output signal is also made one by a proper selection of R_L, R_E, and h_{fe}. In this way, this circuit provides no gain but results in two opposite phase signals required for the two stages of the push-pull amplifier. The disadvantage of this circuit is that the two opposite phased output signals do not remain equal under load conditions. This is because the load of the emitter and the collector circuits are different.

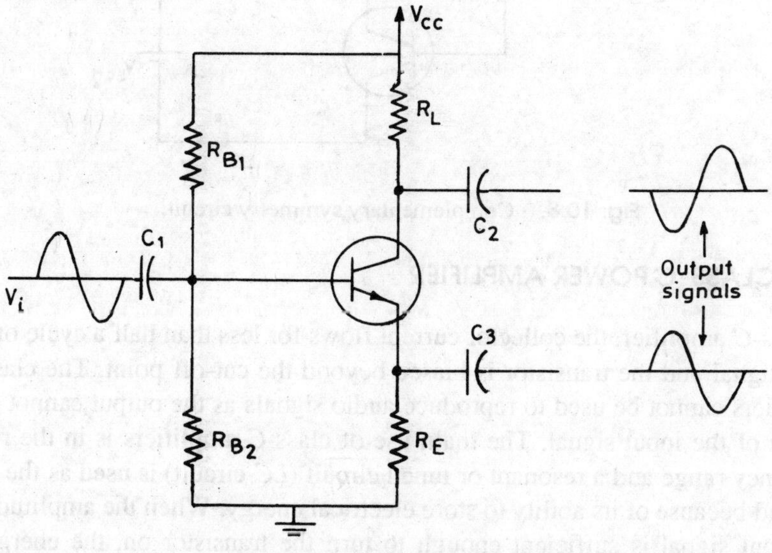

Fig. 10.7. Paraphase circuit.

10.6.2 Complementary Symmetry Circuit to Replace Input and Output Transfomers

Various circuits have been developed to replace not only the input phase inverting transfomer but also to replace the output transfomer so that the circuit becomes completely transfomerless. Figure 10.8 shows one such circuit known as a *complementary symmetry circuit*, in which the complementary transistors (i.e., n-p-n and p-n-p) are used, instead of using the same type. The two transistors conduct on opposite half cycles of the input signal. For example, the positive half-cycle of the input signal derives the n-p-n transistor into conduction while the p-n-p transistor remains off. The negative half-cycle of the input signal makes the n-p-n transistor off and the p-n-p transistor on. Hence, the complete cycle of the signal output is developed across the load. The major disadvantage of this circuit is that it requires two power supplies. Also, some cross-over distortion is developed in the output signal as the exact switching of one transistor to off and the other to on at the zero voltage is not possible; both the transistors may be off or partially conducting simultaneously.

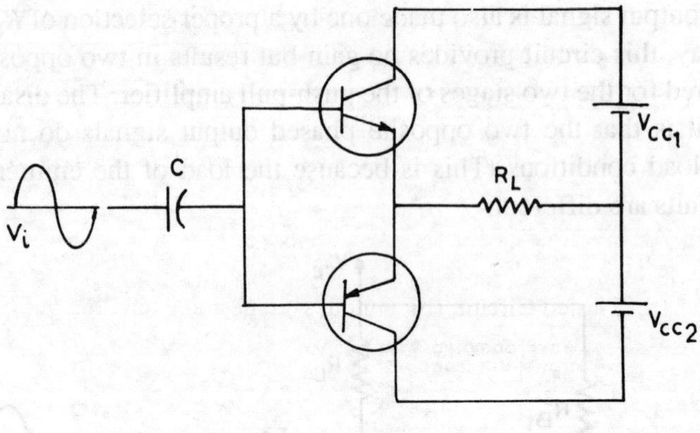

Fig. 10.8. Complementary symmetry circuit.

10.7 CLASS-C POWER AMPLIFIER

In class-*C* amplifier, the collector current flows for less than half a cycle of the input signal and the transistor is biased beyond the cut-off point. The class-*C* amplifiers cannot be used to reproduce audio signals as the output cannot be a replica of the input signal. The main use of class-*C* amplifiers is in the radio frequency range and a resonant or tuned circuit (*LC* circuit) is used as the output load because of its ability to store electrical energy. When the amplitude of the input signal is sufficient enough to turn the transistor on, the energy is supplied to the *LC* circuit from the dc source. The transistor switch is shown by *S* in Fig.10.9. This results in ac signal appearing across the secondary winding L_2, as shown in the figure. Thus, oscillations are produced with frequency equal to the resonant frequency of *LC* circuit owing to the transfer of energy from inductor to capacitor and back. These oscillations decay owing to circuit losses. Hence, in order to maintain oscillations and to get a complete cycle of output current, low circuit losses or a high *Q*-tuned circuit is required in the class-*C* amplifier. The output voltage may also be kept at the same level by injecting excitation pulses with *S* in phase with the output signal. In the class-*C* operation, the current flows for very short periods of time during each cycle and as a result, the power drawn from the dc supply is low. Hence, the efficiency becomes very high, reaching 85 percent.

Figure 10.10 shows the operation of a class-*C* *RF* amplifier using waveform analysis. As the bias is shifted beyond the cut-off value, the conduction angle decreases and the efficiency is increased. Here the conduction angle means the duration of *RF* input cycle during which the collector current flows. The output resonant circuit produces the sine wave output voltage, as shown in Fig. 10.10(d). A typical circuit for a class-*C* amplifier is shown in Fig. 10.11. The efficiency of a turned circuit is defined as

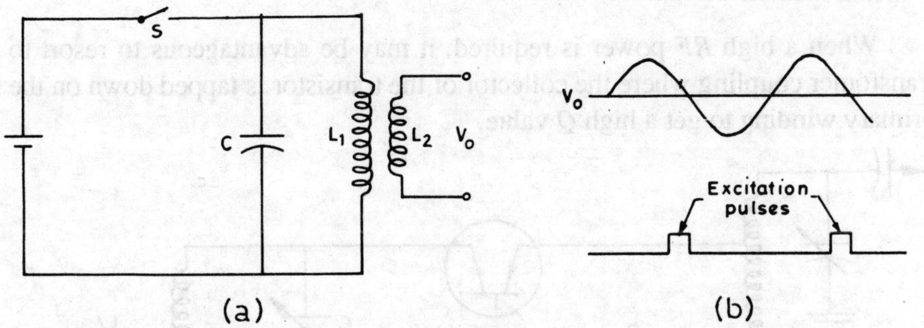

Fig. 10.9. (a) Turned Circuit (b) Output voltage and excitation pulses.

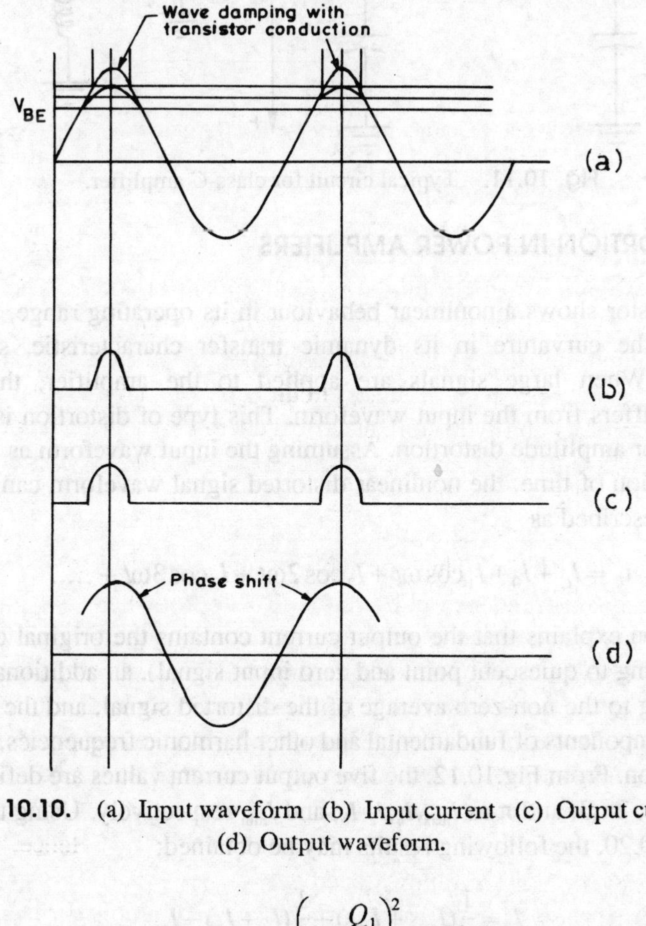

Fig. 10.10. (a) Input waveform (b) Input current (c) Output current (d) Output waveform.

$$\eta_{(LC\,\text{circuit})} = \left(1 - \frac{Q_1}{Q_2}\right)^2 \times 100\% \qquad (10.19)$$

where Q_1 and Q_2 are the Q values in loaded and unloaded conditions. Hence, a 10:1 ratio of Q_2 and Q_1 gives 81 percent efficiency.

When a high RF power is required, it may be advantageous to resort to transfomer coupling where the collector of the transistor is tapped down on the primary winding to get a high Q value.

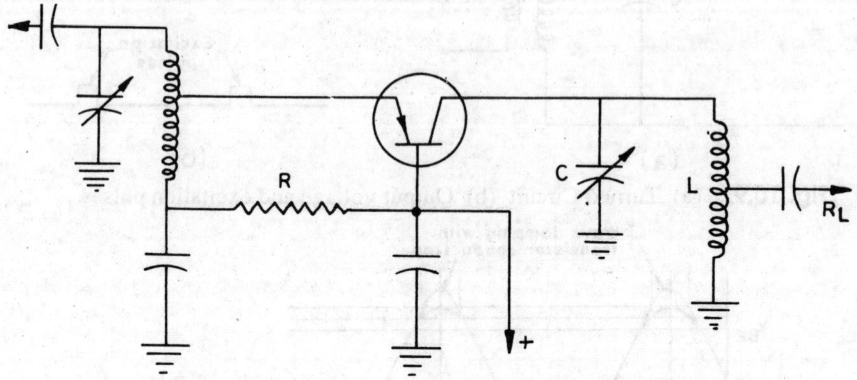

Fig. 10.11. Typical circuit for class-C amplifier.

10.8 DISTORTION IN POWER AMPLIFIERS

Every transistor shows a nonlinear behaviour in its operating range, as is evident from the curvature in its dynamic transfer characteristic, shown in Fig. 10.12. When large signals are applied to the amplifier, the output waveform differs from the input waveform. This type of distortion is called a non-linear, or amplitude distortion. Assuming the input waveform as a simple cosine function of time, the nonlinear distorted signal waveform can approximately be described as

$$i_C = I_C + I_0 + I_1 \cos \omega t + I_2 \cos 2\omega t + I_3 \cos 3\omega t + \ldots.. \qquad (10.20)$$

This equation explains that the output current contains the original current I_C (corresponding to quiescent point and zero input signal), an additional dc current I_0 owing to the non-zero average of the distorted signal, and the distorted ac signal components of fundamental and other harmonic frequencies, viz. $2\omega t$, $3\omega t$, and so on. From Fig.10.12, the five output current values are defined at $\omega t = 0$, $\pi/3$, $\pi/2$, $2\pi/3$, and π as I_{max}, I_a, I_c, I_b, and I_{min} respectively. Using these values in Eq. 10.20, the following results may be obtained:

$$I_0 = \frac{1}{6}(I_{max} + I_{min}) + \frac{1}{3}(I_a + I_b) - I_c$$

$$I_1 = \frac{1}{3}(I_{max} - I_{min}) + \frac{1}{3}(I_a - I_b)$$

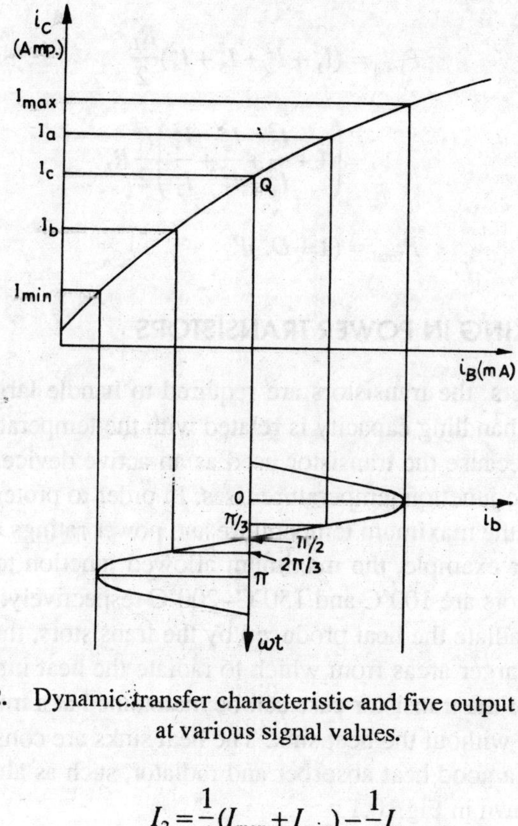

Fig. 10.12. Dynamic transfer characteristic and five output current values at various signal values.

$$I_2 = \frac{1}{4}(I_{max} + I_{min}) - \frac{1}{2}I_c$$

$$I_3 = \frac{1}{6}(I_{max} - I_{min}) - \frac{1}{3}(I_a - I_b)$$

$$I_4 = \frac{1}{12}(I_{max} + I_{min}) - \frac{1}{3}(I_a - I_b) + \frac{1}{2}I_c \tag{10.21}$$

By definition, the 2nd, 3rd, and 4th order harmonic distortions are defined as

$$\%D_2 = \left(\frac{I_2}{I_1}\right) \times 100, \%D_3 = \left(\frac{I_3}{I_1}\right) \times 100, \%D_4 = \left(\frac{I_4}{I_1}\right) \times 100 \tag{10.22}$$

The total distortion when only up to the 4th harmonics are present is given by,

$$D = \sqrt{D_2^2 + D_3^2 + D_4^2} \tag{10.23}$$

In the presence of distortion, the output power owing to the fundamental component and the total output power owing to all the harmonics are expressed as

$$P_1 = \frac{I_1^2}{2}R_L \tag{10.24}$$

and
$$P_{\text{total}} = (I_1^2 + I_2^2 + I_3^2 + I_4^2)\frac{R_L}{2}$$

$$= \left(1 + \frac{I_2^2}{I_1^2} + \frac{I_3^2}{I_1^2} + \frac{I_4^2}{I_1^2}\right)\frac{I_1^2}{2}R_L$$

or
$$P_{\text{total}} = (1 + D^2)P_1 \tag{10.25}$$

10.9 HEAT SINKING IN POWER TRANSISTORS

In power amplifiers, the transistors are required to handle large powers. The maximum power handling capacity is related with the temperature of the transistor junctions because the transistor used as an active device dissipates heat and as a result, the junction temperature rises. In order to protect the transistor from destruction, the maximum temperature and power ratings are provided by manufacturer. For example, the maximum allowed junction temperatures for Ge and Si transistors are 100°C and 150°C–200°C respectively.

In order to radiate the heat produced by the transistors, they are mounted on heat sinks of larger areas from which to radiate the heat into the air. Thus, the temperature of the transistor case can be maintained at a much lower value than would result without the heat sink. The heat sinks are constructed in various shapes using a good heat absorber and radiator, such as aluminium. Some heat sinks are shown in Fig.10.13

Fig. 10.13. Shapes of power sinks.

As it is not possible to keep the transistor case temperature at the room temperature even with good heat sink, it is advisable to derate the maximum power rating as a function of increased transistor temperature. Fig. 10.14 shows a typical power derating characteristics of a Si transistor. The transistor can dissipate 100W if the case temperature is not greater than 75°C. Above 75°C, the transistor power dissipation capacity must be reduced so that at higher case temperature, the power handling capacity is reduced

10.10 THERMAL ANALOGY OF A POWER TRANSISTOR

The power handling capacity of a transistor depends upon its thermal charac-

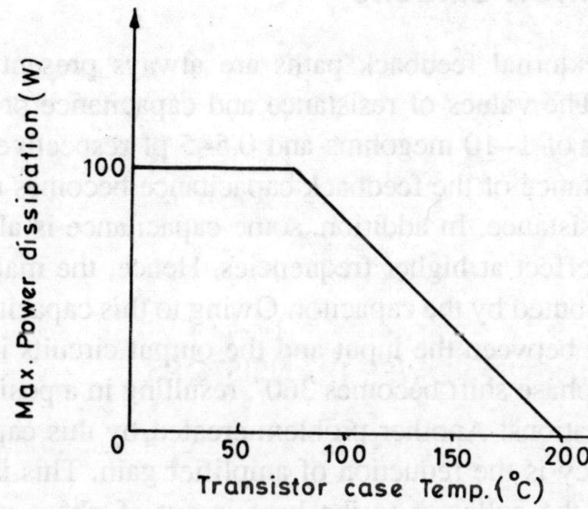

Fig. 10.14. Power derating curve for a Si transistor.

teristics. Figure 10.15(a) shows a thermal equivalent of an electrical circuit, where the temperature difference $T_1 - T_2$ is equivalent to the potential difference $V_1 - V_2$ and the power dissipated P is analogous to the current I. The equivalent of the electrical resistance R is known as the thermal resistance and is defined as

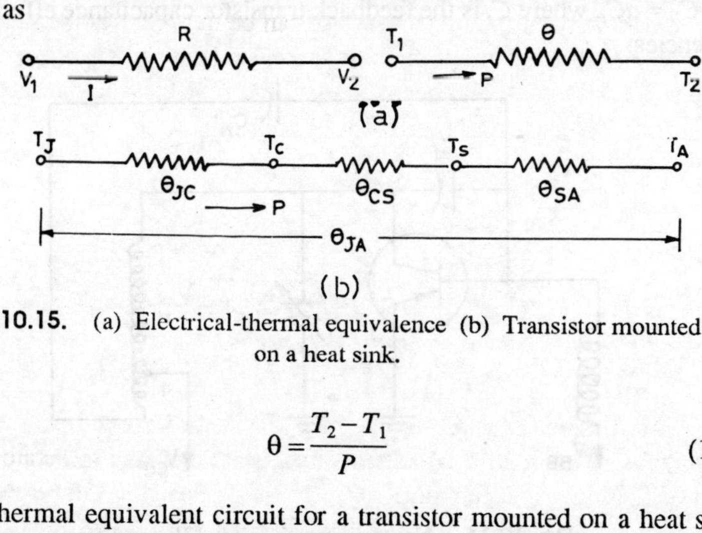

Fig. 10.15. (a) Electrical-thermal equivalence (b) Transistor mounted on a heat sink.

$$\theta = \frac{T_2 - T_1}{P} \tag{10.26}$$

The thermal equivalent circuit for a transistor mounted on a heat sink is represented in Fig.10.15(b), where T_J, T_C, T_S and T_A are the temperatures of the transistor junction, case, heat sink, and room (ambient) respectively. The various thermal resistances: junction-to ambient θ_{JA}, junction-to-case θ_{JC}, case-to-heat sink θ_{CS}, and heat sink-to-ambient θ_{SA} can be defined by Eq.10.26. Using the thermal-electrical analogy, the total thermal resistance.

$$\theta_{JA} = \theta_{JC} + \theta_{CS} + \theta_{SA} \tag{10.27}$$

10.11 NEUTRALIZATION CIRCUITS

Some internal and external feedback paths are always present between the transistor elements. The values of resistance and capacitance providing feedback are of the order of 1–10 megohms and 0.5–5 pf respectively. At higher frequencies, the reactance of the feedback capacitance becomes much smaller than the feedback resistance. In addition, some capacitance is also developed owing to the Miller effect at higher frequencies. Hence, the major portion of the feedback is contributed by the capacitor. Owing to this capacitive feedback, sufficient phase shift between the input and the output circuits is created and sometimes, the total phase shift becomes 360°, resulting in a positive feedback and starting of oscillations. Another problem created by this capacitive feedback at high frequency is the reduction of amplifier gain. This is because the signal coupled from the collector to the base is out of phase with the input signal. In order to compensate for this capacitive feedback, neutralization is used. In this process, signal from the collector is coupled to the base through a capacitance C_N in such a way that the signal through this neutralizing capacitance cancels the signal through the transistor capacitance.

A typical neutralizing circuit is shown in Fig.10.16. The transformer provides the phase inversion of the collector signal. If a is the transformer turns ratio, $C_N = aC_f$, where C_f is the feedback transistor capacitance effective at high frequencies.

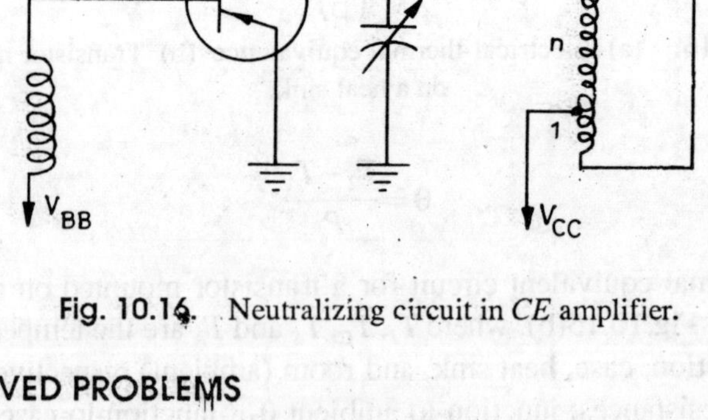

Fig. 10.16. Neutralizing circuit in *CE* amplifier.

10.12 SOLVED PROBLEMS

Problem 1

Calculate the power losses in the circuit if the efficiency of the power amplifier is 25% and the ac output power is 50 W.

Solution

The efficiency of the power amplifier is given by

$$\eta = \frac{P_{O(ac)}}{P_{S(dc)}} \times 100\%$$

$\therefore$ Power delivered by the battery

$$P_{S(dc)} = P_{O(ac)}/\eta \times 100$$

$$= \frac{50}{25} \times 100 = 200 \text{ W}$$

Hence the power losses

$$P_D = P_{S(dc)} - P_{O(ac)}$$

$$= 200 - 50 = 150 \text{ W}$$

Problem 2

Find the various power terms and efficiency of a class-*A* power amplifier if V_{CC} = 15 V and R_L = 15 Ω.

Solution

In a class-*A* power amplifier,

$$P_{O(ac)} = \frac{V_{CC}^2}{8R_L} = \frac{225}{8 \times 15} = 1.9 \text{ W}$$

and

$$P_{S(dc)} = \frac{V_{CC}^2}{2R_L} = \frac{225}{2 \times 15} = 7.5 \text{ W}$$

Hence, the power losses

$$P_D = P_{O(ac)} - P_{S(dc)}$$

$$= 7.5 - 1.9 = 5.6 \text{ W}$$

The efficiency of the circuit is

$$\eta = \frac{P_{O(ac)}}{P_{S(dc)}} \times 100\%$$

$$= \frac{1.9}{7.5} \times 100 = 25\%$$

Problem 3

Calculate the effective impedance seen into the primary of a transformer with turns ratio 20 connected to a speaker of 8 Ω.

Solution

The effective load seen into the primary is

$$R_L' = a^2 R_L$$

$$= (20)^2 \times 8$$

$$= 3.2\ \text{k}\Omega$$

Problem 4

Design a transformer capable of matching an 8Ω speaker load with an amplifier of output impedance as 5 kΩ.

Solution

The effective load seen into the primary is given by

$$R_L' = a^2 R_L$$

or

$$a^2 = R_L'/R_L = \frac{5000}{8} = 625$$

∴

$$a = 25$$

Hence, the required transformer should have turns ratio as 25:1.

Problem 5

Calculate the power efficiency of a class-*A* transformer amplifier if $V_{CC} = 15$ V, $R_L = 15\ \Omega$, and $a = 25$.

Solution

The power output is given by

$$P_{O(ac)} = \frac{V_{CC}^2}{2R_L} = \frac{225}{2 \times 15} = 7.5\ \text{W}$$

and the power input is

$$P_{S(dc)} = \frac{V_{CC}^2}{R_L} = \frac{225}{15} = 15\ \text{W}$$

Hence, the power efficiency is

$$\eta = \frac{P_{O(ac)}}{P_{S(dc)}} \times 100$$

$$= \frac{7.5}{15} \times 100 = 50\%$$

Problem 6

Find the total distortion, fundamental component, and the total output power for a power amplifier with $D_2 = 0.1$, $D_3 = 0.05$, $D_4 = 0.02$, $I_1 = 2$ A, and $R_L = 15\ \Omega$.

Solution

Total distortion in a power amplifier is defined as in Eq.10.23,

$$D = \sqrt{D_2^2 + D_3^2 + D_4^2}$$

$$= \sqrt{0.01 + 0.0025 + 0.0004} \cong 0.11$$

The fundamental power component,

$$P_1 = \frac{I^2 R_L}{2} = \frac{4 \times 15}{2} = 30 \text{ W}$$

Total power as in Eq.10.25

$$P_{total} = (1 + D^2) P_1$$

$$= [1 + (0.11)^2] \times 30$$

$$= 30.36 \text{ W}$$

Problem 7

Calculate various power components and the efficiency of a class-*B* push-pull power amplifier with the following parameters:
$R_L = 8\ \Omega$, $V_{cc} = 24$ V, and $V_i = 12\ V_{r.m.s}$.

Solution

The peak value of the input signal

$$V_i = \sqrt{2} \times 12 = 17 \text{ V}$$

As the gain of the amplifier is unity (emitter follower), the output voltage

$$V_0 = 17 \text{ V}$$

Hence, the output power

$$P_{0(ac)} = \frac{V_0^2}{2R_L} = \frac{(17)^2}{2 \times 8} = 18.1 \text{ W}$$

and the load current

$$I_0 = \frac{V_0}{R_L} = \frac{17}{8} = 2.13 \text{ A}$$

The dc current drawn from the battery

$$I_{dc} = \frac{2}{\pi} \cdot I_0 = \frac{2 \times 7}{22} \times \frac{17}{8} = 1.35 \text{ A}$$

So, the dc power input is

$$P_{S(\mathrm{dc})} = V_{CC} \cdot I_C = 24 \times 1.35$$

$$= 32.4 \text{ W}$$

Hence, the efficiency of the amplifier

$$\eta = \frac{P_{0(\mathrm{ac})}}{P_{S(\mathrm{dc})}} \times 100$$

$$= \frac{18.1}{32.4} \times 100 = 55.9\%$$

Power losses

$$P_D = 32.4 - 18.1 = 14.3 \text{ W}$$

Hence, the power losses by each transistor

$$= \frac{14.3}{2} = 7.15 \text{ W}$$

Problem 8

Calculate the efficiency of a tuned circuit if Q values are in the ratio of 10 : 1 for loaded and unloaded conditions.

Solution

The efficiency of a tuned circuit is given by

$$\eta = \left(1 - \frac{Q_1}{Q_2}\right)^2 \times 100$$

$$= \left(1 - \frac{1}{10}\right)^2 \times 100$$

$$= 81\%$$

Problem 9

If the thermal resistances of a power transistor are:

(a) Junction-to-case $\theta_{JC} = 0.6°\text{C/W}$;

(b) case-to-heat sink $\theta_{CS} = 0.5°\text{C/W}$; and

(c) heat sink-to-ambient $\theta_{SA} = 1.2°\text{C/W}$;

find the thermal resistance of the transistor and maximum power dissipated for the temperature difference of 100°C.

Solution

The thermal resistance of a power transistor is given by

$$\theta_{JA} = \theta_{JC} + \theta_{CS} + \theta_{SA}$$

$$= 0.6 + 0.5 + 1.2$$

$$= 2.3° \text{ C/W}$$

and the maximum power dissipated,

$$P = \frac{T_2 - T_1}{\theta_{JA}}$$

$$= \frac{100}{2.3} = 43.48 \text{ W}$$

Questions and Problems

1. Describe the various classes of power amplifier in terms of their operating points.
2. Compare the efficiencies in case of *A*, *B* and *C* classes of amplifiers.
3. What do you mean by a nonlinear amplifier?
4. Describe a class-*B* push-pull amplifier.
5. Define efficiency of a power amplifier.
6. Find an expression for the efficiency of a class-*A* power amplifier .
7. What is the role of a transformer in power amplifiers?
8. How does a transformer improve the efficiency of a power amplifier?
9. Distinguish between a linear and nonlinear amplifier.
10. What is a phase splitter? Explain it.
11. Describe a complementary symmetry circuit used in power amplifiers.
12. Describe a class-*C* power amplifier.
13. Calculate the distortion in power amplifiers.
14. What is the need of heat sinking in power amplifier?
15. What are the derating characteristics of a power transistor?
16. Establish a thermal analogy of a power transistor.
17. What do you mean by neutralization circuits? Why are they required in power amplifiers?
18. Should a good electrical conductor be used for good heat sinks?
19. Find the efficiency of a power amplifier if the input and output powers are 200W and 1 kW respectively.
20. Calculate the power loss in a power amplifier with efficiency of 50% and input power 1 kW.
21. Calculate the efficiency of a class-*A* power amplifier.
22. Find the turns ratio of a transformer matching the 7.2 kΩ load to 8 Ω.

23. Find the power loss in a class *A* power amplifier if $R_L = 10\ \Omega$ and $V_{CC} = 9$ V.
24. Calculate the power input, power output, and efficiency of a class-*A* transformer amplifier if $V_{CC} = 9$ V, $R_L = 8\ \Omega$, and $a = 25$.
25. Repeat Problem 9 (solved) if $\theta_{JC} = 0.5$°C/W, $\theta_{CS} = 0.7$°C/W and $\theta_{SA} = 1.0$°C/W.
26. Find the various power components and the efficiency of a class-*B* push-pull amplifier if $R_L = 8\ \Omega$, $V_{cc} = 15$ V, and $V_i = 10\ V_{r.m.s.}$

11

Tuned and Wideband (R.F.) Amplifiers

The radiofrequency (R.F) amplifiers are generally classified into two categories: the untuned and tuned amplifiers. When the aim is to amplify broad frequency range, the untuned amplifier is used. The tuned amplifier is used when very high amplification is required to amplify high frequency signals at one frequency, or a small range of frequencies. Thus, the function of a tuned amplifier is to amplify and select the wanted signals from a broad band frequency spectrum. In addition, the tuned amplifier provides a good signal-to-noise ratio so that weak signals are also received. The selectivity in tuned amplifiers is obtained by using a parallel tuned (L–C) circuit, which resonates at the frequency to be amplified and provides high impedance at this frequency.

11.1 THE PARALLEL TUNED CIRCUIT

A parallel resonant circuit is shown in Fig. 11.1. The two possibilities of the presence of resistive component (owing to L, C, or the transistor of the amplifier) in the circuit are represented in Fig. 11.1(a) and (b), where R_s and R_p are the resistive elements shown in the series and parallel form respectively. The resonant condition is that the magnitude of the capacitive and the inductive reactances are equal at resonant frequency $\omega_o\left(=\frac{1}{2\pi f_o}\right)$, i.e.,

$$\omega_0 L = \frac{1}{\omega_0 C}$$

or

$$\omega_0 = \frac{1}{\sqrt{LC}}$$

or

$$f_0 = \frac{1}{2\pi\sqrt{LC}} \tag{11.1}$$

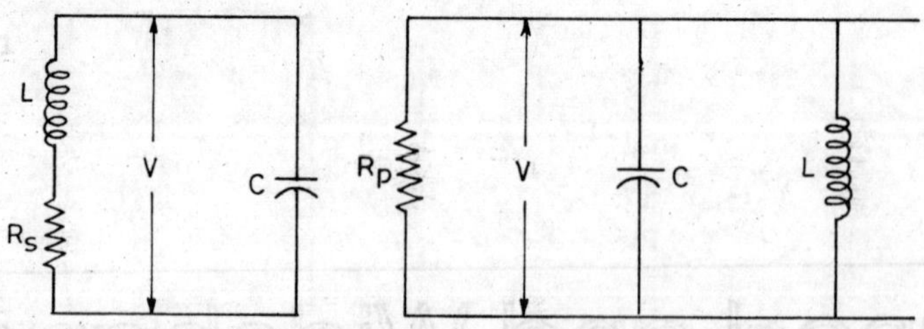

Fig. 11.1. Parallel tuned circuit with resistive component (a) in series (b) in parallel.

The impedance of the tuned circuit shown in Fig. 11.1(a) is written as

$$Z = \frac{(R_s + j\omega L)\frac{1}{j\omega C}}{R_s + j\omega L + \frac{1}{j\omega C}}$$

If R_S is very small in comparison to ωL,

$$Z = \frac{L/C}{R_s + j\left(\omega L - \frac{1}{\omega C}\right)}$$

and at resonance when $\omega L = \frac{1}{\omega C}$,

$$Z = \frac{L}{CR_s} = R_D \tag{11.2}$$

Here, R_D is the dynamic resistance of a parallel tuned circuit at resonance and is purely resistive in nature. This circuit at resonance ($\omega = \omega_o$) gives a current amplification Q defined as,

$$Q = \frac{\text{Amplitude of the osillatory current through the capacitor}}{\text{Amplitude of the current supplied by source}}$$

$$= \frac{V/\frac{1}{\omega_0 C}}{V/\frac{L}{CR_s}}$$

$$= \frac{L\omega_0}{R_s} \tag{11.3}$$

Similarly, if the circuit in Fig. 11.1(b) is considered, the impedance at resonance (dynamic resistance) is

$$Z = R_p = \frac{L}{CR_s} \tag{11.4}$$

Using Eq. 11.4, the current magnification from Eq. 11.3 becomes

$$Q = \omega_0 C R_p \tag{11.5}$$

The tuned circuit is used as load in the amplifier, therefore, the gain of the amplifier at resonance becomes

$$A_V = -g_m R_p$$

$$= -g_m \frac{Q}{\omega_0 C}$$

$$= -g_m Q \omega_0 L \quad \left(\text{as } \frac{1}{\omega_0 C} = \omega_0 L \right) \tag{11.6}$$

11.2 SINGLE-TUNED AMPLIFIER: BANDWIDTH

The impedance of the tuned circuit shown in Fig. 11.1(b) is written as,

$$Z = \frac{R_p \cdot \frac{1}{j\omega C} \cdot j\omega L}{R_p \cdot \frac{1}{j\omega C} + R_p \cdot j\omega L + \frac{1}{j\omega C} \cdot j\omega L}$$

$$= \frac{LR_p/C}{L/C + j(\omega L R_p - R_p/\omega C)} \tag{11.7}$$

If d is taken as the dissipation factor and equal to $1/Q$, then

$$d = \frac{1}{Q} = \frac{1}{\omega_0 C R_p} = \frac{\omega_0 L}{R_p} = \frac{1}{R_p}\sqrt{L/C} \tag{11.8}$$

Equation 11.7 becomes

$$Z = R_P \frac{\left(\frac{1}{R_p}\right)^2 \cdot L/C}{\left(\frac{1}{(R_p)^2}\right)\left[\frac{L}{C} + j(\omega L R_p - R_p/\omega_C)\right]}$$

$$= R_P \frac{\left[\frac{1}{R_p}\sqrt{L/C}\right]^2}{\left(\frac{1}{R_p}\sqrt{L/C}\right)^2 + j\left(\frac{\omega L}{R_p} - \frac{1}{\omega C R_p}\right)}$$

$$= R_p \frac{d^2}{d^2 + j\left(\frac{\omega L}{R_p} - \frac{1}{\omega c R_p}\right)}$$

$$= R_p \frac{d^2}{d^2 + jd\left(\frac{\omega L R_p}{\omega_0 L R_p} - \frac{\omega_0 C R_p}{\omega C R_p}\right)}$$

$$= R_p \frac{d}{d + j\left(\frac{f}{f_0} - \frac{f_0}{f}\right)} \tag{11.9}$$

where f_o is the resonance frequency and f is any other frequency at which Z is defined by the above equation. This expression shows that the impedance is maximum ($= R_P$) and pure resistive at resonance, i.e. $f = f_o$ and is reduced on either side of the resonant frequency. The same behaviour is observed for the amplifier gain as $A_v = -g_m Z$ as defined in Eq. 11.6. This gain variation with frequency is shown in Fig. 11.2. The bandwidth as defined in Sec 8.4 of Chapter 8, is the range of frequencies between two points at which the gain drops to $\frac{1}{\sqrt{2}}$ times the gain at resonance. From Eq. 11.9, this condition is valid when

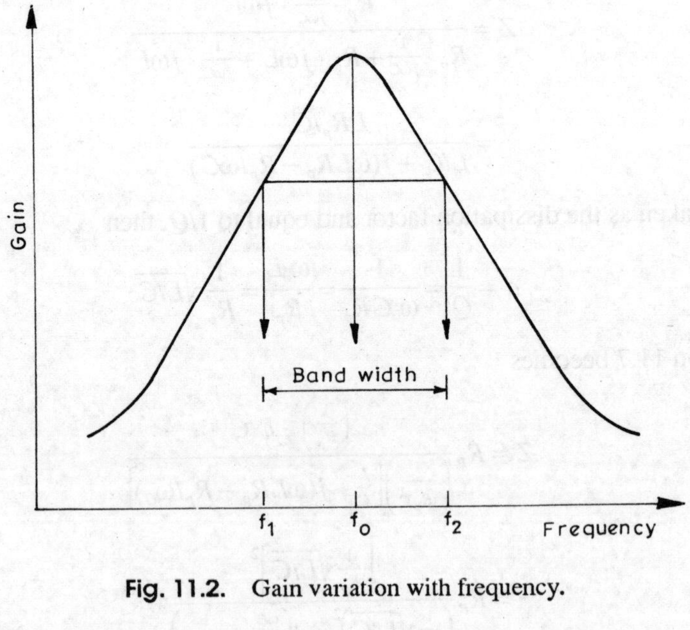

Fig. 11.2. Gain variation with frequency.

$$\frac{f}{f_0}-\frac{f_0}{f}=\pm\, d$$

or
$$\frac{f^2-f_0^2}{f_0 f}=d$$

For the tuned circuits of low dissipation factor, f will be closer to f_0. Thus

$$\frac{f^2-f_0^2}{f_0^2}=d$$

or
$$\frac{f-f_0}{f_0}\cdot\frac{f+f_0}{f_0}=d$$

or
$$\frac{f-f_0}{f_0}\cdot 2=d \quad \left(\text{as } \frac{f+f_0}{f_0}\cong 2\right)$$

or
$$\delta=\frac{f_0 d}{2} \tag{11.10}$$

where $\delta=f-f_0$. Hence, the lower- and upper-band frequencies may be defined as $f_o\pm\delta$ or

$$f_1=f_0-\frac{f_0 d}{2} \quad \text{and} \quad f_2=f_0+\frac{f_0 d}{2}$$

Therefore, the bandwidth is

$$B=f_2-f_1=f_0 d$$

$$=f_0\cdot\frac{1}{2\pi f_0 C R_P} \quad \text{(using Eq. 11.8)}$$

or
$$B=f_2-f_1=\frac{1}{2\pi C R_P}=\frac{f_0}{Q} \tag{11.11}$$

As the gain at resonance defined by Eq. 11.6 is $g_m\, R_P$, the gain-bandwidth product is $g_m/2\pi C$, which is a constant. Therefore, it is possible to increase the bandwidth only at the cost of gain and vice-versa.

11.3 CASCADED SINGLE-TUNED AMPLIFIERS

In the higher-frequency range, the bandwidth of an amplifier is sometimes required to be very broad, in the order of several megahertz. As discussed in Sec. 11.2, the gain bandwidth product in an amplifier remains constant. This dictates that the bandwidth may be increased only by reducing the gain.

Therefore, in order to fulfil the higher gain and broadband requirements simultaneously, a number of identical single tuned amplifier stages are used in cascade. If the number of amplifier stages used is n and the gain and bandwidth of single stage are $A^{\frac{1}{n}}$ and B respectively, then

$$\text{over-all gain} = A \qquad [11.12(a)]$$

and

$$\text{over-all bandwidth} = \frac{B}{1.2\sqrt{n}} \qquad [11.12(b)]$$

Using Eqs 11.6 and 11.11, the gain-bandwidth product is $g_m/2\pi C$, which remains constant.

Thus

$$A^{1/n} \cdot B = \frac{g_m}{2\pi C}$$

From Eq. 11.12,

$$(\text{over-all gain})^{1/n} \cdot 1.2\sqrt{n}\,(\text{over-all bandwidth}) = \frac{g_m}{2\pi C}$$

or

$$A^{1/n} \cdot B_n = \frac{1}{1.2\sqrt{n}} \cdot \frac{g_m}{2\pi C}$$

or

$$B_n = \frac{1}{1.2\sqrt{n}} \cdot \frac{1}{A^{1/n}} \cdot \frac{g_m}{2\pi C} \qquad (11.13)$$

For obtaining the condition for maximum over-all bandwidth B_n, $\frac{dB_n}{dn} = 0$

Hence

$$B_n = \frac{1}{1.2\sqrt{n}} \cdot \frac{1}{A^{1/n}} \cdot \frac{g_m}{2\pi C}$$

$$= \frac{1}{1.2} \cdot \frac{g_m}{2\pi C}(n^{-1/2} \cdot A^{1/n})$$

Differentiating w.r.t. n

$$\frac{dB_n}{dn} = \frac{g_m}{1.2(2\pi C)}\left[n^{-1/2}A^{1/n}\log_e A\left(\frac{1}{n^2}\right) - \frac{1}{2}n^{-3/2} \cdot A^{-1/n}\right]$$

$$\left(\text{since} \frac{d}{dx}a^x = a^x \log_e a\right)$$

For max $\cdot B_n$, $$\frac{dB_n}{dn}=0$$

or $$\frac{1}{n^2}\cdot\frac{1}{\sqrt{n}}\cdot\log_e A=\frac{1}{2}\cdot\frac{1}{n^{3/2}}$$

or $$1=2\log_e A^{1/n}$$

or $$A^{1/n}=e^{1/2}=1.65 \tag{11.14}$$

This expression shows that the maximum bandwidth is obtained when the single stage gain is 1.65 or 4.34 dB.

11.4 STAGGER-TUNED AMPLIFIERS

In order to obtain a better gain-bandwidth product and sharper cut-off response outside the passband, the stagger-tuned circuits are used. The impedance of a parallel tuned circuit as given in Eq. 11.9 is

$$Z=R_p\frac{d}{d+j(f/f_0-f_0/f)}$$

where the dissipation factor

$$d=\frac{1}{Q}=\frac{B}{f_0}=\frac{\text{Bandwidth of single stage}}{f_0}$$

Equation 11.10 defines

$$\frac{f}{f_0}-\frac{f_0}{f}=2\frac{f-f_0}{f_0}=\frac{2\delta}{f_0}$$

Therefore, the impedance becomes

$$Z=R_p\frac{B/2}{B/2+j\delta}$$

$$=R_p\cdot\frac{1}{1+j\delta} \quad (\text{if} \quad B=2)$$

or $$\left|\frac{Z}{R_p}\right|=\frac{1}{\sqrt{1+\delta^2}} \tag{11.15}$$

Equation 11.15 is plotted in Fig. 11.3(a) and is known as the *normalized selectivity curve*. For an *n*-stage cascaded amplifier Eq. 11.15 may be written as

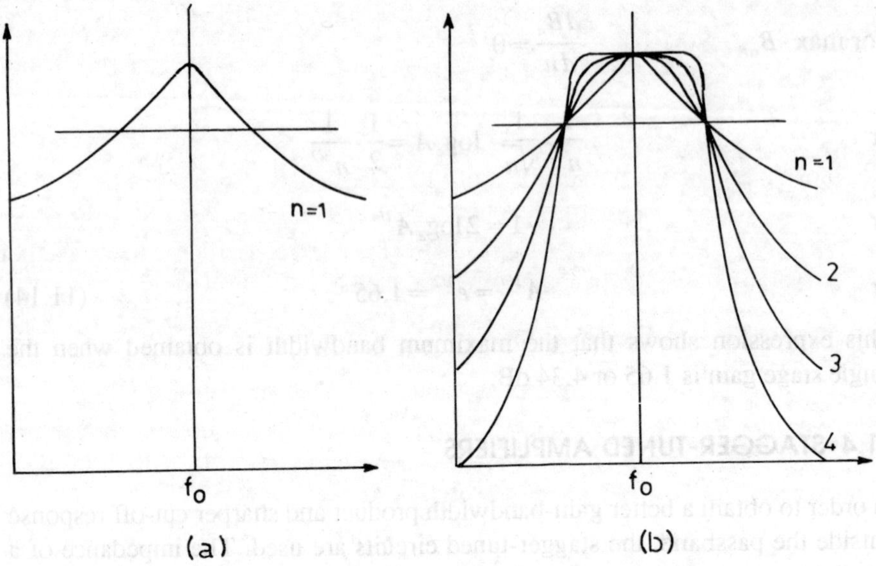

Fig. 11.3. (a) Normalized selectivity curve for $n = 1$

(b) Selectivity curve for various values of n in $\left|\frac{Z}{R_p}\right| = \frac{1}{\sqrt{1+\delta^{2n}}}$

$$\left|\frac{Z}{R_p}\right| = \frac{1}{\sqrt{1+X^{2n}}}$$

which gives the selectivity curves for various values of n as shown in Fig. 11.3(b). As n increase, the bandwidth remains constant but the curve flattens. Also, the cut-off response after the pass-band becomes sharper. The number of stagger-tuned stages required for a particular bandwidth at a resonant frequency may be obtained by using the Butterworth's technique and the results are summarised in Table 11.1.

Table 11.1 Stages of staggered tuned n-stage amplifier

n	Number of single stages (staggered)
2	Two stages at $f_o \pm 0.35\ B_n$ and $B = 0.71\ B_n$
3	Two stages at $f_o \pm 0.43\ B_n$ and $B = 0.5\ B_n$ and one stage at f_o and $B = B_n$
4	Two stages at $f_o \pm 0.46\ B_n$ and $B = 0.38\ B_n$ and Two stages at $f_o \pm 0.19\ B_n$ and $B = 0.92\ B_n$
5	Two stages at $f_o \pm 0.48\ B_n$ and $B = 0.31\ B_n$, Two stages at $f_o \pm 0.29\ B_n$ and $B = 0.81\ B_n$ one stage at f_o and $B = B_n$

where B_n is the over-all bandwidth of the n-stage amplifier and B is the bandwidth of a single stage.

11.5 COUPLING METHODS IN TUNED AMPLIFIERS

In most applications, the tuned amplifiers are used as power amplifiers. Therefore, it becomes essential to use impedance matching devices to assure maximum power transfer from one stage to the next stage. There are many methods for coupling the signal from one stage to another.

11.5.1 Single-tuned Coupling

Figure 11.4 shows two circuits for single-tuned interstage coupling. In Fig. 11.4(a), the source voltage is in series with the tank circuit, while in Fig. 11.4(b), both are in parallel. Hence, the first circuit is called series-fed arrangement, while the second is called the shunt fed arrangement. These single tuned circuits with high Q give good selectivity, but the bandwidth is narrow, which can be made broad by reducing Q as B.W. $= f_o/Q$.

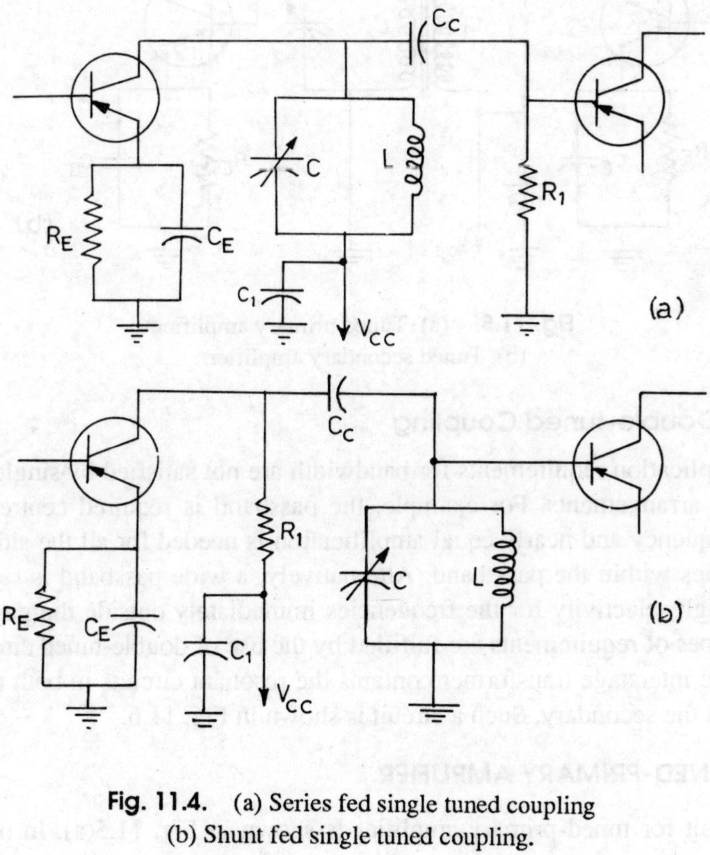

Fig. 11.4. (a) Series fed single tuned coupling
(b) Shunt fed single tuned coupling.

11.5.2 Tuned Primary and Tuned Secondary Coupling

In Figure 11.5, two arrangements of single tuned transformer coupling are shown. Figure 11.5(a) shows a tuned-primary untuned-secondary coupling, while Fig. 11.5(b) shows untuned-primary tuned secondary coupling.

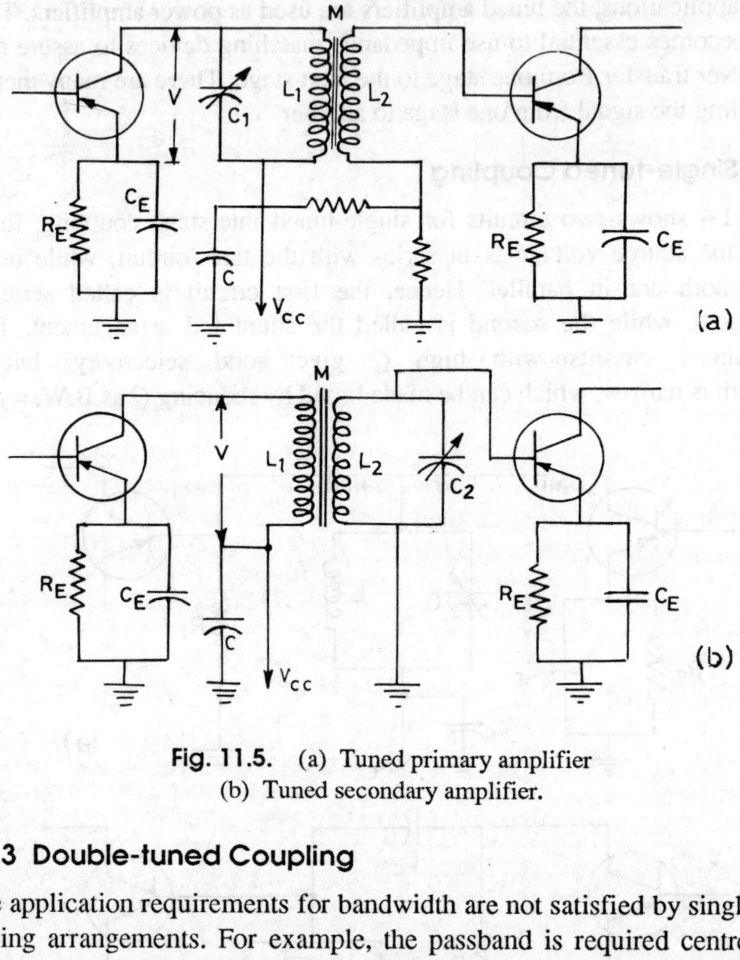

Fig. 11.5. (a) Tuned primary amplifier
(b) Tuned secondary amplifier.

11.5.3 Double-tuned Coupling

Some application requirements for bandwidth are not satisfied by single-tuned coupling arrangements. For example, the passband is required centred on a fixed frequency and nearly equal amplification is needed for all the side band frequencies within the pass band. Alternatively, a wide passband is required, with a high selectivity for the frequencies immediately outside the passband. These types of requirements are fulfilled by the use of double-tuned circuits in which the interstage transformer contains the resonant circuits in both the primary and the secondary. Such a circuit is shown in Fig. 11.6.

11.6 TUNED-PRIMARY AMPLIFIER

The circuit for tuned-primary amplifier is shown in Fig. 11.5(a). In order to

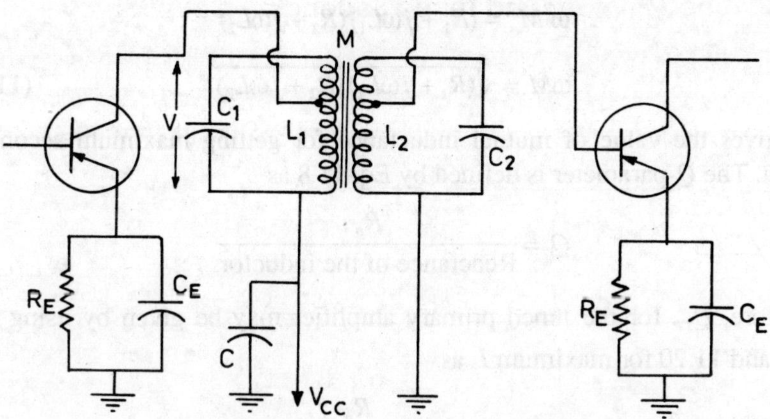

Fig. 11.6. (a) Double tuned amplifier.

provide impedance matching, the inductive coupling having tuned circuit in the primary is used to transfer signal form one stage to the second stage of the amplifier. The mutual inductance present between the primary and the secondary coils is $M = K\sqrt{L_1L_2}$, where k is the coefficient of coupling. The resistance of L_1 and L_2 are R_1 and R_2 respectively.

The impedance coupled from the secondary to the primary $= \dfrac{\omega^2M^2}{R_2 + j\omega L_2}$ (11.16)

The equivalent primary impedance

$$= (R_1 + j\omega L_1) + \frac{\omega^2M^2}{R_2 + j\omega L_2} \tag{11.17}$$

The primary current will be,

$$I_1 = \frac{V}{(R_1 + j\omega L_1) + \frac{\omega^2M^2}{R_2 + j\omega L_2}} \tag{11.18}$$

The voltage induced in the secondary $= -j\,\omega MI_1$

and the secondary current

$$I_2 = -\frac{j\omega MI_1}{R_2 + j\omega L_2} = \frac{-j\omega MV}{(R_1 + j\omega L_1)(R_2 + j\omega L_2) + \omega^2M^2} \tag{11.19}$$

To get the condition for maximum secondary current, Eq. 11.19 is differentiated with respect to M and putting $dI_2/dM = 0$ yields,

$$\omega^2M^2 = (R_1 + j\omega L_1)(R_2 + j\omega L_2)$$

or
$$\omega M = \sqrt{(R_1 + j\omega L_1)(R_2 + j\omega L_2)} \tag{11.20}$$

This gives the value of mutual inductance for getting maximum secondary current. The Q-parameter is defined by Eq. 11.8 as

$$Q = \frac{R_p}{\text{Reactance of the inductor}}$$

Therefore, Q_{TP} for the tuned primary amplifier may be given by using Eqs. 11.17 and 11.20 for maximum I_2 as

$$Q_{TP} = \frac{R_p}{(R_1 + j\omega L_1) + \frac{\omega^2M^2}{R_2 + j\omega L_2}}$$

$$= \frac{R_2}{2(R_1 + j\omega L_1)}$$

$$\cong \frac{R_p}{2\,|\,\omega L_1\,|} \quad (\text{as} \quad R_1 \ll \omega L_1)$$

$$= \frac{Q}{2} \tag{11.21}$$

where Q is the current amplification for a simple resonant circuit. The bandwidth of the tuned-primary amplifier is given by

$$(\text{B.W.})_{TP} = \frac{f_0}{Q_{TP}}$$

$$= \frac{2f_0}{Q}$$

$$= 2(\text{B.W.})_{\text{Resonant curcuit}} \tag{11.22}$$

This shows that the tuned-primary circuit has twice the bandwidth of a simple resonant circuit when I_2 is made maximum.

11.7 TUNED SECONDARY AMPLIFIER

In the tuned secondary amplifier, the tuned circuit is present in the secondary of the transformer, as shown in Fig. 11.5(b). The impedance coupled from the secondary into the primary

$$= \frac{\omega^2 M^2}{R_2 + j\left(\omega L_2 - \frac{1}{\omega C_2}\right)}$$

$$= \frac{\omega_0^2 M^2}{R_2} \quad \text{at resonance} \tag{11.23}$$

The equivalent primary impedance

$$= (R_1 + j\omega_0 L_1) + \frac{\omega_0^2 M^2}{R_2} \tag{11.24}$$

The primary current will be,

$$I_1 = \frac{V}{(R_1 + j\omega_0 L_1) + \frac{\omega_0^2 M^2}{R_2}} \tag{11.25}$$

The voltage induced in the secondary

$$= -j\omega_0 M I_1$$

and the secondary current,

$$I_2 = -\frac{j\omega_0 M I_1}{R_2} \quad \text{at resonance}$$

$$= \frac{-j\omega_0 M V}{(R_1 + j\omega_0 L_1)R_2 + \omega_0^2 M^2} \tag{11.26}$$

For maximum secondary current $\frac{dI_2}{dM} = 0$ yields,

$$\omega_0^2 M^2 = (R_1 + j\omega_0 L_1)R_2$$

or

$$\omega_0 M = \sqrt{(R_1 + j\omega_0 L_1)R_2} \tag{11.27}$$

The Q_{TS} for tuned-secondary amplifier will be given by using Eqs. 11.24 and 11.27 as

$$Q_{TS} = \frac{R_p}{(R_1 + j\omega_0 L_1) + \frac{\omega_0^2 M^2}{R_2}}$$

$$= \frac{R_p}{2(R_1 + j\omega_0 L_1)}$$

$$= \frac{R_p}{2\omega_0 L_1} \quad (\text{as} \quad R_1 << \omega_0 L_1)$$

$$=\frac{Q}{2} \tag{11.28}$$

which gives

$$(\text{B.W})_{TS} = 2(\text{B.W})_{\text{Resonant circuit}} \tag{11.29}$$

Equation 11.29 shows that the tuned secondary amplifier has also twice the bandwidth of a simple resonant circuit as in the case of the tuned primary amplifier.

11.8 DOUBLE TUNED AMPLIFIER

In a double-tuned amplifier, the tuned circuits are used in both the primary and the secondary of the transformer. For a high gain, L should be large compared with C. But the L/C ratio is limited because of the frequency stability requirements. As the circuit is to operate at a fixed frequency, it becomes essential that no detuning should take place owing to any reason. This is achieved by using permeability tuning, in which the inductor is made tunable rather than capacitor. Hence, the capacitor C_1 comes in series with L_1, as shown in Fig. 11.6. The impedance coupled from the secondary to the primary

$$=\frac{\omega^2 M^2}{R_2 + j\left(\omega L_2 - \frac{1}{\omega C_2}\right)}$$

$$=\frac{\omega_0^2 M^2}{R_2} \quad \text{at resonance} \tag{11.30}$$

The equivalent primary impedance

$$= R_1 + j\left(\omega L_1 - \frac{1}{\omega C_1}\right) + \frac{\omega_o^2 M^2}{R_2} \tag{11.31}$$

The primary current will be,

$$I_1 = \frac{V}{R_1 + j\left(\omega L_1 - \frac{1}{\omega C_1}\right) + \frac{\omega_0 M^2}{R_2}} \tag{11.32}$$

The voltage induced in the secondary

$$= -j\omega_0 M \cdot I_1$$

and the secondary current

$$I_2 = -\frac{j\omega_0 M I_1}{R_2}$$

$$= \frac{-j\omega_0 MV}{R_2\left[R_1 + j\left(\omega L_1 - \frac{1}{\omega C_1}\right)\right] + \omega_0^2 M^2} \tag{11.33}$$

The $\frac{dI_2}{dM} = 0$ gives the value of mutual inductance for maximum I_2 as

$$\omega_0^2 M^2 = R_2\left[R_1 + j\left(\omega L_1 - \frac{1}{\omega C_1}\right)\right]$$

or

$$\omega_0 M = \sqrt{R_2\left[R_1 + j\left(\omega L_1 - \frac{1}{\omega C_1}\right)\right]} \tag{11.34}$$

The Q_{DT} for double tuned amplifier will be given by using Eqs. 11.31 and 11.34 as

$$Q_{DT} = \frac{R_p}{R_1 + j\left(\omega L_1 - \frac{1}{\omega C_1}\right) + \frac{\omega_0^2 M^2}{R_2}}$$

$$= \frac{R_p}{2\left[R_1 + j\left(\omega L_1 - \frac{1}{\omega C_1}\right)\right]}$$

$$= \frac{R_p}{2\sqrt{\omega^2 L_1^2 + \frac{1}{\omega^2 C_1^2}}} \quad (\text{as} \quad R_1 \ll \omega L_1)$$

At resonance in the primary, $\omega_0 L_1 = \frac{1}{\omega_0 C_1}$.

Hence,

$$Q_{DT} = \frac{R_p}{2\sqrt{2}\,\omega_0 L_1}$$

$$\cong \frac{1}{3} \cdot \frac{R_p}{\omega_0 L_1}$$

$$= \frac{Q}{3} \tag{11.30}$$

which gives

$$(\text{B.W.})_{DT} = 3(\text{B.W.})_{\text{Resonant circuit}} \tag{11.31}$$

Equation 11.31 shows that the double tuned amplifier has thrice the bandwidth of a simple resonant circuit.

11.9 SOLVED PROBLEMS

Problem 1

Design a tuned circuit for an amplifier of gain 200 at a frequency of 15 kHz having bandwidth 10 kHz. The transconductance of the transistor is 2×10^{-3} mhos.

Solution

The gain of an amplifier with tuned circuit as a load is given by

$$A_v = g_m R_p$$

$$\therefore \qquad R_p = \frac{A_v}{g_m} = \frac{200}{2\times10^{-3}} = 100\ \text{k}\Omega$$

and the bandwidth of the tuned circuit is

$$f_2 - f_1 = \frac{1}{2\pi C R_p}$$

or

$$C = \frac{1}{2\pi R_p (f_2 - f_1)}$$

$$= \frac{7}{2\times22\times100\times10^3\times10\times10^3}$$

$$= 159\ \text{pF}$$

The resonant frequency of the tuned circuit is given by

$$f_0 = \frac{1}{2\pi\sqrt{LC}}$$

or

$$L = \frac{1}{4\pi^2 f_0^2 C} = \frac{7\times7\times10^{12}}{4\times22\times22\times(15\times10^3)^2\times159}$$

$$= 707\ \text{mH}$$

Hence, the tuned circuit will have $R = 100\ \text{k}\Omega$, $L = 707$ mH, and $C = 159$ pF.

Problem 2

Calculate the maximum bandwidth of a cascaded single-tuned amplifier with gain of 50 dB. Given: $g_m = 2\times10^3$ mhos and $C = 330\,\text{pF}$.

Solution

The maximum bandwidth is obtained when the single-stage gain is 4.34 dB. Therefore, the number of stages to get the 50 dB gain,

$$n = \frac{50}{4.34} = 12$$

According to Eq. 11.13, the gain-bandwidth product is given by

$$A^{1/n} \cdot B_n = \frac{1}{1.2\sqrt{n}} \cdot \frac{g_m}{2\pi C}$$

$$= \frac{1}{1.2\sqrt{12}} \cdot \frac{2 \times 10^{-3} \times 7}{2 \times 22 \times 330 \times 10^{-12}}$$

$$= \frac{1400}{6035.85} \times 10^6$$

$$= 232 \text{ kHz}$$

The gain of one stage $A^{1/n} = 1.65$ as in Eq. 11.14. Hence, the maximum bandwidth

$$B_n = \frac{232}{1.65} = 140.6 \text{ kHz}$$

Problem 3

Find the various stages for a stagger-tuned amplifier with normalized selectivity curve of the form $1/\sqrt{1+x^{10}}$. The overall gain is 50 dB and the maximum bandwidth is 10 MHz at 50 MHz.

Solution

In a stagger-tuned amplifier, the selectivity curve is of the form $1/\sqrt{1+x^{2n}}$ where n is the number of stages. Hence, in the present problem, the number of stages $n = 5$ and the gain per stage $= \frac{50}{5} = 10$ dB $= 3.16$.

For $n = 5$, from the Table 11.1 we get,

(i) one stage at 50 MHz with $B = 10$ MHz.
(ii) one stage at $50 + 0.48\, B_n = 54.8$ MHz with $B = 3.1$ MHz.
(iii) one stage at $50 - 0.48\, B_n = 45.2$ MHz with $B = 3.1$ MHz
(iv) one stage at $50 + 0.29\, B_n = 52.9$ MHz with B = 8.1 MHz.
(v) one stage at $50 - 0.29\, B_n = 47.1$ MHz with $B = 8.1$ MHz.

Problem 4

How many stages are required for getting gain 100 from a tuned amplifier with bandwidth of 500 kHz at 10 MHz. C = 100 pF and $g_m = 2 \times 10^{-3}$ mhos.

Solution

The gain-bandwidth product is given by,

$$A^{1/n} \cdot B_n = \frac{1}{1.2\sqrt{n}} \cdot \frac{g_m}{2\pi C}$$

If $n = 1$

$$A^{1/n} \cdot B = \frac{2 \times 10^{-3} \times 7}{1.2 \times 2 \times 22 \times 100 \times 10^{-12}}$$

$$= \frac{14}{24 \times 22} \times 10^{8} = 2.65 \text{ MHz}$$

Therefore, gain of one stage

$$A^{1/n} = \frac{2.65 \times 10^{6}}{500 \times 10^{3}} = 5.3$$

Hence, for getting a gain of 100, we need three stages (total gain = 5.3 × 5.3 × 5.3)

Problem 5

Calculate the resonant frequency, bandwidth, and Q of the tuned transformer amplifier with $R_L = 10$ kΩ, $C = 30$ pF, $L = 10$ mH, and $a = 10$.

Solution

The resonant frequency is

$$f_0 = \frac{1}{2\pi\sqrt{LC}} = \frac{7}{2 \times 22\sqrt{10 \times 10^{-3} \times 30 \times 10^{-12}}}$$

$$= \frac{70}{241} \times 10^{6} = 0.29 \text{ MHz}$$

$$= 290 \text{ kHz}$$

The impedance reflected into the primary is

$$R_p = a^2 R_L = 100 \times 10 \text{ k}\Omega$$

$$= 1 \text{ M}\Omega$$

The bandwidth is given by

$$\text{B.W} = \frac{1}{2\pi C R_p} = \frac{7}{2 \times 22 \times 30 \times 10^{-12} \times 10^{6}}$$

$$= 5.3 \text{ kHz}$$

and the quality factor

$$Q = \frac{f_o}{\text{B.W}} = \frac{290 \times 10^3}{5.3 \times 10^3}$$

$$= 55$$

Questions and Problems

1. Describe a parallel tuned circuit. Calculate the resonant frequency for it.
2. How do you compare the two tuned circuits?
3. Find the bandwidth of a single-tuned amplifier.
4. Discuss the role of frequency in impedance and gain of a single-tuned amplifier.
5. What is the relationship between gain and bandwidth of a tuned amplifier?
6. How do impedance and gain vary with frequency in a tuned amplifier?
7. Why are the cascaded amplifiers required?
8. Describe a cascaded single-tuned amplifier.
9. What is the effect of cascading on gain and bandwidth of a tuned amplifier?
10. What is the impedance of stagger-tuned amplifier? Describe it.
11. What are the various coupling methods used to couple many stages in tuned amplifiers?
12. Describe a tuned primary amplifier. How is its bandwidth related that of a simple resonant circuit?
13. Describe a tuned secondary amplifier.
14. Discuss a double-tuned amplifier.
15. Compare the bandwidths of tuned primary, tuned secondary, and double-tuned amplifiers
16. Find L, C and R for a tuned circuit to be used in an amplifier of gain 100 at a frequency of 20 kHz having bandwidth 10 kHz. $g_m = 2 \times 10^{-3}$ mhos.
17. Find the gain-bandwidth product of a cascaded single-tuned amplifier with gain of 40 dB and $g_m = 2 \times 10^{-3}$. $C = 200$ pF.
18. Establish the various stages of a stagger-tuned amplifier with selectivity curve governed by $1/\sqrt{1+X^8}$. The overall gain is 40 dB and maximum bandwidth is 10 MHz at 40 MHz.
19. Find the quality factor of a tuned transformer amplifier with $R_L = 15$ kΩ, $C = 50$ pF, $L = 10$ mH, and $a = 12$.

12

Oscillators

An oscillator is an important device for many electronic circuit applications and its prime function is to generate waveforms at a constant amplitude and desired frequency. Basically, an oscillator is an electronic circuit which converts the dc supply voltage to an output waveform of some frequency. The oscillator circuit must also be capable of producing sustained oscillations. The oscillators are classified into two basic catagories: sinusoidal and nonsinusoidal. If the waveform generated looks like sine or cosine wave, the circuit is called a sinusoidal oscillator and the circuits producing all other waveforms are called nonsinusoidal oscillators. Sometimes, the oscillators are also classified on the basis of frequency of the generated waveform, viz. audio frequency, radio-frequency, and ultrahigh frequency oscillators. In this chapter, some important sinusoidal oscillators are discussed.

12.1 POSITIVE FEEDBACK AND OSCILLATIONS: BERKHAUSEN CRITERION

In order to generate sustained oscillations, a feedback circuit is used which couples energy of the proper amount and phase from the output stage to the input circuit. This may be achieved through the positive feedback shown in Fig.12.1. As discussed in Sec. 9.2, the overall gain of the amplifier using positive feedback may be written as

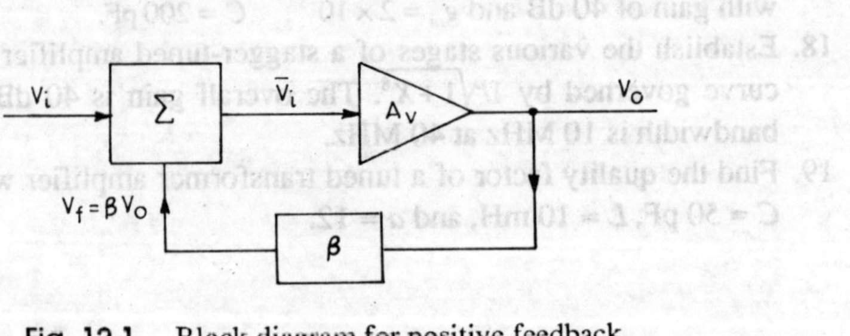

Fig. 12.1. Block diagram for positive feedback.

$$A_{vf} = \frac{A_v}{1 - A_v \cdot \beta} \tag{12.1}$$

where A_v is the open loop gain of the basic amplifier and β is the feedback factor. If the denominator in Eq.12.1 is set to be zero, the amplifier gain becomes infinitely large and an infinitesimal voltage (even noise) can provide a measurable output voltage. The circuit behaves as an oscillator even without an input signal. Hence the condition for oscillation known as Berkhausen criterion is

$$1 - A_v \cdot \beta = 0$$

or

$$A_v\beta = 1 \tag{12.2}$$

This condition is obtained assuming feedback to be positive, i.e., the voltage fed back must be in phase with the input voltage. In practical circuits, $A_v\,\beta$ is set greater than one and the system starts oscillating by amplifying even the noise signals. The saturation and cut-off characteristics of the circuit settle the value of $A_v\beta = 1$.

12.2 BASIC RESONANT CIRCUIT OSCILLATOR

The basic circuit for resonant oscillator is shown in Fig.12.2. The feedback is achieved through the impedance network consisting of Z_1, Z_2 and Z_3. The total impedance of this feedback network is

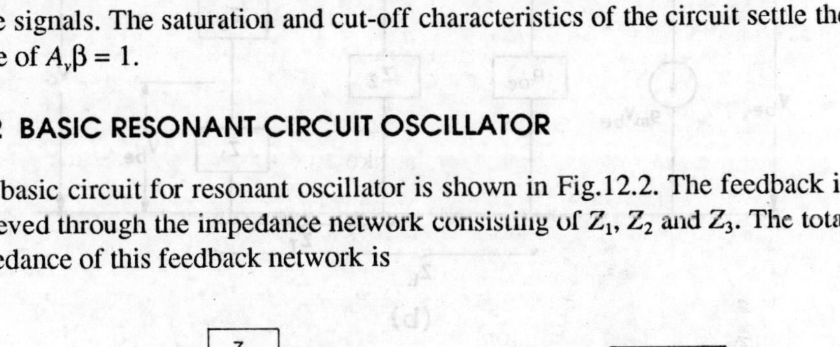

Fig. 12.2. Basic circuit for resonant oscillator.

$$Z_T = \frac{Z_2(Z_1 + Z_3)}{Z_1 + Z_2 + Z_3} \tag{12.3}$$

The transconductance model for resonant oscillator using BJT is shown in Fig.12.3, where a parallel combination of R_{oe} and Z_T makes the total load Z_L given by

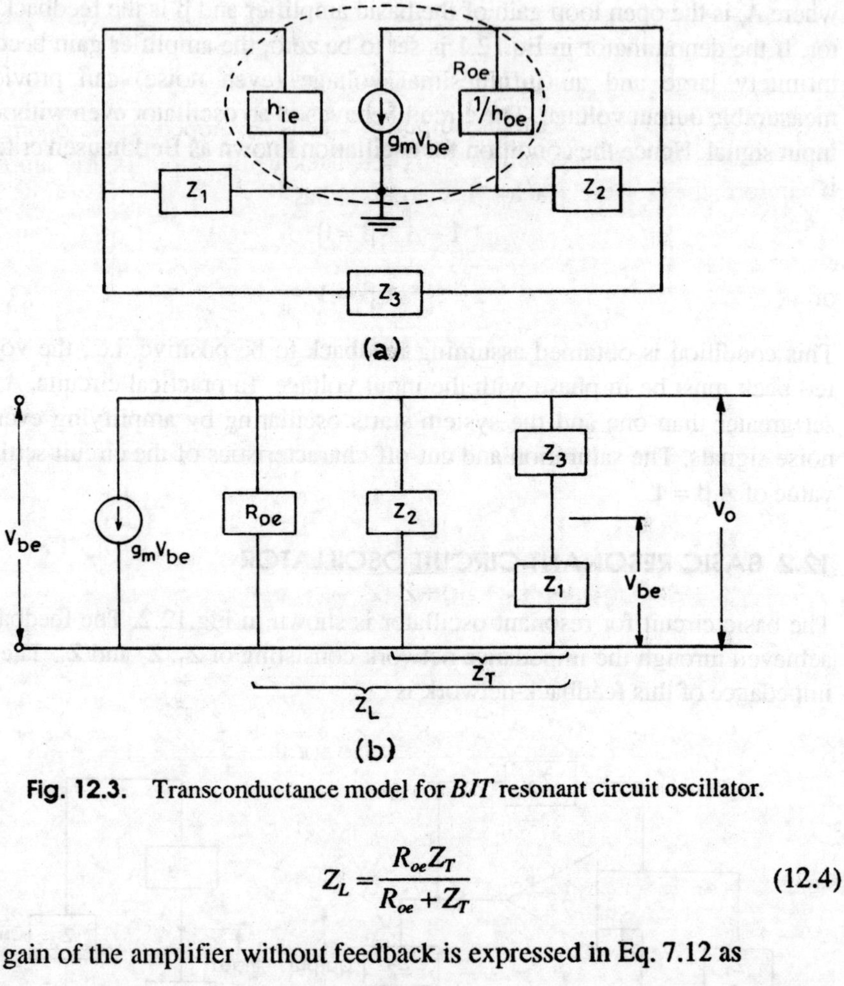

Fig. 12.3. Transconductance model for *BJT* resonant circuit oscillator.

$$Z_L = \frac{R_{oe} Z_T}{R_{oe} + Z_T} \tag{12.4}$$

The gain of the amplifier without feedback is expressed in Eq. 7.12 as

$$A_v = -g_m Z_L \tag{12.5}$$

which shows that the amplifier is introducing a phase shift of 180°. Substituting the value of Z_L from Eq.12.4,

$$A_v = -\frac{g_m R_{oe} Z_T}{R_{oe} + Z_T}$$

Substituting the value of Z_T from Eq.12.3,

$$A_v = \frac{-g_m R_{0e} \frac{Z_2(Z_1+Z_3)}{Z_1+Z_2+Z_3}}{R_{0e} + \frac{Z_2(Z_1+Z_3)}{Z_1+Z_2+Z_3}}$$

$$= \frac{-g_m R_{0e} Z_2(Z_1+Z_3)}{R_{0e}(Z_1+Z_2+Z_3)+Z_2(Z_1+Z_3)} \tag{12.6}$$

In the circuit of Fig.12.3(b), the output voltage is developed across Z_2 or across $Z_1 + Z_3$ and the voltage across Z_1 is fed back to the input of the amplifier. Therefore, the feedback factor β is expressed as

$$\beta = \frac{\text{Voltage fed back (across } Z_1)}{\text{Total voltage (across } Z_1+Z_3)}$$

$$= \frac{Z_1}{Z_1+Z_3} \tag{12.7}$$

The Berkhausen, criterion for oscillation as in Eq.12.2 requires that

$$A_v\beta = 1$$

or

$$\frac{-g_m R_{0e} Z_2(Z_1+Z_3)}{R_{0e}(Z_1+Z_2+Z_3)+Z_2(Z_1+Z_3)} \cdot \frac{Z_1}{Z_1+Z_3} = 1$$

or

$$\frac{-g_m R_{0e} Z_2 Z_1}{R_{0e}(Z_1+Z_2+Z_3)+Z_2(Z_1+Z_3)} = 1 \tag{12.8}$$

If the impedances Z_1, Z_2 and Z_3 are assumed for the moment to be purely reactive, i.e.,

$$Z_1 = jX_1, \quad Z_2 = jX_2, \quad \text{and} \quad Z_3 = jX_3$$

where $j^2 = -1$, Eq.12.8 becomes

$$\frac{g_m R_{0e} X_1 X_2}{jR_{0e}(X_1+X_2+X_3)-X_2(X_1+X_3)} = 1 \tag{12.9}$$

According to the Berkhausen criterion, $A_v\beta$ should be equal to one and real and therefore, the j terms in Eq.12.9 must be zero, i.e

$$X_1+X_2+X_3 = 0 \tag{12.10}$$

as R_{0e} is not zero. Hence, Eq.12.9 becomes,

$$g_m R_{0e} \frac{X_1}{X_1 + X_3} = -1$$

or using Eq.12.10 $$g_m R_{0e} = \frac{X_2}{X_1} \qquad (12.11)$$

where g_m R_{0e} represents the internal gain of the BJT without load. The above Eq. 12.11 shows that the reactance of X_1 and X_2 should be of the same sign. Therefore, X_1 and X_2 both must be either inductors or capacitors. Equation 12.10 represents that if X_1 and X_2 are inductors, X_3 must be capacitor and this oscillator is known as a Hartley oscillator. And if X_1 and X_2 are capacitors, X_3 must be an inductor and this oscillator is known as the Colpitts oscillator.

12.3 COLPITTS OSCILLATOR

The circuit configuration for a BJT Colpitts oscillator is shown in Fig.12.4. The reactances of C_1 and C_2 are $1/j\omega C_1$ and $1/j\omega C_2$ respectively, where $j^2 = -1$. In the circuit of Fig.12.4(b), the Kirchoff's equation for the loop of current i_3 may be written as

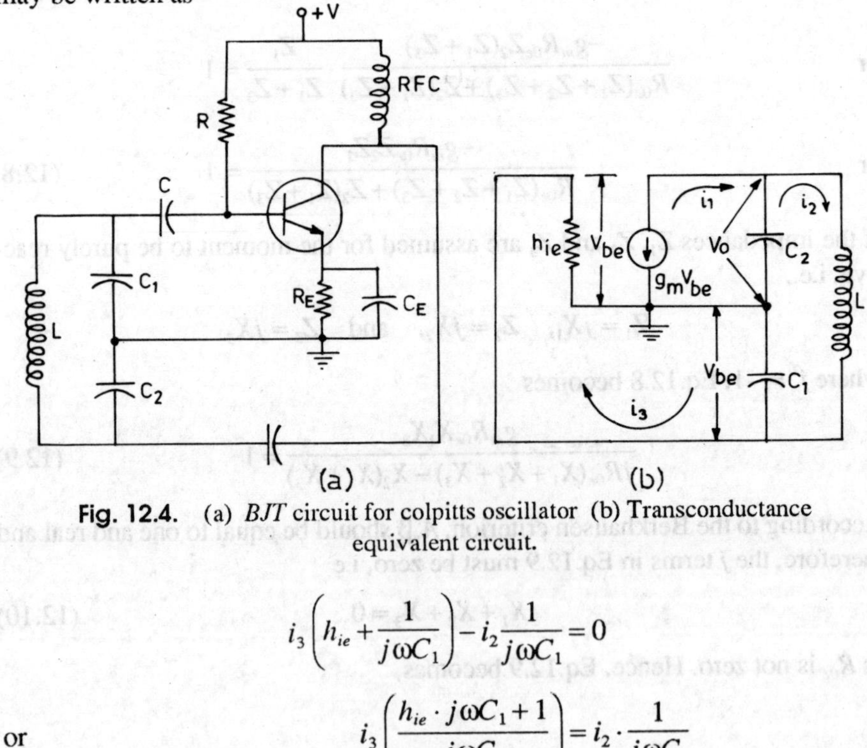

Fig. 12.4. (a) *BJT* circuit for colpitts oscillator (b) Transconductance equivalent circuit.

$$i_3 \left(h_{ie} + \frac{1}{j\omega C_1} \right) - i_2 \frac{1}{j\omega C_1} = 0$$

or $$i_3 \left(\frac{h_{ie} \cdot j\omega C_1 + 1}{j\omega C_1} \right) = i_2 \cdot \frac{1}{j\omega C_1}$$

or $$\frac{i_3}{i_2}=\frac{1}{1+j\omega C_1\cdot h_{ie}} \tag{12.12}$$

Similarly, Kirchoff's equation for the loop of current i_2 may be written as,

$$i_2\left(\frac{1}{j\omega C_1}+\frac{1}{j\omega C_2}+j\omega L\right)-i_1\frac{1}{j\omega C_2}-i_3\frac{1}{j\omega C_1}=0$$

or $$i_2\left(\frac{C_2+C_1-\omega^2LC_1C_2}{j\omega C_1C_2}\right)-\frac{i_1}{j\omega C_2}-\frac{1}{j\omega C_1}\cdot\frac{i_2}{1+j\omega C_1\cdot h_{ie}}=0$$

(using Eq. 12.12)

or $$i_2\left[\frac{C_1+C_2-\omega^2LC_1C_2}{j\omega C_1C_2}-\frac{1}{j\omega C_1(1+j\omega C_1h_{ie})}\right]=\frac{i_1}{j\omega C_2}$$

or $$i_2\left[\frac{(C_1+C_2-\omega^2LC_1C_2)(1+j\omega C_1h_{ie})-C_2}{j\omega C_1C_2(1+j\omega C_1h_{ie})}\right]=\frac{i_1}{j\omega C_2}$$

or $$\frac{i_2}{i_1}=\frac{C_1(1+j\omega C_1\cdot h_{ie})}{(C_1+C_2-\omega^2LC_1C_2)(1+j\omega C_1h_{ie})-C_2}$$

$$=\frac{C_1(1+j\omega C_1\cdot h_{ie})}{C_1+C_2-\omega^2LC_1C_2+j\omega C_1^2h_{ie}+j\omega C_1C_2h_{ie}-j\omega^3LC_1^2C_2h_{ie}-C_2}$$

$$=\frac{(1+j\omega C_1h_{ie})}{[1-\omega^2LC_2+j\omega C_1h_{ie}+j\omega C_2h_{ie}-j\omega^3LC_1C_2h_{ie}]}$$

or $$\frac{i_2}{i_1}=\frac{1+j\omega C_1\cdot h_{ie}}{1+j\omega h_{ie}(C_1+C_2)-\omega^2LC_2-j\omega^3LC_1C_2h_{ie}} \tag{12.13}$$

Multiplying Eqs 12.12 and 12.13,

$$\frac{i_3}{i_1}=\frac{1}{1+j\omega h_{ie}(C_1+C_2)-\omega^2LC_2-j\omega^3L_1L_2h_{ie}} \tag{12.14}$$

From the circuit of Fig.12.4(b), the output voltage across C_2 may be written as,

$$V_0=\frac{1}{j\omega C_2}(i_1-i_2)$$

$$=\frac{i_1}{j\omega C_2}\left(1-\frac{i_2}{i_1}\right) \tag{12.15}$$

where $i_1 = -g_m V_{be}$. Therefore,

$$V_0 = \frac{-g_m V_{be}}{j\omega C_2}\left(1 - \frac{i_2}{i_1}\right)$$

Hence the gain is

$$\frac{V_0}{V_{be}} = A_v = \frac{-g_m}{j\omega C_2}\left(1 - \frac{i_2}{i_1}\right) \tag{12.16}$$

The feedback parameter β may be written as,

$$\beta = \frac{V_{be}}{V_0} = \frac{h_{ie} \cdot i_3}{i_1 / j\omega C_2\left(1 - \frac{i_2}{i_1}\right)} \tag{12.17}$$

The Berkhausen criterion for oscillation requires $A_v\beta = 1$
Substituting the values of A_v and β from Eqs.12.16 and 12.17,

$$\frac{-g_m}{j\omega C_2}\left(1 - \frac{i_2}{i_1}\right) \cdot \frac{h_{ie} i_3}{\frac{i_1}{j\omega C_2}\left(1 - \frac{i_2}{i_1}\right)} = 1$$

or

$$\frac{-g_m}{j\omega C_2} \cdot \frac{h_{ie} i_3}{i_1 / j\omega C_2} = 1$$

or

$$-g_m h_{ie} \cdot \frac{i_3}{i_1} = 1$$

Substituting the value of $\frac{i_3}{i_1}$ from Eq.12.14,

$$-g_m h_{ie} \cdot \frac{1}{1 + j\omega h_{ie}(C_1 + C_2) - \omega^2 L C_2 - j\omega^3 L C_1 C_2 h_{ie}} = 1$$

or

$$\frac{-g_m h_{ie}}{1 - \omega^2 L C_2 + j\omega h_{ie}(C_1 + C_2 - \omega^2 L C_1 C_2)} = 1 \tag{12.18}$$

According to the Berkhausen criterion, $A_v\beta$ should be 1 and real and therefore, the j term in Eq.12.18 must be zero, i.e.,

$$C_1 + C_2 - \omega^2 L C_1 C_2 = 0 \tag{12.19}$$

as ωh_{ie} is not zero. Hence,

$$\omega^2 = \frac{C_1+C_2}{LC_1C_2} = \frac{1}{L\cdot\frac{C_1C_2}{C_1+C_2}}$$

or
$$\omega = \sqrt{\frac{1}{LC_{eff}}}$$

Hence, the frequency of oscillation for the Colpitts oscillator is written as

$$f = \frac{1}{2\pi\sqrt{LC_{eff}}} \quad (12.20)$$

where
$$C_{eff} = \frac{C_1C_2}{C_1+C_2}$$

Using Eq.12.19 in Eq.12.18,

$$-g_m h_{ie} = 1-\omega^2 LC_2$$

$$= 1 - \frac{C_1+C_2}{LC_1C_2}\cdot LC_2$$

$$= \frac{C_1-C_1-C_2}{C_1} = -\frac{C_2}{C_1}$$

or
$$h_{fe} = \frac{C_2}{C_1} \quad (12.21)$$

This gives the value of current gain for the Colpitts oscillator. Thus, the minimum value of transconductance must be $C_2/C_1\,h_{ie}$.

12.4 HARTLEY OSCILLATOR

The circuit configuration for the Hartley oscillator using a BJT is shown in Fig.12.5. The inductors L_1 and L_2 have a mutual coupling M so that the effective values of these inductors become $L_1 + M$ and $L_2 + M$ respectively. In the transconductance equivalent circuit shown in Fig.12.5(b), the Kirchoff's equation for the loop of current i_3 may be written as

$$j\omega(L_1+M)i_2 - h_{ie}i_3 - j\omega(L_1+M)i_3 = 0$$

or
$$\frac{i_3}{i_2} = \frac{j\omega(L_1+M)}{j\omega(L_1+M)+h_{ie}} \quad (12.22)$$

Similarly, the Kirchoff's equation for the loop having current i_2 may be written as

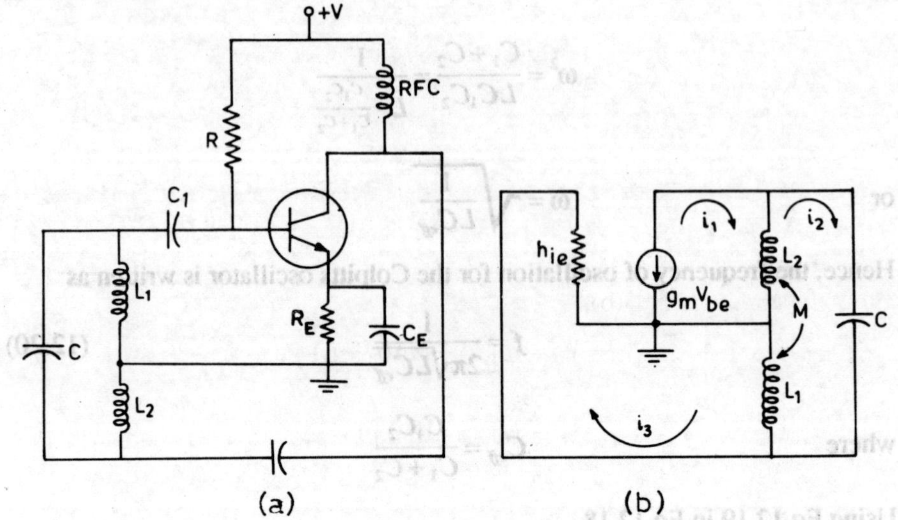

Fig. 12.5. (a) *BJT* circuit for Hartley oscillator (b) Transconductance equivalent circuit.

$$\frac{1}{j\omega C} i_2 + j\omega(L_1 + M)i_2 + j\omega(L_2 + M)i_2$$

$$-j\omega(L_1 + M)i_3 - j\omega(L_2 + M)i_1 = 0$$

Substituting the value of i_3 from Eq.12.22,

$$i_2\left[\frac{1}{j\omega C} + j\omega(L_1 + L_2 + 2M)\right]$$

$$-j\omega(L_1 + M)\cdot\frac{j\omega(L_1 + M)}{j\omega(L_1 + M) + h_{ie}} i_2 = j\omega(L_2 + M)i_1$$

or $$i_2\left[\frac{1 - \omega^2 C(L_1 + L_2 + 2M)}{j\omega C} - \frac{\omega^2(L_1^2 + M^2 + 2L_1 M)}{j\omega(L_1 + M) + h_{ie}}\right] = j\omega(L_2 + M)i_1$$

or $$\frac{i_2[j\omega L_1 + j\omega M + h_{ie} - \omega^2 C(L_1 + L_2 + 2M)(j\omega L_1 + j\omega M + h_{ie}) + j\omega^3 C(L_1^2 + M^2 + 2L_1 M)]}{j\omega C[j\omega(L_1 + M) + h_{ie}]}$$

$$= j\omega(L_2 + M)i_1$$

or $$\frac{i_2}{i_1} = \frac{-\omega^2 C(L_2 + M)[j\omega(L_1 + M) + h_{ie}]}{j\omega L_1 + j\omega M + h_{ie} - \omega^2 C(L_1 + L_2 + 2M)\times(j\omega L_1 + j\omega M + h_{ie}) + j\omega^3 C(L_1^2 + M^2 + 2L_1 M)} \qquad (12.23)$$

Multiplying Eq.12.22 with Eq.12.23,

$$\frac{i_3}{i_1} = \frac{i_3}{i_2} \cdot \frac{i_2}{i_1}$$

$$= \frac{-j\omega^3 C(L_1+M)(L_2+M)}{j\omega L_1 + j\omega M + h_{ie} - \omega^2 C(L_1+L_2+2M) \times (j\omega L_1 + j\omega M + h_{ie}) + j\omega^3 C(L_1^2+M^2+2L_1M)} \tag{12.24}$$

From the circuit of 12.5(b),

$$V_0 = j\omega(L_2+M)(i_1 - i_2)$$

$$= j\omega(L_2+M)i_1\left(1 - \frac{i_2}{i_1}\right)$$

$$= j\omega(L_2+M)(-g_m V_{be})\left(1 - \frac{i_2}{i_1}\right)$$

since $i_1 = -g_m V_{be}$

Therefore, gain is

$$A_v = \frac{V_0}{V_{be}} = -g_m \cdot j\omega(L_2+M)\left(1 - \frac{i_2}{i_1}\right) \tag{12.25}$$

The feedback parameter β may be written as

$$\beta = \frac{V_{be}}{V_0} = \frac{h_{ie} i_3}{j\omega(L_2+M)i_1\left(1 - \frac{i_2}{i_1}\right)} \tag{12.26}$$

The Berkhausen criterion for oscillations requires $A_v\beta = 1$. Substituting the values of A_v and β from Eqs.12.25 and 12.26,

$$-g_m \cdot j\omega(L_2+M)\left(1 - \frac{i_2}{i_1}\right) \cdot \frac{h_{ie} i_3}{j\omega(L_2+M)i_1\left(1 - \frac{i_2}{i_1}\right)} = 1$$

or

$$-g_m h_{ie} \cdot \frac{i_3}{i_1} = 1$$

From Eq.12.24, the value of $\frac{i_3}{i_1}$ is substituted in the above equation to get,

$$\frac{g_m h_{ie} \cdot j\omega^3 C(L_1+M)(L_2+M)}{j\omega(L_1+M) + h_{ie} - j\omega^3 C(L_1+M)(L_1+L_2+2M) - \omega^2 C h_{ie}(L_1+L_2+2M) + j\omega^3 C(L_1^2+M^2+2L_1M)} = 1 \tag{12.27}$$

Equating real terms,

$$h_{ie} - \omega^2 C h_{ie}(L_1 + L_2 + 2M) = 0$$

or
$$\omega^2 C (L_1 + L_2 + 2M) = 1 \tag{12.28}$$

or
$$\omega = \frac{1}{\sqrt{C(L_1 + L_2 + 2M)}}$$

Hence, the frequency of oscillation for the Hartley oscillator is

$$f = \frac{1}{2\pi\sqrt{C(L_1 + L_2 + 2M)}}$$

$$= \frac{1}{2\pi\sqrt{L_{eff} \cdot C}} \tag{12.29}$$

where $L_{eff} = L_1 + L_2 + 2M$

The Berkhausen criterion dictates that $A_v\beta$ must be real and therefore, the j terms in Eq.12.27 must be zero, i.e.,

$$\omega(L_1 + M) - \omega^3 C(L_1 + M)(L_1 + L_2 + 2M) + \omega^3 C(L_1 + M)^2 - g_m h_{ie} \omega^3 C(L_1 + M)(L_2 + M) = 0$$

or
$$\omega - \omega^3 C(L_1 + L_2 + 2M) + \omega^3 C(L_1 + M) - g_m h_{ie} \omega^3 C(L_2 + M) = 0$$

Substituting the value of ω from Eq.12.28,

$$\omega - \omega + \frac{\omega(L_1 + M)}{L_1 + L_2 + 2M} - g_m h_{ie} \frac{\omega(L_2 + M)}{L_1 + L_2 + 2M} = 0$$

or
$$g_m h_{ie} = \frac{L_1 + M}{L_2 + M}$$

or
$$g_m = \frac{1}{h_{ie}} \cdot \frac{L_1 + M}{L_2 + M} \tag{12.30}$$

or
$$h_{fe} = \frac{L_1 + M}{L_2 + M} \tag{12.31}$$

Equations 12.30 and 12.31 define the minimum transconductance and current gain requirements for the Hartley oscillator.

12.5 RESISTANCE-CAPACITANCE OSCILLATORS

The resistance-capacitance oscillators are generally used to produce sine waveforms up to 10 MHz because the resistances and the capacitances are more convenient and cheap in comparison to the inductors. In the *R-C* oscillators, the required phase shift of 180° is achieved through an *R-C* network, whereas another 180° phase difference is introduced by the amplifier so that the total phase shift of 360° is obtained. There are two types of *R-C* oscillators, viz. Phase-shift oscillator and Wien-bridge oscillator.

12.5.1 *R-C* Phase Shift Oscillator

The circuit configuration of a typical *R-C* phase shift oscillator is shown in Fig.12.6, where three identical *R-C* combinations are connected to achieve a phase shift of 180°. The values of *R* and *C* are so selected that a phase shift of 60° per section is obtained at a particular frequency. In Fig. 12.6(b), the loop equations for the loops of current i_2, i_3, and i_4 may be written as,

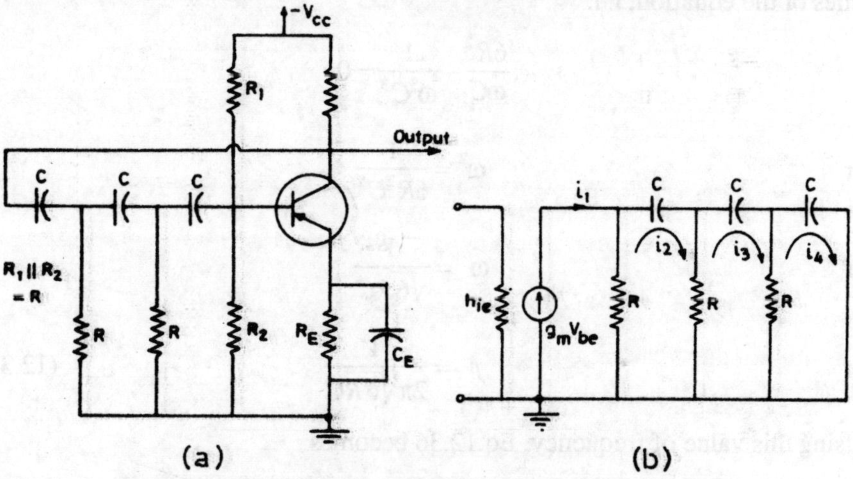

Fig. 12.6. (a) *BJT* circuit for *R-C* phase shift oscillator
(b) Transconductance equivalent circuit.

$$i_1R - i_2R - i_2\cdot\frac{1}{j\omega C} - i_2R + i_3R = 0 \tag{12.32}$$

$$i_2R - i_3R - i_3\cdot\frac{1}{j\omega C} - i_3R + i_4R = 0 \tag{12.33}$$

$$i_3R - i_4R - i_4\cdot\frac{1}{j\omega C} = 0 \tag{12.34}$$

Substituting the value of i_2 from Eq.12.33 in Eq.12.32,

$$i_1R = \frac{\left(2R + \frac{1}{j\omega C}\right)^2}{R}\cdot i_3 - \left(2R + \frac{1}{j\omega C}\right)i_4 - i_3R \tag{12.35}$$

Substituting the value of i_3 from Eq.12.34 in Eq.12.35,

$$i_1 R = i_4 \left[\frac{\left(2R+\frac{1}{j\omega C}\right)^2 \left(R+\frac{1}{j\omega C}\right)}{R^2} - \left(2R+\frac{1}{j\omega C}\right) - \left(R+\frac{1}{j\omega C}\right) \right]$$

or

$$\frac{i_1}{i_4} = \frac{\left(2R+\frac{1}{j\omega C}\right)^2 \left(R+\frac{1}{j\omega C}\right) - \left(3R+\frac{1}{j\omega C}\right) R^2}{R^3}$$

$$= \frac{R^3 - j \cdot \frac{6R^2}{\omega C} - \frac{5R}{\omega^2 C^2} + \frac{j}{\omega^3 C^3}}{R^3} \qquad (12.36)$$

In order to find the frequency of oscillation, the j terms are equated on both the sides of the equation, i.e.

$$\frac{6R^2}{\omega C} - \frac{1}{\omega^3 C^3} = 0$$

or

$$\omega^2 = \frac{1}{6R^2C^2}$$

or

$$\omega = \frac{1}{\sqrt{6}RC}$$

or

$$f = \frac{1}{2\pi\sqrt{6}RC} \qquad (12.37)$$

Using this value of frequency, Eq.12.36 becomes

$$\frac{i_1}{i_4} = \frac{R^3 - \frac{5R}{C^2} \cdot 6R^2C^2}{R^3}$$

$$= -\ 29 \qquad (12.38)$$

The above equation shows that the phase shift between i_1 and i_4 is 180° and the loss in the network at the frequency given in Eq.12.37 is 1/29. Therefore, the minimum current gain of the transistor must be 29 in order to obtain oscillations.

12.5.2 Wien-Bridge Oscillator

The circuit configuration for a typical Wien-bridge oscillator is shown in Fig.12.7. This involves an *RC* bridge circuit in which a frequency adjusting network is constructed of series combination of R_1 and C_1, connected from the output of the operational amplifier (OP.AMP) to the noninverting input of the OP.AMP (shown as positive) and of a parallel combination of R_2 and C_2 from the noninverting input of the OP.AMP to the ground connection. The resistances R_3 and R_4 make the feedback path and determine the gain of the amplifier. The bridge circuit causes a phase shift of 360° between the output and the input of OP.AMP.

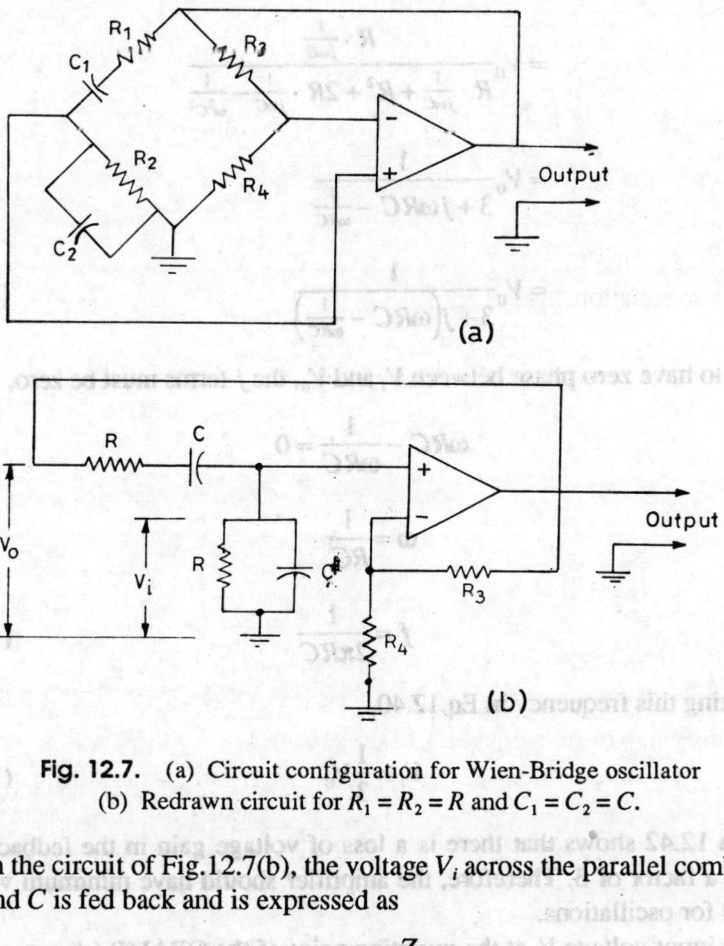

Fig. 12.7. (a) Circuit configuration for Wien-Bridge oscillator
(b) Redrawn circuit for $R_1 = R_2 = R$ and $C_1 = C_2 = C$.

In the circuit of Fig.12.7(b), the voltage V_i across the parallel combination of R and C is fed back and is expressed as

$$V_i = V_0 \frac{Z_1}{Z_1 + Z_2} \tag{12.39}$$

where
$$Z_1 = R \parallel C = \frac{R \cdot \frac{1}{j\omega C}}{R + \frac{1}{j\omega C}}$$

and $Z_2 = R$ in series with $C = R + \dfrac{1}{j\omega C}$

Hence,

$$V_i = \frac{V_0 \dfrac{R \cdot \frac{1}{j\omega C}}{R + \frac{1}{j\omega C}}}{\dfrac{R \cdot 1/j\omega C}{R + 1/j\omega C} + R + \dfrac{1}{j\omega C}}$$

$$= V_0 \frac{R \cdot \frac{1}{j\omega C}}{R \cdot \frac{1}{j\omega C} + R^2 + 2R \cdot \frac{1}{j\omega C} - \frac{1}{\omega^2 C^2}}$$

$$= V_0 \frac{1}{3 + j\omega RC - \frac{j}{\omega RC}}$$

$$= V_0 \frac{1}{3 + j\left(\omega RC - \frac{1}{\omega RC}\right)} \tag{12.40}$$

In order to have zero phase between V_i and V_0, the j-terms must be zero, i.e.,

$$\omega RC - \frac{1}{\omega RC} = 0$$

or

$$\omega = \frac{1}{RC}$$

or

$$f = \frac{1}{2\pi RC} \tag{12.41}$$

Substituting this frequency in Eq.12.40,

$$V_i = \frac{1}{3} V_0 \tag{12.42}$$

Equation 12.42 shows that there is a loss of voltage gain in the fedback network by a factor of 3. Therefore, the amplifier should have minimum voltage gain as 3 for oscillations.

The input voltage V_i at the inverting point of the OPAMP (shown a negative) may be written as

$$V_i = V_0 \frac{R_4}{R_3 + R_4} \tag{12.43}$$

Equating Eqs. 12.42 and 12.43,

$$\frac{R_4}{R_3 + R_4} = \frac{1}{3}$$

or

$$R_3 = 2R_4 \tag{12.44}$$

Hence, for a stable operation of the oscillator; R_3 is selected slightly larger than $2R_4$ so that the minimum gain requirement is fulfilled.

12.6 CRYSTAL OSCILLATOR : FREQUENCY STABILITY

A crystal oscillator is also a resonant frequency oscillator in which a piezo-electric crystal is used as a tuned/resonant circuit because a properly cut crystal shows the characteristics of a resonant circuit. The crystal oscillator provides excellent frequency stability.

12.6.1 Characteristics of a Piezo-Electric Crystal

When a piezo-electric crystal is compressed or streched in a specific direction, a potential difference is developed across the opposite faces of the crystal. Conversely, when a potential difference is applied across one set of opposite faces of the crystal, the crystal expands or contracts. This relationship between the electrical and mechanical properties of a crystal is known as the piezo-electric effect. The most suitable piezo-electric crystals used as resonant circuits are quartz, Rochelle salt, and tourmaline. The quartz crystals are generally used in crystal oscillators because they are inexpensive, mechanically rugged, and the effect of heat on them is very little. It has a form of hexagonal prism as shown in Fig. 12.8(a) where ideally symmetrical hexagonal cross section is shown in Fig.12.8(b). The line joining the points at each end of the crystal is referred as the optic or *Z*-axis. No piezo-electric effect is observed if the stress along the *z*-axis is applied.

The line parallel to any face of the crystal and at right angles to the *z*-axis is known as the *x*-axis or *electrical axis*. The line perpendicular to both the *x*-and *z*-axes is known as the *y*-axis or *mechanical axis*. The crystal is generally shown by the circuit symbol as in Fig.12.8(c).

At resonant frequency, the crystal action is represented by an equivalent resonant circuit shown in Fig.12.8(d). The shunt capacitance C_m represents the capacitance of the mechanical mounting. The inductor *L* represents the electrical equivalent of the crystal mass. The capacitor *C* represents the electrical equivalent of the mechanical elasticity, and the resistance *R* represents the friction during vibration.

The variation of the crystal impedance/reactance with frequency is shown in Fig.12.9, where the presence of two resonant frequencies is observed. In the first resonant condition, the series resonance impedance is very low (in the order of *R*), and the inductive and the capacitive reactances of *RLC* circuit are equal. The second resonant condition shows very high impedance when the reactance of *RLC* circuit becomes equal to the reactance of C_m.

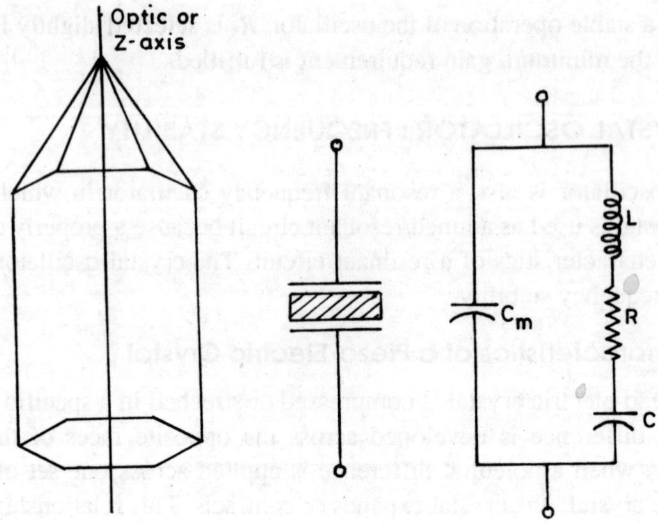

Fig. 12.8. (a) Piezoelectric quartz crystal (b) Cross-section of quartz crystal (c) Circuit symbol for quartz crystal (d) Equivalent circuit of quartz crystal.

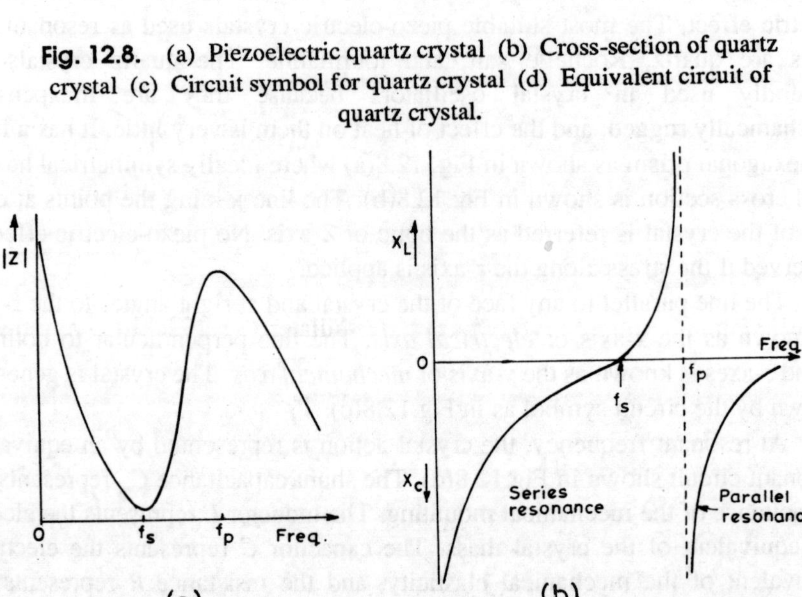

Fig. 12.9. (a) Variation of crystal impedance with frequency. f_s and f_p are series and parallel resonant frequencies (b) Variation of crystal reactance in series and parallel resonances.

12.6.2 Crystal Oscillator in Series Resonant Mode

A typical four-resistor circuit configuration of a crystal oscillator in series-resonant mode is shown in Fig.12.10(a), where the crystal is connected as a series element in the feedback path. The *RFC* coil is used to get dc bias while stopping any ac signal. At the resonant frequency, the crystal impedance is minimum and voltage fed back is maximum. The circuit frequency is held stabilized by the crystal and there is no effect of changes in the supply voltage and the transistor parameters. This circuit is also called a Pierce crystal controlled oscillator. The series resonant frequency may be expressed as

$$f_s = \frac{1}{2\pi\sqrt{LC}} \tag{12.45}$$

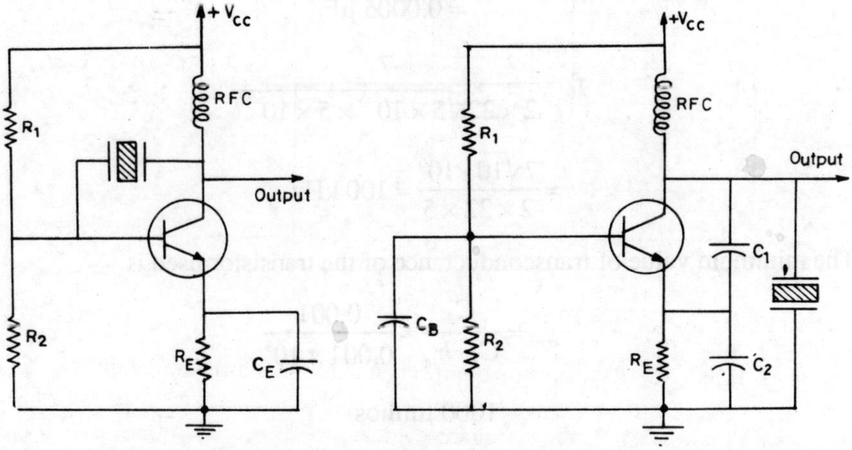

Fig. 12.10. (a) Circuit for crystal oscillator in series-resonant mode
(b) Circuit for crystal oscillator in parallel-resonant mode.

12.6.3 Crystal Oscillator in Parallel Resonant Mode

A typical circuit configuration of a crystal oscillator in parallel resonant mode is shown in Fig.12.10(b), where the crystal is connected as a shunt. At the parallel-resonant frequency, the crystal shows the highest inductive reactance and the maximum voltage is developed across the crystal. The parallel resonant frequency may be given by

$$f_p = \frac{1}{2\pi\sqrt{L\left(\frac{C \cdot C_m}{C + C_m}\right)}} \tag{12.46}$$

which is slightly higher than the series frequency.

12.7 SOLVED PROBLEMS

Problem 1

Calculate the frequency of a Colpitts oscillator if the feedback network has $C_1 = C_2 = 0.001$ μF and $L = 5$ mH. What is the minimum value of g_m if h_{ie} = 1 kΩ?

Solution

The frequency of oscillations for the Colpitts oscillator is given by

$$f_0 = \frac{1}{2\pi\sqrt{LC_{eff}}}.$$

where
$$C_{eff} = \frac{C_1 C_2}{C_1 + C_2} = \frac{0.001 \times 0.001}{0.001 + 0.001}$$

$$= 0.0005\ \mu\text{F}$$

$\therefore$
$$f_0 = \frac{7}{2 \times 22\sqrt{5 \times 10^{-3} \times 5 \times 10^{-10}}}$$

$$= \frac{7\sqrt{10} \times 10^6}{2 \times 22 \times 5} = 100\ \text{kHz}$$

The minimum value of transconductance of the transistor used is

$$g_m = \frac{C_2}{C_1 \cdot h_{ie}} = \frac{0.001}{0.001 \times 10^3}$$

$$= 1000\ \mu\text{mhos}$$

Problem 2

Find the frequency of a Hartley oscillator with feedback network as $L_1 = 4$ mH, $L_2 = 5$ mH, $C = 0.001$ μF, and $M = 0.05$ mH. If $h_{ie} = 1$ kΩ, find the minimum value of transconductance required.

Solution

In a Hartley oscillator, the frequency of oscillation is given by

$$f_0 = \frac{1}{2\pi\sqrt{(L_1 + L_2 + 2M)C}}$$

$$= \frac{7}{2 \times 22\sqrt{(4 + 5 + 0.1) \times 10^{-3} \times 0.001 \times 10^{-6}}}$$

$$= \frac{7 \times 10^6}{44 \times \sqrt{9.1}} = 52.7\ \text{kHz}$$

and the minimum value of transconductance

$$g_m = \frac{1}{h_{ie}} \cdot \frac{L_1 + M}{L_2 + M}$$

$$= \frac{(4 + 0.05) \times 10^{-3}}{1 \times 10^3 (5 + 0.05) \times 10^{-3}}$$

$$= 802\ \mu\text{mhos}.$$

Problem 3

In an *R-C* phase-shift oscillator of frequency 5 kHz, R = 1 kΩ and g_m = 1000 μmhos. Find the value of C and load resistance R_L if gain is 50.

Solution

In a phase-shift oscillator, the frequency of oscillation is given by

$$f_0 = \frac{1}{2\pi RC\sqrt{6}}$$

or

$$C = \frac{1}{2\pi R f_0 \sqrt{6}}$$

$$= \frac{7}{2\times 22\times 1\times 10^3\times 5\times 10^3\times\sqrt{6}}$$

$$= 0.013\ \mu\text{F}$$

The gain of the amplifier is given by

$$A = g_m R_L$$

$$\therefore \qquad R_L = \frac{A}{g_m} = \frac{50}{1000\times 10^{-6}}$$

$$= 50\ \text{k}\Omega$$

Problem 4

Find the frequency of oscillation for a Wien-bridge oscillator if R = 100 kΩ and C = 200 pF.

Solution

The frequency of oscillation for a Wien-bridge oscillator is given by

$$f_0 = \frac{1}{2\pi RC}$$

$$= \frac{7}{2\times 22\times 100\times 10^3\times 200\times 10^{-12}}$$

$$= 8\ \text{kHz}$$

Problem 5

Find the series and parallel resonant frequencies for a crystal oscillator if L = 50 mH, C = 0.005 pF, and C_m = 0.5 pF.

Solution

In a crystal oscillator, the series resonant frequency is

$$f_s = \frac{1}{2\pi\sqrt{LC}} = \frac{7}{2\times 22\sqrt{50\times 10^{-3}\times 5\times 10^{-15}}}$$

$$= 10.06 \text{ MHz}$$

and the parallel resonant frequency is

$$f_p = \frac{1}{2\pi\sqrt{L\left(\frac{C\cdot C_m}{C+C_m}\right)}}$$

$$= \frac{7}{2\times 22\sqrt{\frac{50\times 10^{-3}\times 5\times 10^{-15}\times 500\times 10^{-15}}{(5\times 10^{-15}+500\times 10^{-15})}}}$$

$$= \frac{7\times 10^{9}}{220\sqrt{\frac{5000}{505}}} = 10.11 \text{ MHz}$$

Questions and Problems

1. What do you mean by oscillators? Discuss the Berkhausen criterion for oscillations.
2. Why is positive feedback required for oscillators?
3. What are the circuit requirements for an oscillator?
4. Derive the expression for frequency of a Colpitts oscillator. What is the minimum value of transconductance of the transistor used in this case?
5. Describe a Hartley oscillator.
6. State a difference between the Colpitts and Hartley oscillators.
7. What is an *R-C* oscillator? How is the Berkhausen criterion fulfilled in this oscillator?
8. Describe an *R-C* phase oscillator. What is the minimum current gain required for the transistor used for this oscillator?
9. Derive an expression for the frequency of oscillations for a Wien-bridge oscillator. How is the minimum gain requirement fulfilled in this oscillator?
10. What are the advantages of a crystal oscillator over an *LC* oscillator?
11. What is the Piezo-electric effect and how is it used in oscillators?
12. Derive the expressions for frequency of crystal oscillator for the series and parallel resonant modes.
13. Find the frequency of a Colpitts oscillator if $C_1 = 0.001$ μF, $C_2 = 0.005$ μF, and $L = 10$ mH.

14. Calculate the frequency of a Hartley oscillator if $L_1 = L_2 = 5$ mH, $C = 0.002$ µF, and $M = 0.02$ mH.
15. Find the minimum values of transconductance in case of the above Problems 13 and 14 if $h_{ie} = 2$ kΩ.
16. Find the frequency of oscillations of a *R-C* phase-shift oscillator if $R = 2$ kΩ and $C = 0.05$ µF.
17. Find the frequency of oscillations for a Wien-bridge oscillator if $R = 150$ kΩ and $C = 100$ pF.
18. Calculate the series and parallel resonant frequencies for a crystal oscillator if $L = 40$ mH, $C = 0.002$ pF, and $C_m = 0.1$ pF.

13

Multivibrators

The multivibrators, also called relaxation oscillators, are the electronic circuits producing nonsinusoidal waveforms. The term 'multi' represents the superposition of many harmonics of different frequencies, amplitudes, and compositions. More specifically, the multivibrators are the oscillators which produce rectangular waves. These are widely used as switching circuits in digital and computer electronics. Basically, a multivibrator represents a closed-loop regenerative circuit with a two-stage amplifier in which the output of each stage is coupled back to the input of the other stage in positive feedback or in the regenerative mode. There are mainly three types of multivibrators, viz. bistable multivibrator, monostable multivibrator, and astable multivibrator. The function of each of these multivibrators is based upon the switching characteristics of transistors, which is discussed in the following section.

13.1 SWITCHING ACTION OF A TRANSISTOR

A bipolar junction transistor (BJT) may be operated in one of the three regions, viz. saturation region, active region, and cut-off region. In the saturation region, the BJT conducts heavily even with small voltages and it is referred to as being on or closed. In the cut-off region, the BJT normally does not conduct even with large voltages and it is referred to as being off or open. Therefore, a BJT may be used as a switch if it is driven from saturation to the cut-off region or vice versa. Figure 13.1(a) shows a transistor used as a switch in the off position because its base-emitter junction is made reverse-biased by connecting V_{BB}. In this condition, only the leakage current flows in the collector.

The transistor may be driven to saturation by applying an input pulse waveform v_i to the base as shown in Fig.13.2(a). At $v_i = 0$ the transistor is cut-off and at $v_i = V$ the transistor is in saturation. The variation of collector current in response to the input waveform v_i is shown in Fig.13.2(b). When the signal becomes V, there is a *delay time* involved in the response of the collector current because:

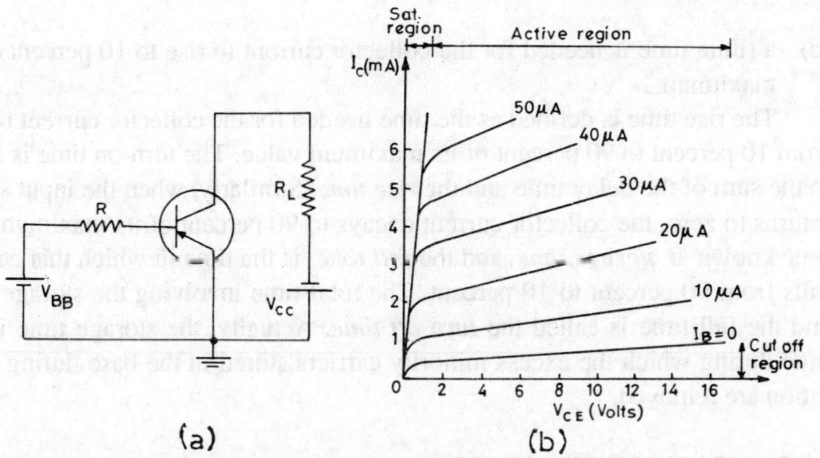

Fig. 13.1. (a) Transistor used as a switch (b) Output characteristics in common emitter configuration.

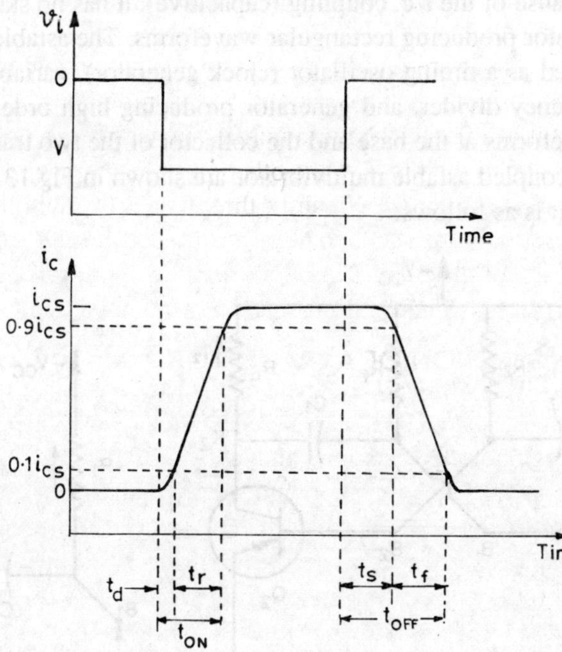

Fig. 13.2. (a) Input signal waveform (b) Response of collector current to the input signal where, t_d = Delay time; t_r = Rise time; t_{on} = Turn-on-Time; t_s = Storage time; t_f = Fall time; t_{off} = Turn-off-Time.

(a) a finite time is needed to charge up the junction capacitances;

(b) a finite time is needed for the minority charge carriers to cross the base region; and

(c) a finite time is needed for the collector current to rise to 10 percent of its maximum.

The rise time is defined as the time needed for the collector current to rise from 10 percent to 90 percent of its maximum value. The turn-on time is equal to the sum of the delay time and the *rise time*. Similarly, when the input signal returns to zero, the collector current decays to 90 percent of its maximum in a time known as *storage time* and the *fall time* is the time in which this current falls from 90 percent to 10 percent. The total time involving the storage time and the fall time is called the *turn-off time*. Actually, the storage time is the time during which the excess minority carriers stored in the base during saturation are removed.

13.2 ASTABLE MULTIVIBRATOR

The circuit configuration of a collector-coupled astable multivibrator is shown in Fig.13.3. Because of the a.c. coupling (capacitive), it has no stable state and acts as an oscillator producing rectangular waveforms. The astable multivibrator is widely used as a timing oscillator (clock generator), variable frequency oscillator, frequency divider, and generator producing high order harmonics. The typical waveforms at the base and the collector of the two transistors used in the collector coupled astable multivibrator are shown in Fig.13.4. The function of the circuit is as follows:

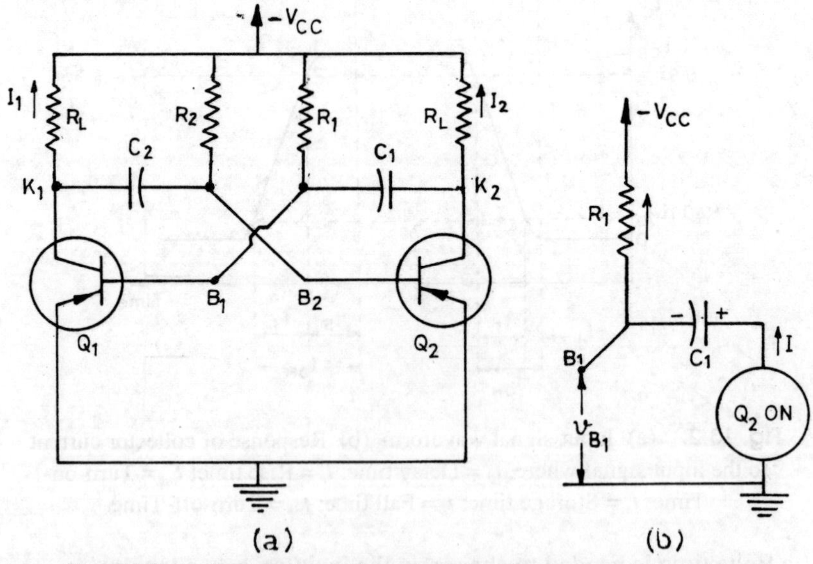

Fig. 13.3. (a) Circuit configuration of collector-coupled astable multivibrator (b) The effective circuit for charging C_1 when Q_1 is OFF and Q_2 is ON.

(a) When the circuit is switched on, it may be possible that $I_1 = I_2$ but a minute fluctuation, say increase in I_1 due to any random cause, causes a decrease in the voltage at k_1. This decrease in the voltage is passed to B_2 by the capacitor C_2. The decrease in the voltage at B_2 causes an increase in the voltage at K_2. This increase in the voltage is passed to B_1 by the capacitor C_1 and B_1 becomes more positive and as a result the current I_1 increases further. This cycle of transitions repeats itself and the current I_1 continues to increase, whereas the current I_2 continues to decrease and the circuit finally reaches the condition that I_1 becomes maximum and I_2 becomes zero. In other words, the transistor Q_1 is in saturation (on) and Q_2 is below the cut-off (off). Suppose this is the condition before time $t = 0$ as shown in Fig.13.4. Hence, at $t < 0$, v_{B_2} is positive, $v_{k_2} = -V_{CC}$, $v_{B_1} = V_{BE(sat)}$, and $v_{k_1} = V_{CE(sat)}$.

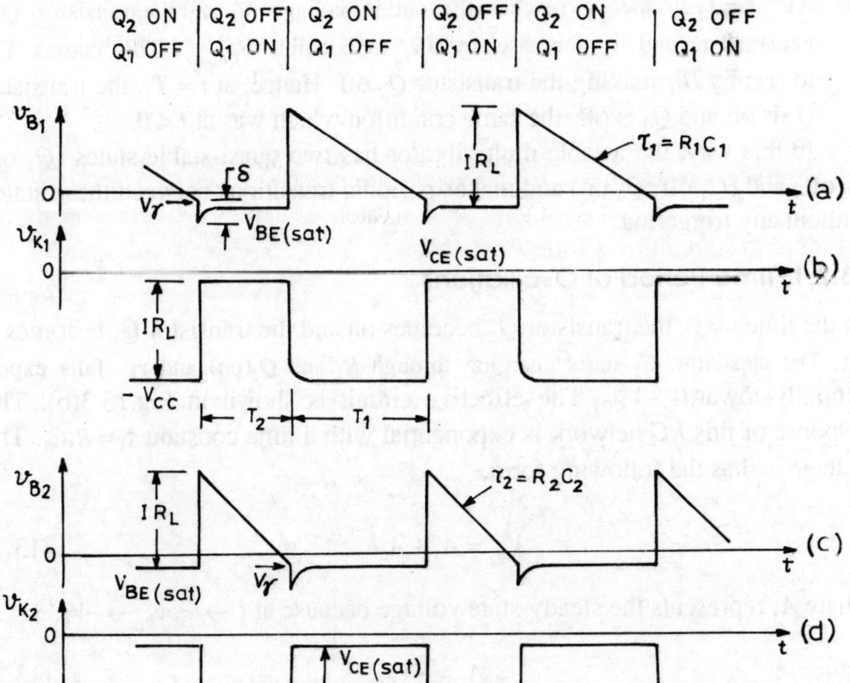

Fig. 13.4. Voltage waveforms for collector-coupled astable multivibrator.

(b) As the transistor Q_1 is on, the capacitor C_2 charges through R_2 and Q_1, and v_{B_2} falls exponentially towards $-V_{CC}$ as shown in Fig. 13.4(c).

(c) At time $t = 0$, the v_{B_2} reaches the cut-in voltage $V\gamma$ and the transistor Q_2 becomes on and the voltage v_{k_2} increases by IR_L to become $v_{CE(sat)}$ shown in Fig.13.4(d). This causes v_{B_1} to rise by IR_L since the two are capacitively coupled, as shown in Fig.13.4(a).

(d) The increase in v_{B_1} makes the transistor Q_1 off and v_{k_1} falls towards $-V_{CC}$.

(e) The fall in v_{k_1} is coupled through C_2 to the base of Q_2, causing the under shoot δ in v_{B_2}, shown in Fig. 13.4(c). The voltage v_{k_1} also shows an abrupt fall by δ and then reaches $-V_{CC}$.

(f) At $t = 0+$, v_{B_1} is positive, $v_{k_1} = -V_{CC}$, $v_{B_2} = V_{BE(sat)}$, and $v_{k_2} = V_{CE(sat)}$.

(g) As the transistor Q_2 is on for $t = 0+$, the capacitor C_1 charges through R_1 and Q_2 and v_{B_1} falls exponentially towards $-V_{CC}$.

(h) At $t = T_1$, the v_{B_1} reaches the cut in voltage V_γ and the transistor Q_1 becomes on and v_{k_1} increases by IR_L to become $V_{CE(sat)}$. This causes V_{B_2} to rise by IR_L making the transistor Q_2 off. Hence, at $t = T_1$, the transistor Q_1 is on and Q_2 is off, the same condition which was at $t < 0$.

In this way, the astable multivibrator has two quasi-stable states (Q_1 on, Q_2 off) and (Q_1 off, Q_2 on) and makes periodic transitions between these states without any triggering.

13.2.1 Time Period of Oscillations

At the time $t = 0$, the transistor Q_2 becomes on and the transistor Q_1 becomes off. The capacitor C_1 starts charging through R_1 and Q_2(on), and v_{B_1} falls exponentially towards $-V_{CC}$. The effective circuit is shown in Fig.13.3(b). The response of this RC network is exponential with a time constant $\tau_1 = R_1C_1$. The voltage v_{B_1} has the following form,

$$V_{B_1} = A_1 + A_2 e^{-t/\tau_1} \tag{13.1}$$

where A_1 represents the steady-state voltage because at $t \to \infty$, $v_{B_1} \to -V_{CC}$

Hence,
$$A_1 = -V_{CC} \tag{13.2}$$

At $t = 0$, Eq, 13.1 gives

$$v_{B_1} = -V_{CC} + A_2$$

or
$$A_2 = V_{CC} + v_{B_1} \tag{13.3}$$

where $v_{B_1} = IR_L - V_{BE(sat)}$ at $t = 0+$, as shown in Fig.13.4(a). Hence, the Eq.13.3 becomes

$$A_2 = V_{CC} + IR_L - V_{BE(sat)} \tag{13.4}$$

Substituting the values of A_1 and A_2 in Eq.(13.1),

$$v_{B_1} = -V_{CC} + [V_{CC} + IR_L - V_{BE(sat)}]e^{-t/\tau_1} \tag{13.5}$$

At $t = T_1$, the base potential of Q_1 reaches the cut in potential $-V_\gamma$, i.e. $v_{B_1} = -V_\gamma$.

Hence at $t = T_1$, Eq.13.5 becomes

$$-V_\gamma = -V_{CC} + [V_{CC} + IR_L - V_{BE(sat)}]e^{-T_1/\tau_1}$$

or
$$e^{-T_1/\tau_1} = \frac{V_{CC} - V_\gamma}{V_{CC} + IR_L - V_{BE(sat)}}$$

or
$$e^{T_1/\tau_1} = \frac{V_{CC} + IR_L - V_{BE(sat)}}{V_{CC} - V_\gamma}$$

or
$$T_1 = \tau_1 l_n \frac{V_{CC} + IR_L - V_{BE(sat)}}{V_{CC} - V_\gamma} \tag{13.6}$$

From Fig.13.4(d), $IR_L = V_{CC} - V_{CE(sat)}$. Hence, Eq.13.6 becomes,

$$T_1 = \tau_1 \ln \frac{V_{CC} + V_{CC} - V_{CE(sat)} - V_{BE(sat)}}{V_{CC} - V_\gamma}$$

$$= \tau_1 \ln \frac{2V_{CC} - [V_{CE(sat)} + V_{BE(sat)}]}{V_{CC} - v_\gamma}$$

$$= \tau_1 \ln 2 + \tau_1 \ln \frac{V_{CC} - \frac{1}{2}[V_{CE(sat)} + V_{BE(sat)}]}{V_{CC} - V_\gamma}$$

At room temperature, for Ge as well as for Si, $V_{CE(sat)} + V_{BE(sat)} = 2V_\gamma$

Hence,

$$T_1 = \tau_1 \ln 2$$

$$= R_1 C_1 \ln 2 \quad \text{as } \tau_1 = R_1 C_1$$

or
$$T_1 = 0.69 R_1 C_1 \quad \text{as } \ln 2 = 0.69 \tag{13.7}$$

Similarly, the time period for the second quasi-stable state (Q_2 off, Q_1 on) with time constant $\tau_2 = R_2 C_2$ may be found as

$$T_2 = 0.69 R_2 C_2 \tag{13.8}$$

The time period of complete cycle is

$$T = T_1 + T_2 = 0.69(R_1C_1 + R_2C_2) \tag{13.9}$$

For a symmetrical circuit, let us assume $R_1 = R_2 = R$ and $C_1 = C_2 = C$. Hence,

$$T = 1.38RC \tag{13.10}$$

13.3 MONOSTABLE MULTIVIBRATOR

The circuit configuration for a collector coupled monostable multivibrator is shown in Fig.13.5. The transistor Q_1 is made off by applying a negative bias $-V_{BB}$ at the base, while the transistor Q_2 is on. Hence, a monostable multivibrator has one stable state and one quasi-stable state. Here C_1 is a speed-up or commutating capacitor. When a triggering signal is applied, a transition from the stable state to the quasi-stable state takes place and after the time period of quasi-stable state, the reverse transition takes place without any triggering signal, as in the case of astable multivibrator. Since only one triggering signal is required to complete one cycle of the transitions, it is also called one-shot, single-cycle, or single-step circuit. Sometimes, it is also known as *univibrator*. The monostable multivibrator is widely used as a pulse generator, counter, delaying circuit, and gating circuit. The typical waveforms at the base and the collector of the two transistors are shown in Fig.13.6. The functioning of the circuit is as follows:

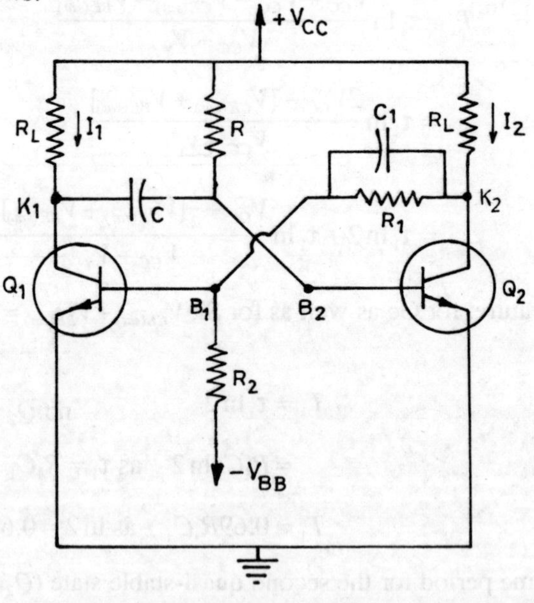

Fig. 13.5. Circuit configuration for collector-coupled monostable multivibrator.

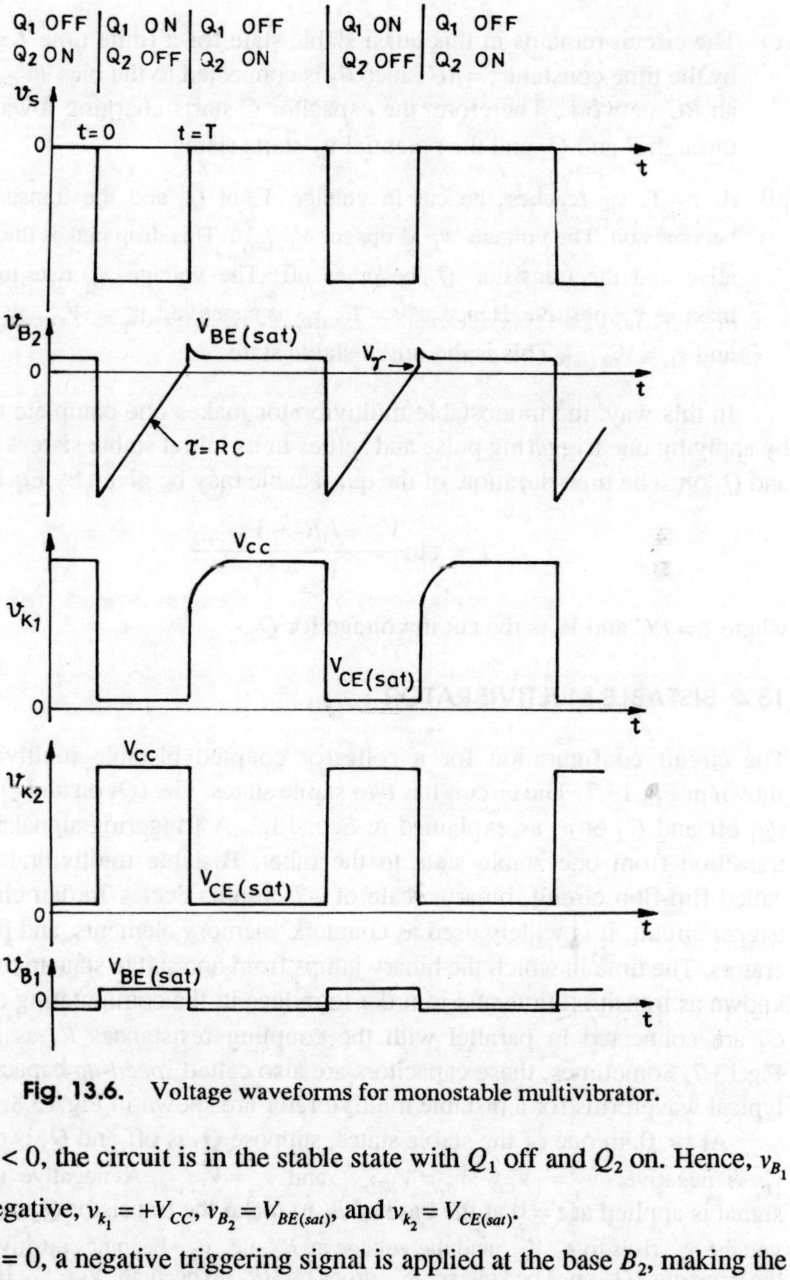

Fig. 13.6. Voltage waveforms for monostable multivibrator.

(a) At $t < 0$, the circuit is in the stable state with Q_1 off and Q_2 on. Hence, v_{B_1} is negative, $v_{k_1} = +V_{CC}$, $v_{B_2} = V_{BE(sat)}$, and $v_{k_2} = V_{CE(sat)}$.

(b) At $t = 0$, a negative triggering signal is applied at the base B_2, making the transistor Q_2 cut-off. The voltage v_{k_2} rises to $+V_{cc}$ and the voltage at B_1, i.e. v_{B_1} becomes positive to make the transistor Q_1 on. The voltage v_{k_1} drops by I_1R_L to become $V_{CE(sat)}$. Hence at $t = 0$, $V_{B_1} = V_{BE(sat)}$, $V_{k_1} = V_{CE(sat)}$, V_{B_2} is negative, and $v_{k_2} = +V_{CC}$. This is the quasi stable state.

(c) The circuit remains in this quasi stable state for a finite time T governed by the time constant $\tau = RC$ since B_2 is connected to the bias $+V_{CC}$ through an RC network. Therefore, the capacitor C starts charging towards $+V_{CC}$ through R and Q_1 and the potential v_{B_2} starts rising.

(d) At $t = T$, v_{B_2} reaches the cut in voltage V_γ of Q_2 and the transistor Q_2 becomes on. The voltage v_{k_2} drops to $V_{CE(sat)}$. This drop makes the v_{B_1} negative and the transistor Q_1 becomes off. The voltage v_{k_1} rises to $+V_{CC}$ making v_{B_2} positive. Hence at $t = T$, v_{B_1} is negative, $v_{k_1} = +V_{CC}$, $v_{B_2} = V_{BE(sat)}$, and $v_{k_2} = V_{CE(sat)}$. This is the initial stable state.

In this way, the monostable multivibrator makes one complete transition by applying one triggering pulse and settles in its initial stable state with Q_1 off and Q_2 on. The time duration of the quasistable may be given by Eq.13.6 as

$$T = \tau \ln \frac{V_{cc} + I_1 R_L - V_{BE(sat)}}{V_{cc} - V_\gamma} \tag{13.11}$$

where $\tau = RC$ and V_γ is the cut in voltage for Q_2.

13.4 BISTABLE MULTIVIBRATOR

The circuit configuration for a collector coupled bistable multivibrator is shown in Fig.13.7. The circuit has two stable states, viz. (Q_1 on and Q_2 off) and (Q_1 off and Q_2 on), as explained in Sec. 13.2. A triggering signal induces a transition from one stable state to the other. Bistable multivibrator is also called flip-flop circuit, binary, scale of -2 circuit, Eccles-Jordan circuit, and trigger circuit. It is widely used as counters, memory elements, and pulse generators. The time in which the binary jumps from one stable state to the other is known as transition time and in order to reduce it, the commutating capacitors C_1 are connected in parallel with the coupling resistances R_1 as shown in Fig.13.7. Sometimes, these capacitors are also called *speed-up* capacitors. The typical waveforms for a bistable multivibrator are shown in Fig.13.8.

At $t < 0$, in one of the stable states, suppose Q_1 is off and Q_2 is on. Hence v_{B_1} is negative, $v_{k_1} = +V_{CC}$, $v_{B_2} = V_{BE(sat)}$, and $v_{k_2} = V_{CE(sat)}$ A negative triggering signal is applied at $t = 0$ at the base of B_2 to make the transistor Q_2 cut-off. The voltage v_{k_2} rises to $+V_{CC}$ and the voltage at B_1, i.e. v_{B_1} becomes positive to make the transistor Q_1 on. The voltage v_{k_1} drops by IR_L to become $V_{CE(sat)}$. Hence, at $t = 0$, $v_{B_1} = V_{BE(sat)}$, $v_{k_1} = V_{CE(sat)}$, v_{B_2} is negative, and $v_{k_2} = +V_{CC}$. This is the second stable state of the binary with Q_1 on and Q_2 off. The flip-flop remains in this stable state indefinitely until a negative triggering signal is applied at the base

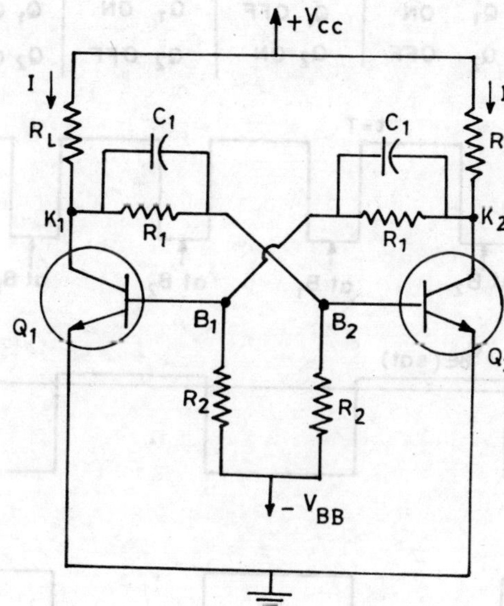

Fig. 13.7. Circuit configuration for collector-coupled bistable multivibrator.

of Q_1 to make it cut-off. This causes v_{k_1} to rise to $+V_{cc}$ and v_{B_2} becomes positive and transistor Q_2 becomes on. The voltage v_{k_2} drops by IR_L and becomes $V_{CE(sat)}$. Hence, at $t = T$ (say) when the second triggering signal is applied at the base of Q_1, the initial stable state with Q_1 off, Q_2 on is reached, which again continues indefinitely until a negative pulse is applied to make Q_2 off.

In this way, the bistable multivibrator completes one cycle after the application of two triggering signals.

13.5 ROLE OF COMMUTATING CAPACITORS

In order to reduce the transition time from one stable state to the other stable/quasi-stable state, a capacitor is connected in parallel to the coupling resistance in multivibrator circuits as shown in Fig.13.5 for a monostable multivibrator and in Fig.13.7 for a bistable multivibrator. In order to induce a transition, when a triggering signal is applied at B_2, the transistor Q_2 becomes cut-off owing to which the voltage v_{k_2} rises to $+V_{cc}$ rapidly. This rapid rise in v_{k_2} is to be transmitted to B_1 in minimum time. But owing to the input characteristic of Q_1, the voltage v_{B_1} rises with the time constant $(R_1R_2/R_1+R_2)C_i$ as $R_1 \parallel R_2$. The speed of rising v_{B_1} may be increased by connecting a capacitance C_1 in parallel to R_1 because the time constant of this rise may even become zero for rapidly varying signal at k_2 owing to a short circuit path $(1/j\omega C_1 = 0)$ provided by C_1 having a large value.

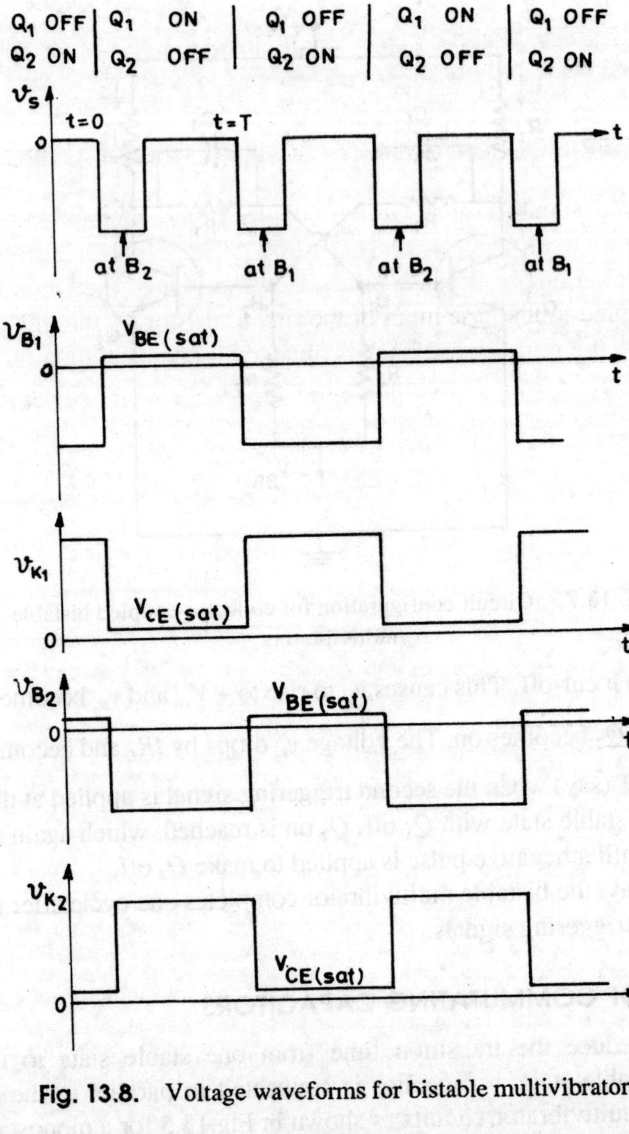

Fig. 13.8. Voltage waveforms for bistable multivibrator.

When a transition from one stable state to the other is induced the binary settles in the new stable state only after the interchange of voltages of the two commutating capacitors. This is because the capacitor with on transistor has voltage = $V_{CE(sat)} - V_{B_1}$ (small value) and the capacitor with off transistor has voltage $V_{CC} - V_{BE(sat)}$ (large value) and this interchange takes place with a time constant $\tau = \dfrac{R_1 R_2}{R_1 + R_2} C_1$ Hence, some minimum time is required to induce the reverse transition and is called the *resolving time* of the binary.

13.6 EMITTER COUPLED BINARY : SCHMITT TRIGGER CIRCUIT

The circuit configuration for an emitter-coupled binary popularly known as the Schmitt trigger circuit is shown in Fig.13.9(a). The circuit is different from a collector coupled binary because the collector output of the second transistor Q_2 is not coupled to the base input of the first transistor Q_1 through an *RC* network. Instead, the positive feedback is obtained by connecting a resistance R_z, as shown in Fig. 13.9(a). The circuit behaviour depends upon the loop gain and is described below:

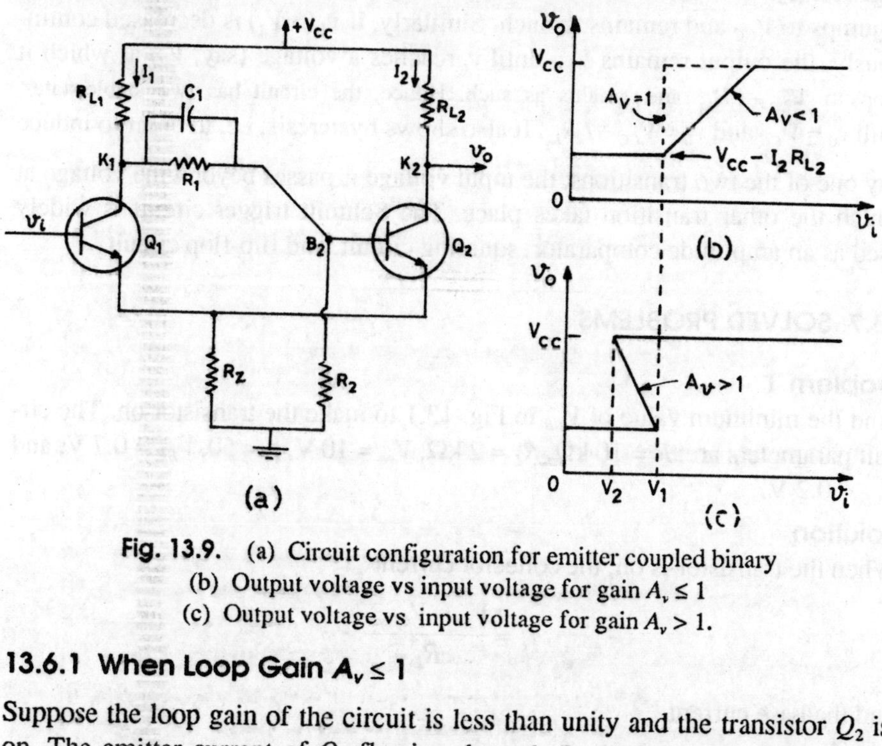

Fig. 13.9. (a) Circuit configuration for emitter coupled binary
(b) Output voltage vs input voltage for gain $A_v \le 1$
(c) Output voltage vs input voltage for gain $A_v > 1$.

13.6.1 When Loop Gain $A_v \le 1$

Suppose the loop gain of the circuit is less than unity and the transistor Q_2 is on. The emitter current of Q_2 flowing through R_z causes some voltage drop across R_z and as a result, the emitter of Q_1 becomes positive. If this value of emitter voltage is more than the input voltage v_i, the transistor Q_1 remains cut-off and if v_i is increased, there is no effect on v_0, which remains at $V_{cc} - I_2 R_{L_2}$. But when increasing v_i reaches the cut-in voltage of Q_1 represented by, say, V_1, the transistor Q_1 becomes on and the collector potential V_{k_1} drops, which causes

v_{B_2} to drop and v_{k_2} to rise. This rise in v_{k_2} (also represented by v_0) with increasing v_i, continues, as shown in Fig.13.9(b). At a particular value of v_i, the transistor Q_2 becomes off and v_0 becomes V_{CC} which now does not depend on increasing v_i (Fig. 13.9(b)).

If the loop gain is increased, the behaviour of the circuit remains similar except that the rise in v_0 with v_i becomes steeper and for $A_v = 1$, the rise becomes vertical as shown in Fig.13.9(b) and its slope becomes infinite.

13.6.2 When Loop Gain $A_v > 1$

The slope of the rising portion of the v_0 *vs* v_i curve increases with gain and when gain becomes more than one, the slope becomes negative, as shown in Fig. 13.9(c). This shows that when increasing v_i reaches V_1, the output voltage v_0 jumps to V_{CC} and remains as such. Similarly, if v_i (> V_1) is decreased continuously, the output remains V_{CC} until v_i reaches a voltage (say, V_2) at which it drops to $V_{CC} - I_2 R_{L_2}$ and remains as such. Hence, the circuit has two stable states with $v_0 = V_{CC}$ and $v_0 = V_{CC} - I_2 R_{L_2}$. It also shows hysteresis, i.e. in order to induce any one of the two transitions, the input voltage v_i passes beyond the voltage at which the other transition takes place. The Schmitt trigger circuit is widely used as an amplitude comparator, squaring circuit, and flip-flop circuit.

13.7 SOLVED PROBLEMS

Problem 1

Find the minimum value of V_{BB} in Fig. 13.1 to make the transistor on. The circuit parameters are: $R = 10\ \text{k}\Omega$, $R_L = 2\ \text{k}\Omega$, $V_{cc} = 10$ V, $\beta = 50$, $V_{BE} = 0.7$ V, and $V_{CE} = 0.2$ V.

Solution

When the transistor is on, the collector current

$$I_c = \frac{V_{cc} - V_{CE\,sat}}{R_L}$$

and the base current

$$I_b = \frac{V_{BB} - V_{BE\,sat}}{R}$$

The transistor β is defined as

$$\beta = \frac{I_C}{I_B} = \frac{V_{cc} - V_{CE\,sat}}{R_L} \cdot \frac{R}{V_{BB} - V_{BE\,sat}}$$

or, $$V_{BB}=\frac{R}{\beta R_L}[V_{cc}-V_{CEsat}]+V_{BEsat}$$

$$=\frac{10}{50\times 2}(10-0.2)+0.7$$

$$=1.78\text{ V}$$

Problem 2

For an astable multivibrator of period 200 μsec, V_{CC} = 10V, h_{fe} = 50, and R_{L_1} = 2 kΩ. Find the values of R_1 and C_1 for the timing circuit.

Solution

The collector current is given by,

$$I_c=\frac{V_{cc}}{R_{L_1}}=\frac{10}{2\times 10^3}=5\text{ mA}$$

and the base current

$$I_b=\frac{I_c}{h_{fe}}=\frac{5\times 10^{-3}}{50}=100\ \mu\text{A}$$

Hence, the resistance of the timing network

$$R_1=\frac{V_{cc}-V_{BE}}{I_b}=\frac{10-0.7}{100\times 10^{-6}}$$

$$=93\text{ k}\Omega$$

The half-time period of astable multivibrator is

$$\tau=0.69R_1C_1$$

or $$100\times 10^{-6}=0.69\times 93\times 10^3\times C_1$$

or $$C_1=1560\text{ pF}$$

Problem 3

In the circuit of monostable multivibrator of Fig.13.5, V_{CC} = 5V, R_L = 1 kΩ, h_{fe} = 25, R_2 = 4 kΩ, $V_{BE_1}=V_{BE_2}$ = 0.7V, $I_{B\max}$ = 1 mA, and V_{BB} = −1.5 V. Find the values of R, C, and R_1 if the frequency of oscillations is 0.2 MHz.

Solution

When Q_1 is on, the current in R_2 will be

$$I_{R_2}=\frac{1.5+0.7}{4\times 10^3}=0.55\text{ mA}$$

and the current in R_1 is

$$I_{R_1} = I_B + I_{R_2} \text{ where}$$

$$I_B = \frac{I_c}{h_{fe}} = \frac{V_{cc}/R_L}{h_{fe}}$$

$$= \frac{5}{1 \times 10^3 \times 25} = 0.2 \text{ mA}$$

Hence

$$I_{R_1} = 0.2 + 0.55 = 0.75 \text{ mA}$$

The voltage drop across the resistor R_1 is

$$V_{R_1} = V_{cc} - I_{R_1} \cdot R_L - V_{BE_1}$$

$$= 5 - 0.75 \times 10^{-3} \times 1 \times 10^3 - 0.7$$

$$= 3.55 \text{ V}$$

Therefore,

$$R_1 = \frac{V_{R_1}}{I_{R_1}} = \frac{3.55}{0.75 \times 10^{-3}} = 4.7 \text{ k}\Omega$$

The minimum value of R will be

$$R = \frac{V_{cc}}{I_{B\max}} = \frac{5}{1 \times 10^{-3}} = 5 \text{ k}\Omega$$

The time period of oscillations,

$$\tau = \frac{1}{f} = 0.69\, RC$$

or

$$C = \frac{1}{0.69Rf} = \frac{1}{0.69 \times 5 \times 10^3 \times 0.2 \times 10^6}$$

$$= 1450 \text{ pF}$$

Problem 4

In Fig.13.7, a circuit for bistable multivibrator is shown. Take the following values of the components:

$V_{CC} = 5$ V, $V_{BB} = -1.5$ V, $R_L = 1$ kΩ, $R_2 = 20$ kΩ, $V_{BE} = 0.7$V, and $h_{fe} = 25$. Find the value of coupling resistor R_1.

Solution

When Q_1 is on, the current through R_2 is

$$I_{R_2} = \frac{1.5 + 0.7}{20 \times 10^3} \cong 0.1 \text{ mA}$$

and the current in R_1

$$I_{R_1} = I_B + I_{R_2} \text{ where}$$

$$I_B = \frac{I_c}{h_{fe}} = \frac{V_{cc}/R_L}{h_{fe}}$$

$$= \frac{5}{1 \times 10^3 \times 25} = 0.20 \text{ mA}$$

Hence

$$I_{R_1} = 0.2 + 0.1 = 0.3 \text{ mA}$$

The voltage drop across R_1 is

$$V_{R_1} = V_{cc} - I_{R_1} \cdot R_L - V_{BE}$$

$$= 5 - 0.3 \times 10^{-3} \times 1 \times 10^3 - 0.7$$

$$= 4 \text{ V}$$

Therefore, the coupling resistor

$$R_1 = \frac{V_{R_1}}{I_{R_1}} = \frac{4}{0.3 \times 10^{-3}} = 13.4 \text{ k}\Omega$$

In order to allow the voltage variations, R_1 may be selected as 10 kΩ.

Questions and Problems

1. Describe the switching action of a transistor.
2. What are multivibrators? Name the various types of multivibrators.
3. What is an astable multivibrator? Describe its action.
4. Draw the voltage and current waveforms of an astable multivibrator.
5. Calculate the frequency of oscillations in an astable multivibrator.
6. Draw and describe the voltage and current waveforms in a monostable multivibrator.
7. Find the expression for the time period of oscillations for a monostable multivibrator.

8. Describe the working of a bistable multivibrator. Why is it also called a binary?
9. Why are commutating capacitors used in a binary?
10. Describe a Schmitt trigger circuit.
11. Calculate h_{fe} of a transistor used as a switch if $R = 5$ kΩ, $R_L = 1$ kΩ, $V_{CC} = 9$ V, $V_{BE\,sat} = 0.6$ V, and $V_{CE\,sat} = 0.2$ V.
12. Design the timing network for an astable multivibrator of period 100 μsec. if $V_{CC} = 9$ V, $h_{fe} = 75$, and $R_{L_1} = 2$ kΩ.
13. Following the circuit of Fig.13.5 for a monostable multivibrator, find the frequency of oscillations if $R_1 = 3.3$ kΩ, $R = 4.7$ kΩ, and $C = 200$ pF.
14. Find h_{fe} for the transistors used in a bistable multivibrator with the following circuit components:
$V_{CC} = 9$ V, $V_{BB} = -1.5$ V, $R_L = 2$ kΩ, $R_2 = 15$ kΩ, $V_{BE} = 0.6$ V, and $R_1 = 10$ kΩ.

14

Basic Logic Gates

The digital circuits have nowadays become more popular than the analogue circuits because of their applications in communication electronics, calculators, computers, automobiles, and other home appliances. The analog circuits use linearly variable voltages, whereas the digital circuitry responds to two signal levels called *logic levels*. These logic levels are represented by various methods as shown below:

(a) upper voltage level, closed, 1, high, on, true and;

(b) lower voltage level, open, 0, low, off, false. Since only two digits are needed for digital circuits, the binary system of numbers is used in place of the commonly used decimal system. Thus, the binary numbers and its mathematics have been explained in this chapter before discussing the basic logic gates which are the building blocks of digital circuits.

14.1 BINARY NUMBERS

In commonly used decimal system of numbers, ten numerals are required to express any number. For example, the number 8527 may be expressed as

$$8527 = 8\times10^3+5\times10^2+2\times10^1+7\times10^0 \qquad (14.1)$$

where 10 is called the *base* or the *radix*. In the above, number 8 is called the most significant digit which has to be multiplied by the 10 with the largest exponent, whereas 7 is the least significant digit which has to be multiplied by the 10 with zero exponent.

As in digital circuitry, only two levels are used, the binary system of numbers requires only two numerals known as *bits (binary digits)*, viz. 0 and 1. The radix is 2 in this system. For example, the number 17 may be expressed as

$$17 = 10001 = 1\times2^4+0\times2^3+0\times2^2+0\times2^1+1\times2^0 \qquad (14.2)$$

(decimal) (binary)

As another example, the number 21 may be expressed as

$$21 = 10101 = 1\times2^4 + 0\times2^3 + 1\times2^2 + 0\times2^1 + 1\times2^0 \quad (14.2)$$

(decimal) (binary)

The first digit at the left of the number is the most significant number (MSB) and the first digit at the right is called the least significant digit (LSB)

14.1.1 Conversion of Binary to Decimal Numbers

In order to convert any binary number to the equivalent decimal number, the following procedure is followed:

(a) The bit at the extreme right is multiplied by $2^0 = 1$.
(b) The next bit, i.e. second from the right, is multiplied by $2^1 = 2$.
(c) The next, third bit from the right, is multiplied by $2^2 = 4$.
(d) The next, fourth bit from the right, is multiplied by $2^3 = 8$
(e) In this way, the procedure is shifted towards the left digits one by one, adding one number to the exponent of 2 for every next left digit.
(f) All the numbers obtained in steps (a) to (e) above are added to get the equivalent decimal number, as shown in the following example.

Binary number	Procedure to convert to decimal number	Decimal number
00000	$0\times2^4 + 0\times2^3 + 0\times2^2 + 0\times2^1 + 0\times2^0$	0
00001	$0\times2^4 + 0\times2^3 + 0\times2^2 + 0\times2^1 + 1\times2^0$	1
00010	$0\times2^4 + 0\times2^3 + 0\times2^2 + 1\times2^1 + 0\times2^0$	2
00011	$0\times2^4 + 0\times2^3 + 0\times2^2 + 1\times2^1 + 1\times2^0$	3
00100	$0\times2^4 + 0\times2^3 + 1\times2^2 + 0\times2^1 + 0\times2^0$	4
00101	$0\times2^4 + 0\times2^3 + 1\times2^2 + 0\times2^1 + 1\times2^0$	5
00110	$0\times2^4 + 0\times2^3 + 1\times2^2 + 1\times2^1 + 0\times2^0$	6
00111	$0\times2^4 + 0\times2^3 + 1\times2^2 + 1\times2^1 + 1\times2^0$	7
01000	$0\times2^4 + 1\times2^3 + 0\times2^2 + 0\times2^1 + 0\times2^0$	8

14.1.2 Conversion of Decimal to Binary Numbers

In order to convert a decimal number into a binary number, the decimal number is divided by 2 and the remainder (1 or 0) is recorded in a column. The quotient is again divided by 2 and the remainder is recorded in the column below the first remainder. The procedure is repeated until there is nothing to divide. The remainder recorded in the column in reverse order is the equivalent binary number. The procedure may be explained in the following example:

Decimal number	Procedure to convert into binary number	Remainder
85	$\frac{85}{2}=42$	1
	$\frac{42}{2}=21$	0
	$\frac{21}{2}=10$	1
	$\frac{10}{2}=5$	0
	$\frac{5}{2}=2$	1
	$\frac{2}{2}=1$	0
	$\frac{1}{2}=0$	1

So the binary equivalent of 85 is

$$1010101 = 1\times2^6+0\times2^5+1\times2^4+0\times2^3+1\times2^2+0\times2^1+1\times2^0$$

$$= 85$$

14.1.3 Addition of Binary Numbers

The rules for addition of binary numbers are as follows:

(a) $0+0=0$

(b) $1+0=1$

(c) $0+1=1$

(d) $1+1=10$

where 10 represents 0 and a carry of 1. The following examples explain the addition of binary numbers,

Example 1

```
      1  1  0  1  1  1
    + 0  1  1  1  0  1
    ------------------
      1  0  1  0  1  0   sum
    + 1     1     1      carry
    ------------------
      0  0  0  0  0  0   sum
  + 1     1     1        carry
  --------------------
    1  0  1  0  1  0  0
```

The LSB of the two quantities to be added are 1 and 1, the binary sum of which is 10. This 10 is written as 0 in the sum row and first column and 1 in the carry row and second column. The second bits from the right in the quantities are 1 and 0, whose sum is 1, which is written in the sum row and second column.

The third bits from the right are 1 and 1, whose sum is 10, which is written as 0 in the sum row and third column and 1 in the carry row and fourth column. The fourth bits from the right are 0 and 1, whose sum is 1 which is written in the sum row and fourth column. The fifth bits from the right are 1 and 1, whose sum is 10, which is written as 0 in the sum row and fifth column and 1 in the carry row and sixth column. The sixth bits (MSB) are 1 and 0, whose sum is 1, written in the sum row and sixth column.

Now, the sum and carry rows are added, following the above procedure, to get the final result. The second example shown below helps further in an understanding of the addition procedure.

Example 2

```
    1 0 1 0 1
  +   1 0 1 1
  -----------
    1 1 1 1 0   sum
  +       1     carry
  -----------
    1 1 1 0 0   sum
  +     1       carry
  -----------
    1 1 0 0 0   sum
  +   1         carry
  -----------
    1 0 0 0 0   sum
  + 1           carry
  -----------
  1 0 0 0 0 0
  -----------
```

14.1.4 Subtraction of Binary Numbers

The rules for the subtraction of two binary numbers are as follows:

(a) $0-0=0$

(b) $1-0=1$

(c) $10-1=1$

(d) $1-1=0$

The following examples explain the subtraction procedure for binary numbers

Example 1

	Binary						Decimal
	1	0	0	0	0	← Minuend	16
	(1)	(1)	(1)	(1)			
–	0	1	0	1	1	← Subtrahend	–11
	0	0	1	0	1	← Result	5

Starting from LSB (extreme right bits), the 1 in the subtrahend is to be subtracted from 0 in the minuend, which is not possible. Therefore, 1 is borrowed from the second bit of the minuend and is depicted as circled 1 written below the second bit of minuend. This makes LSB of minuend to be 10, out of which 1 of the subtrahend is subtracted to get 1, which is written as the first right bit in the result. But 1 is not available in the second bit of minuend, so 1 is further borrowed from the third bit and depicted as circled 1 below this. But still 1 is not available in the third bit of minuend also, so 1 is further borrowed from the fourth bit and depicted as circled 1 below it. However, 1 is still not available in the fourth bit of the minuend also. Now, in the process of borrowing from the left, 1 is borrowed from the fifth digit in the minuend, which is available and is depicted as circled 1 below it. Now, subtract the second, third, and fourth bits of the subtrahend from the second, third, and fourth bits of the borrowed 1's (marked as circled 1's) to get 0, 1, and 0 respectively. Finally, subtract the borrowed 1 in the fifth column from the fifth bit of the minuend to get 0. The following example further helps in understanding the subtraction procedure for binary numbers.

Example 2

	Binary						Decimal
	1	1	0	0	1	← Minuend	25
		(1)	(1)			← Borrow	
–	1	0	0	1	1	← Subtrahend	–19
	0	0	1	1	0	← Result	6

Here, the LSB of the subtrahend 1 is subtracted from the LSB of minuend 1 and it is feasible to get 0 written in the first column of the result row. Then, for the second bits onwards, the procedure explained in the first example is followed.

14.1.5 Multiplication of Binary Numbers

The rules of multiplication of binary numbers are as follows:

(a) $1 \times 0 = 0 = 0 \times 1$

(b) $1 \times 1 = 1$

The procedure is the same as followed in the decimal numbers and may be understood by the following examples:

Example 1

Binary	Decimal
1 0 0	4
× 1 1 1	×7
1 0 0	28 Result
1 0 0	
1 0 0	
1 1 1 0 0 Result	

Example 2

Binary	Decimal
1 0 1 0 1	21
× 1 0 0 1 1	× 19
1 0 1 0 1	189
1 0 1 0 1	21
0 0 0 0 0	399 Result
0 0 0 0 0	
1 0 1 0 1	
1 1 0 0 0 1 1 1 1 Result	

14.1.6 Division of Binary Numbers

The process of division of binary numbers is the same as followed in the decimal numbers. Some examples follow:

Example 1

Binary		Decimal
1 1 0	←Result →	6
1 1 1) 1 0 1 0 1 0		7) 4 2
1 1 1		4 2
− 1 1 1		0
0 0 1 1 1		
− 1 1 1		
0 0 0		

Example 2

Binary		Decimal
1 1	←Result →	3
1 1 0) 1 0 0 1 0		6) 1 8
1 1		−1 8
− 1 1 0		0
0 0 1 1 0		
− 1 1 0		
0 0 0		

14.2 BOOLEAN ALGEBRA

The conventional mathematics is not used in logic circuits. Instead, a different type of manupulations known as Boolean algebra is applied to logic circuits. In Boolean algebra, the variables may have only two possible values denoted by 1 or 0. The plus sign (+) between two quantities represents the *OR* function, i.e. the presence of either quantity at the input of a logic gate is sufficient to open the gate. The dot or blank sign (.) or ( ) between the two quantities represents the *AND* operation. i.e. the presence of both the quantities simultaneously at the input of a logic gate may open the gate. The prime or bar (′) or (–) denotes the *NOT* operation, i.e.. the presence of the input means absence of the output and vice-versa.

In Boolean algebra, a variable cannot be added, divided, multiplied, subtracted, or squared; and other similar manupulations of conventional algebra are also not possible. If A, B and C are the three quantities having the values 1 or 0, the various rules applicable in Boolean algebra are as follows:

(a) $A + B = B + A$

(b) $A.B = B.A$

(c) $A\,(BC) = B(AC) = C\,(AB)$

(d) $A + (B + C) = B + (A + C) = C + (A + B)$

(e) $A\,(B + C) = (AB) + (AC)$

(f) $A + (B + C) = (A + B)\,(A + C)$

(g) $A + 0 = A$

(h) $A\,.\,1 = A$

(i) $A + \overline{A} = 1$

(j) $A\,.\,\overline{A} = 0$

(k) $A + A = A$

(l) $A.A = A$

(m) $A.0 = 0$

(n) $A + 1 = 1$

(o) $A + AB = A$

(p) $A\,(A + B) = A$

(q) $A + \overline{A}B = A + B$

14.3 DE MORGAN'S LAWS

There are two theorems given by De Morgan for the Boolean algebra as,

(a) $\overline{A + B + C} = \overline{A} \cdot \overline{B} \cdot \overline{C}$

(b) $\overline{A \cdot B \cdot C} = \overline{A} + \overline{B} + \overline{C}$

14.4 TYPES OF LOGIC

There are two types of logic defined as follows:

(a) Positive Logic

In the positive logic, the more positive voltage is considered as the logic–1 state, whereas less positive voltage is assigned as the logic–0 state.

(b) Negative Logic

In the negative logic, the more negative voltage is assigned as the logic–1 state, whereas less negative voltage is assigned as the logic–0 state.

14.5 TRUTH TABLE

Boolean algebraic equations are represented in the form of truth tables. As already described, Boolean algebra involves only two values (0 or 1) of variables. Hence a truth table represents all possible values of the output (dependant variable) owing to various combinations of values assigned to the independent variables, as shown in Figs.14.2 and 14.4.

14.6 DIODE *AND* GATE

The *AND* gate is defined as the circuit which provides the output in logic–1 state, only when all the inputs are in logic–1 states. The Boolean expression for the three input *AND* gate is

$$L = ABC$$

which reads as, "the output *L* is in the logic–1 state, only when *A* and *B* and *C* are in the logic–1 state". A simple *AND* gate using mechanical switches is shown in Fig. 14.1(a). The circuit clearly explains that the bulb (output) will be *ON* (logic–1 state) only when switches *A*, *B* and *C* are closed (logic–1 state). The symbol of the *AND* gate is shown in Fig. 14.1(b).

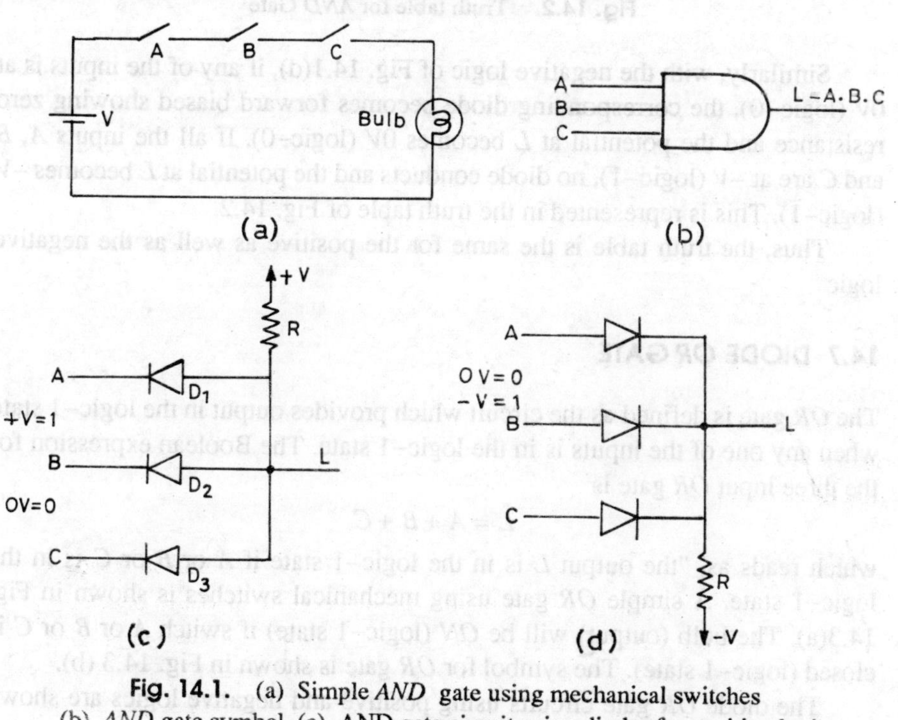

Fig. 14.1. (a) Simple *AND* gate using mechanical switches (b) *AND* gate symbol (c) AND gate circuit using diodes for positive logic (d) AND gate circuit using diodes for negative logic

The diode *AND* gate circuits using positive and negative logics are shown in Figs.14.1(c) and 14.1(d) respectively. In Fig. 14.1(c) for positive logic, if any of the three inputs is at 0V (logic - 0), the corresponding diode becomes forward biased or conducting showing zero resistance and hence the voltage at *L* becomes zero. If all the inputs *A*, *B* and *C* are at +*V* (logic–1), no diode conducts and the voltage at *L* will be +*V* (logic–1). This is described with the help of the truth table shown in Fig. 14.2.

A	B	C	L
0	0	0	0
0	0	1	0
0	1	0	0
0	1	1	0
1	0	0	0
1	0	1	0
1	1	0	0
1	1	1	1

Fig. 14.2. Truth table for *AND* Gate

Similarly, with the negative logic of Fig. 14.1(d), if any of the inputs is at 0*V* (logic–0), the corresponding diode becomes forward biased showing zero resistance and the potential at *L* becomes 0*V* (logic–0). If all the inputs *A*, *B* and *C* are at –*V* (logic–1), no diode conducts and the potential at *L* becomes –*V* (logic–1). This is represented in the truth table of Fig. 14.2.

Thus, the truth table is the same for the positive as well as the negative logic.

14.7 DIODE *OR* GATE

The *OR* gate is defined as the circuit which provides output in the logic–1 state when any one of the inputs is in the logic–1 state. The Boolean expression for the three input *OR* gate is

$$L = A + B + C$$

which reads as, "the output *L* is in the logic–1 state if *A* or *B* or *C* is in the logic–1 state. A simple *OR* gate using mechanical switches is shown in Fig. 14.3(a). The bulb (output) will be *ON* (logic–1 state) if switch *A* or *B* or *C* is closed (logic–1 state). The symbol for *OR* gate is shown in Fig. 14.3 (b).

The diode *OR* gate circuits using positive and negative logics are shown in Figs.14.3(c) and 14.3(d) respectively. In Fig. 14.3 (c), using the positive logic, if input *A*, *B* or *C* is at +*V* (logic–1) the corresponding diode conducts

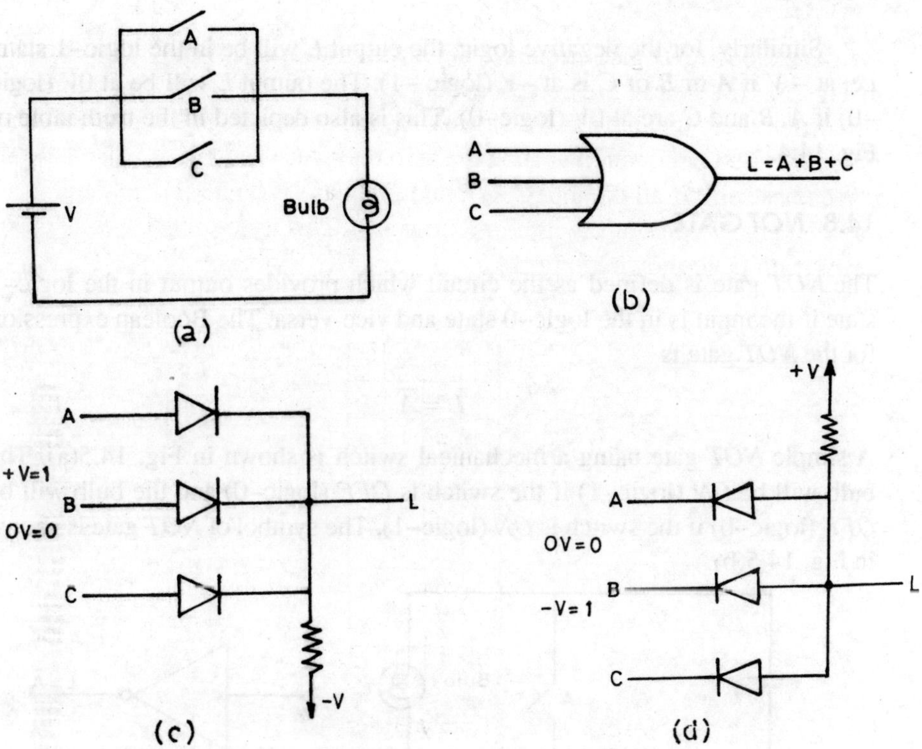

Fig. 14.3. (a) Simple *OR* gate using mechanical switches (b) *OR* gate symbol (c) *OR* gate circuit using diodes for positive logic (d) *OR* Gate circuit using diodes for negative logic

and the potential at *L* becomes +*V* (logic–1). If all the inputs are at 0*V* (logic – 0), all the diodes conduct and the potential at *L* will be 0*V* (logic–0). If any of the three inputs is at + *V* (logic–1), the output will be at + *V* (logic–1). This function of the *OR* gate is depicted in the truth table of Fig. 14.4.

A	B	C	L
0	0	0	0
0	0	1	1
0	1	0	1
0	1	1	1
1	0	0	1
1	0	1	1
1	1	0	1
1	1	1	1

Fig. 14.4. Truth table for *OR* gate

Similarly, for the negative logic, the output L will be in the logic–1 state, i.e. at $-V$ if A or B or C is at $-V$ (logic –1). The output L will be at $0V$ (logic –0) if A, B and C are at $0V$ (logic–0). This is also depicted in the truth table of Fig. 14.4.

14.8 *NOT* GATE

The *NOT* gate is defined as the circuit which provides output in the logic–1 state if the input is in the logic–0 state and vice-versa. The Boolean expression for the *NOT* gate is

$$L = \overline{A}$$

A simple *NOT* gate using a mechanical switch is shown in Fig. 14.5(a). The bulb will be *ON* (logic–1) if the switch is *OFF* (logic–0) and the bulb will be *OFF* (logic–0) if the switch is *ON* (logic–1). The symbol of *NOT* gate is shown in Fig. 14.5(b).

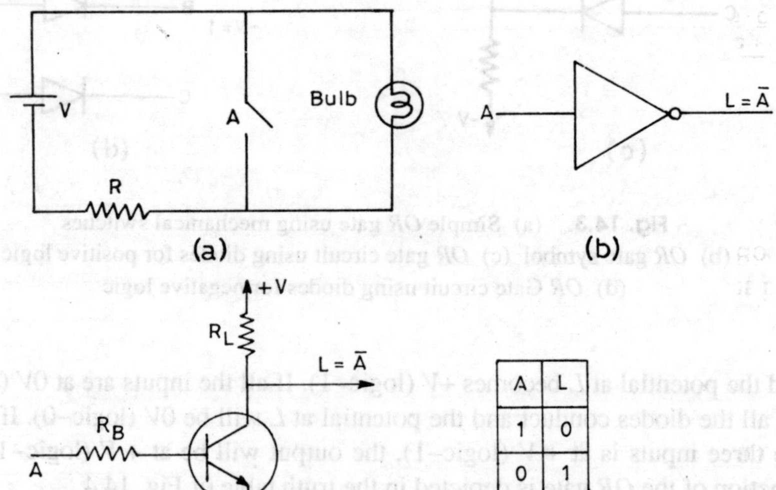

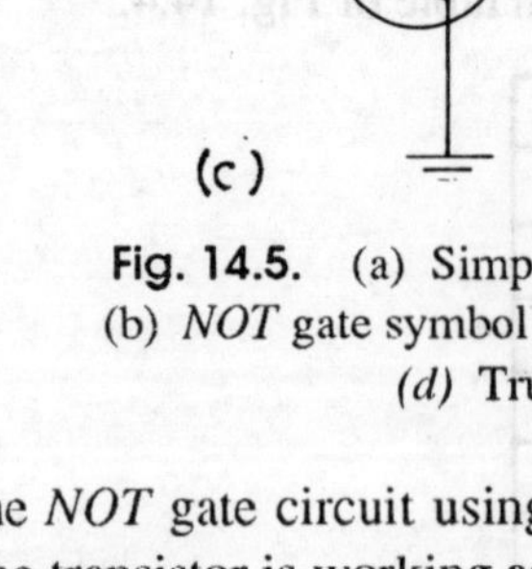

A	L
1	0
0	1

(d)

Fig. 14.5. (a) Simple *NOT* gate using mechanical switch (b) *NOT* gate symbol (c) *NOT* gate circuit using transistor (*d*) Truth table for *NOT* gate.

The *NOT* gate circuit using an n-p-n transistor is shown in Fig. 14.5(c). Here, the transistor is working as a switch, i.e. either it is in the saturation or in cut-off position. When the input is $+V$ (logic–1), the transistor is in saturation and the collector potential (output) will be $0V$ (logic–0), and when the input is

$0V$ (logic–0), the transistor is cut-off, raising the collector potential to be $+V$ (logic–1). This function of the *NOT* gate is depicted in the truth table shown in Fig. 14.5(d).

14.9 *NAND* GATE

The *NAND* gate is basically an *AND* gate driving a *NOT* gate. In other words, a *NOT-AND* combination is called *NAND* gate or a negated *AND* gate. The Boolean expression for a *NAND* gate is

$$L = \overline{ABC} = \overline{A} + \overline{B} + \overline{C}$$

which means as, "the output will be in the logic–0 if *A*, and *B* and *C* are in the logic–1 state". The simple circuit using mechanical switches is shown in Fig. 14.6(a). The symbol for a *NAND* gate is depicted in Fig. 14.6(b). The actual electronic circuit using *AND* and *NOT* gates is shown in Fig. 14.6(c). The function of the *NAND* is explained in the truth table of Fig. 14.6(d).

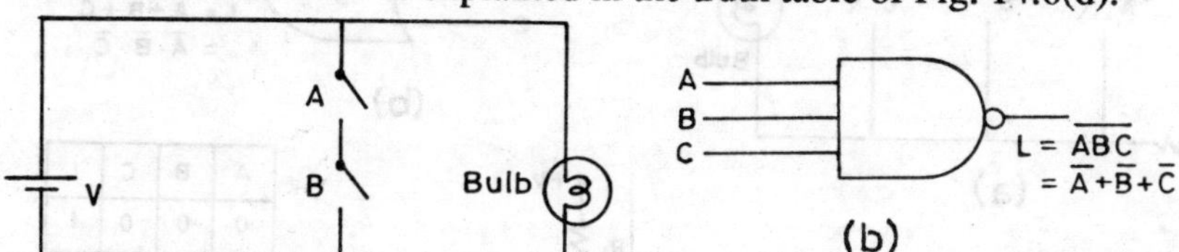

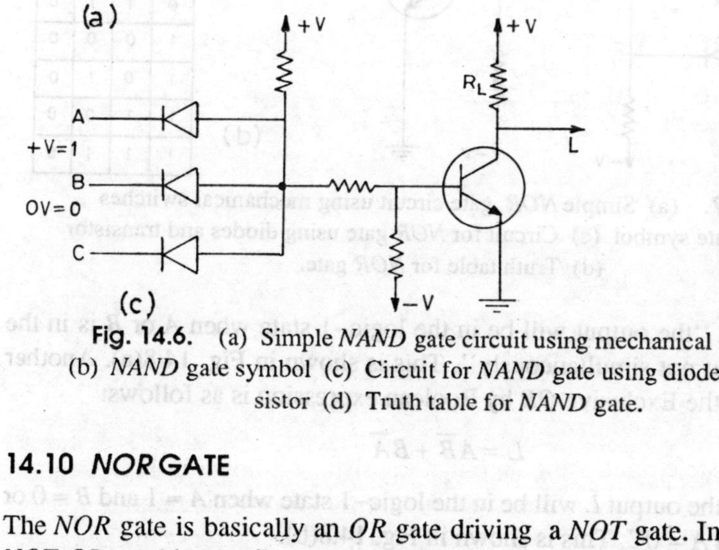

A	B	C	L
0	0	0	1
0	0	1	1
0	1	0	1
0	1	1	1
1	0	0	1
1	0	1	1
1	1	0	1
1	1	1	0

(d)

Fig. 14.6. (a) Simple *NAND* gate circuit using mechanical switches (b) *NAND* gate symbol (c) Circuit for *NAND* gate using diodes and transistor (d) Truth table for *NAND* gate.

14.10 *NOR* GATE

The *NOR* gate is basically an *OR* gate driving a *NOT* gate. In other words, a *NOT-OR* combination is called a *NOR* gate or a negated *OR* gate. The Boolean expression for a *NOR* gate is

$$L = \overline{A + B + C} = \overline{A} \cdot \overline{B} \cdot \overline{C}$$

which spells that, "the output will be in the logic–0 state if *A*, *B* or *C* is in the logic–1 state. The simple circuit using mechanical switches is shown in Fig. 14.7(a). The symbol for a *NOR* gate is depicted in Fig. 14.7(b). The electronic circuit using *OR* and *NOT* gates is shown in Fig. 14.7(c). The function of the *NOR* gate is explained in the truth table of Fig. 14.7(d).

14.11 EXCLUSIVE *OR* GATE

The exclusive *OR* gate is defined as the circuit which provides output in the logic–1 state when one and only one input is at the logic–1 state. The Boolean expression for Exclusive *OR* gate is

$$L = (A + B) \cdot \overline{(AB)}$$

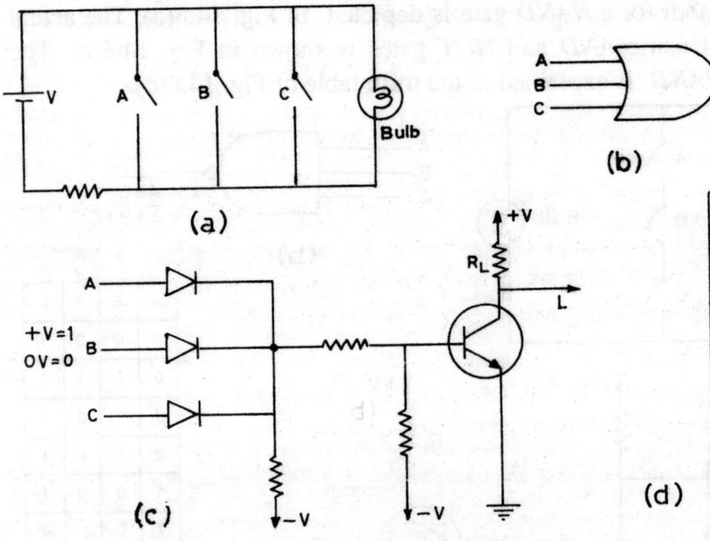

A	B	C	L
0	0	0	1
0	0	1	0
0	1	0	0
0	1	1	0
1	0	0	0
1	0	1	0
1	1	0	0
1	1	1	0

Fig. 14.7. (a) Simple *NOR* gate circuit using mechanical switches (b) *NOR* gate symbol (c) Circuit for *NOR* gate using diodes and transistor (d) Truth table for *NOR* gate.

which reads as, "the output will be in the logic–1 state when *A* or *B* is in the logic–1 state but not simultaneously". This is shown in Fig. 14.8(a). Another way to express the Exclusive *OR* by Boolean expression is as follows:

$$L = A\overline{B} + B\overline{A}$$

which states, "the output *L* will be in the logic–1 state when $A = 1$ and $B = 0$ or when $B = 1$ and $A = 0$". This is shown in Fig. 14.8(b).

The simple circuit using mechanical switches is shown in Fig. 14.9(a). The symbol for Exclusive *OR* is shown in Fig. 14.9 (b). The function of the circuit is explained in the truth table of Fig. 14.9(c).

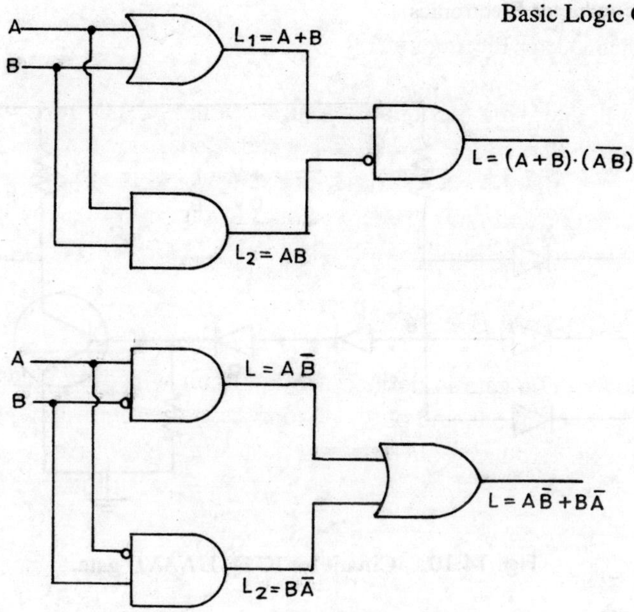

Fig. 14.8. Logic circuits for exclusive *OR* gate.

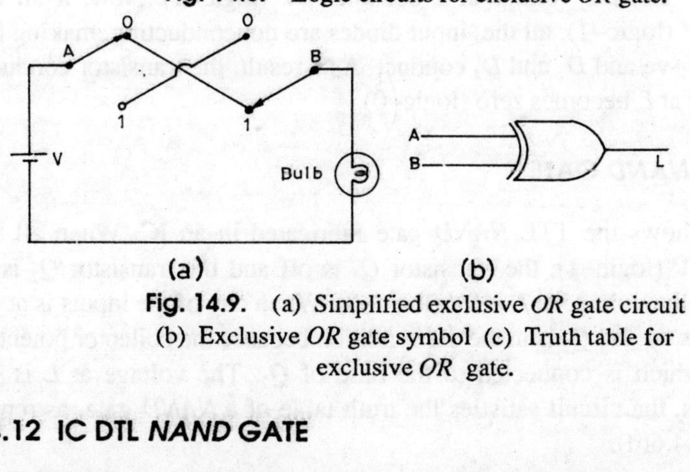

(a) (b)

A	B	L
0	0	0
1	0	1
0	1	1
1	1	0

(c)

Fig. 14.9. (a) Simplified exclusive *OR* gate circuit (b) Exclusive *OR* gate symbol (c) Truth table for exclusive *OR* gate.

14.12 IC DTL *NAND* GATE

A diode-transistor *NAND* gate has been explained in Sec.14.9. In case of a *DTL NAND* gate to be fabricated in an IC, the circuit of Fig. 14.6(c) has to be modified as shown in Fig. 14.10, where high resistances are replaced by diodes and a single power supply is used. This is because the resistors and capacitors of high values are avoided in IC fabrication as discussed in Chapter 4, and therefore, it is advisable to design a circuit for IC fabrication in such a way that it does not involve high values of resistors and capacitors. The circuit of Fig. 14.10 satisfies the truth table of a *NAND* gate shown in Fig. 14.6(d) as discussed below.

If any of the three inputs is at 0*V* (logic–0), the corresponding input diode becomes forward-biased, showing zero resistance and hence, the voltage

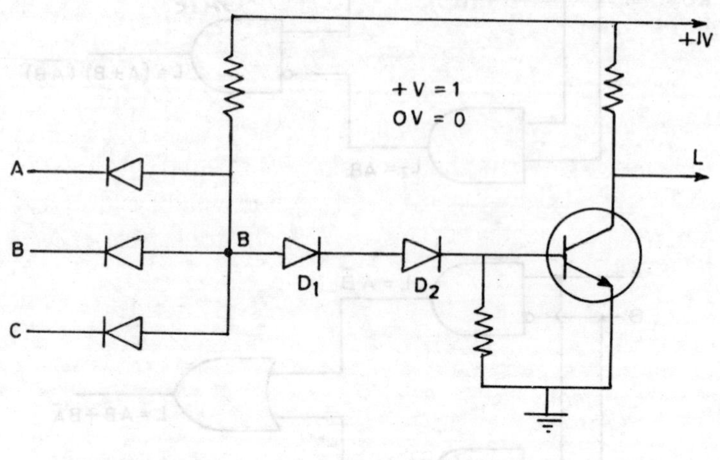

Fig. 14.10. Circuit for IC DTL *NAND* gate.

at B is zero and diode D_1 and D_2 become off. As a result, the transistor becomes nonconducting and the voltage at L becomes $+V$ (logic–1). Now, if all the inputs are at $+V$ (logic–1), all the input diodes are nonconducting, making the voltage at B as +ve and D_1 and D_2 conduct. As a result, the transistor conducts and the voltage at L becomes zero (logic–0).

14.13 IC TTL *NAND* GATE

Figure 14.11 shows the TTL *NAND* gate fabricated in an IC. When all the inputs are at $+V$ (logic–1), the transistor Q_1 is off and the transistor Q_2 is in saturation, and the voltage at L is $0V$ (logic–0). When any of the inputs is at $0V$ (logic–0), Q_1, is on and Q_2 is in the off condition because the collector potential of Q_1 is $0V$, which is connected to the base of Q_2. The voltage at L is $+V$ (logic–1). Thus, the circuit satisfies the truth table of a *NAND* gate, as represented in Fig. 14.6(d).

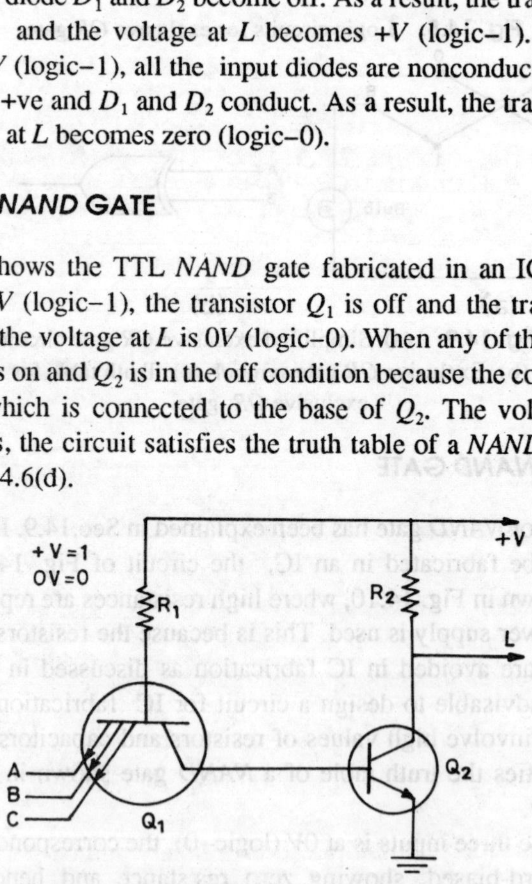

Fig. 14.11. Circuit for IC TTL *NAND* gate.

14.14 RESISTOR-TRANSISTOR LOGIC *NOR* GATE

If we build the various gates (*OR, AND, NOT, NAND*, and *NOR*) using resistors and transistors, these are called *RTL* gates. A circuit for a RTL *NOR* gate is shown in Fig. 14.12. If any input is at +*V* (logic–1), the corresponding transistor will be in saturation and the voltage at *L* will be 0*V* (logic–0). If both the inputs are at 0*V* (logic–0), both the transistors will be cut-off and the voltage at *L* will be +*V* (logic–1). This satisfies the truth table of *NOR* gate as shown in Fig. 14.7(d).

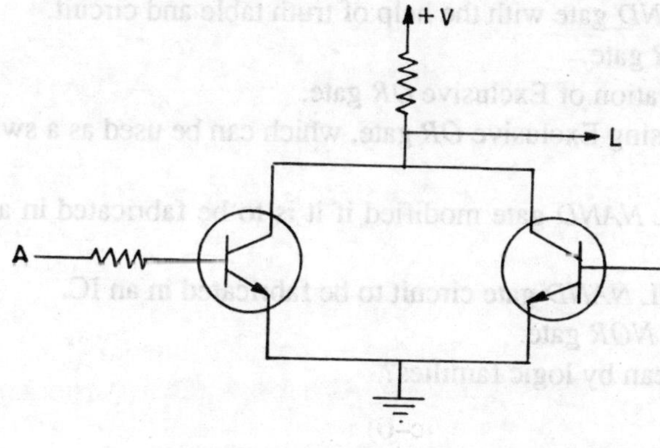

Fig. 14.12. Circuit for RTL *NOR* gate.

Questions and Problems

1. What is the meaning of digital circuits? Why is a digital system more dependable than an analog system?
2. What are binary numbers? Compare them with decimal numbers.
3. Describe the procedure of converting a binary number into an equivalent decimal number.
4. Write the corresponding decimal number for 00101 and 01000 binary numbers.
5. How can a decimal number be converted into an equivalent binary number?
6. Write the binary numbers for the decimal numbers 60, 95, 4000 and 9832.
7. What are the rules for addition of binary numbers. Add 100101 and 110111.
8. Write down the rules for subtraction of binary numbers. Subtract 10000 from 11001.
9. Write down the rules of multiplication of binary numbers. Multiply 00101 and 00110.

10. Divide 11100 by 111.
11. What do you mean by Boolean algebra? Write various rules applicable in this algebra.
12. Write De-Morgan's laws for the Boolean algebra.
13. What is the meaning of positive logic, negative logic, and truth table?
14. Describe diode *AND* gate along with its circuit and truth table.
15. Describe diode *OR* gate along with its circuit and truth table.
16. What is *NOT* gate? Draw the circuit for positive *NOT* gate and explain its operation.
17. Describe the *NAND* gate with the help of truth table and circuit.
18. Discuss the *NOR* gate.
19. Explain the operation of Exclusive *OR* gate.
20. Draw a circuit using Exclusive *OR* gate, which can be used as a switch by three persons.
21. How is the DTL *NAND* gate modified if it is to be fabricated in an integrated circuit?
22. Describe the TTL *NAND* gate circuit to be fabricated in an IC.
23. Describe a RTL *NOR* gate.
24. What do you mean by logic families?

15

Modulation and Demodulation

In order to transmit various forms of intelligence lying in the low frequency range such as speech, music, video, or telemetry to longer distances, *radiofrequency waves* are used. The frequency of these radiowaves, also called *carrier waves*, may be from 100 kHz to many GHz. In the process of transmitting intelligence with the help of radiowaves, the information to be transmitted is superimposed on the radiowave, causing some of its features to change according to the information. This process of mixing information with the carrier is called *modulation* and the resultant wave is called *modulated wave*.

The carrier along with the information in the form of a modulated wave is received at the receiving end and the process known as *demodulation* or *detection* is followed to convert the modulated wave into usable intelligence. Or in other words, demodulation is the process in which the superimposed information is extracted from the modulated wave. In this chapter, the basic aspects and circuits of modulation and demodulation are discussed.

15.1 MODULATION

The radiofrequency wave (carrier) may be represented as

$$v_C = V_C \cos(\omega_C t + \phi) \tag{15.1}$$

where the information may be superimposed on the amplitude V_C, on the frequency ω_C, or on the phase ϕ and accordingly three types of modulation may be achieved, viz.

(a) Amplitude Modulation (*AM*) in which the amplitude varies with the information while the frequency and the phase remain constant.

(b) Frequency Modulation (*FM*) in which the frequency varies with the information while the amplitude and the phase remain constant.

(c) Phase Modulation (*PM*) in which the phase varies with the information while the amplitude and the frequency remain constant.

15.2 AMPLITUDE MODULATION

The carrier wave (with zero phase) and the modulating wave carrying the information may be represented as:

$$v_C = V_C \cos \omega_C t \tag{15.2}$$

and

$$v_m = V_m \cos \omega_m t \tag{15.3}$$

In the amplitude modulation, the amplitude of the modulated wave varies with the information and hence, the equation for the amplitude modulated wave may be written as

$$v_a = (V_C + V_m \cos \omega_m t) \cos \omega_C t$$

$$= V_C \left(1 + \frac{V_m}{V_C} \cos \omega_m t \right) \cos \omega_C t$$

or

$$v_a = V_C (1 + m_a \cos \omega_m t) \cos \omega_C t \tag{5.4}$$

where $m_a = \dfrac{V_m}{V_C}$ and is called the modulation factor or modulation index and represents the extent of modulation. If $V_m = V_C$, $m_a = 1$ or 100 percent modulation is achieved. Figure 15.1 shows the carrier wave, modulating wave or information, and the amplitude modulated wave for 50 percent and 100 percent modulation.

15.2.1 Radiofrequencies in Amplitude Modulated Wave

The amplitude modulated wave is expressed by Eq.15.4 as

$$v_a = V_C (1 + m_a \cos \omega_m t) \cos \omega_C t$$

$$= V_C \cos \omega_C t + V_C m_a \cos \omega_m t \ \cos \omega_C t \tag{15.5}$$

Substituting the value of

$$\cos \omega_m t \cos \omega_C t = \frac{1}{2} [\cos(\omega_C + \omega_m)t + \cos(\omega_C - \omega_m)t]$$

$$v_a = V_C \cos \omega_C t + \frac{V_C m_a}{2} \cos(\omega_C + \omega_m)t + \frac{V_C m_a}{2} \cos(\omega_C - \omega_m)t \tag{15.6}$$

This expression shows that the amplitude modulated wave consists of three types of radio frequencies viz.

(a) the carrier frequency, ω_C;

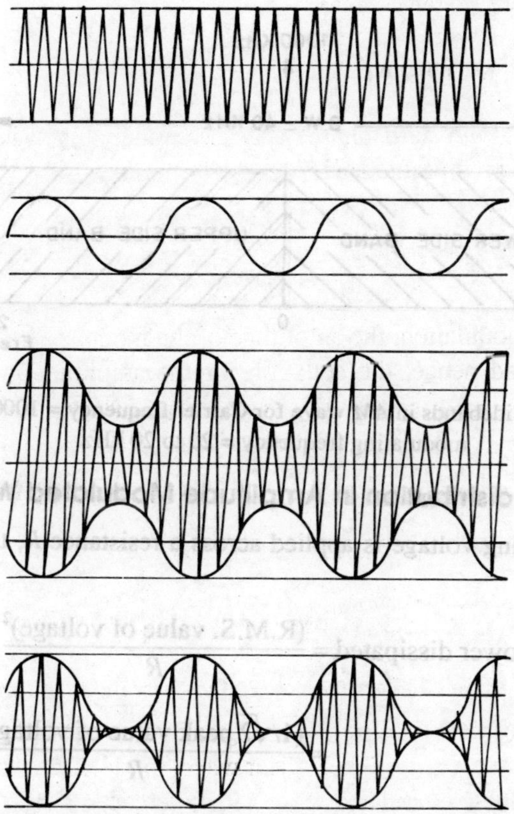

Fig. 15.1 (a) Radio frequency carrier wave (b) Modulating wave or information (c) Amplitude modulated wave for $m_a = 50\%$ (d) Amplitude modulated wave for $m_a = 100\%$

(b) the sum of carrier and modulating frequencies, which is known as the upper side-frequency, $\omega_C + \omega_m$; and

(c) the difference of carrier and modulating frequencies $(\omega_C - \omega_m)$, which is known as the lower-side-frequency.

As the information frequency ω_m involves many frequencies, e.g. audio signal frequency may vary up to 20 kHz, $(\omega_C + \omega_m)$ and $(\omega_C - \omega_m)$ represent a large number of frequencies present in the amplitude modulated wave and are called the upper side band and the lower side band respectively. Figure 15.2 shows the upper and the lower side bands for a carrier of 1000 kHz and information of 20 to 20 kHz. The bandwidth of the amplitude modulated wave of Fig. 15.2 is 40 kHz, which is twice the maximum modulating frequency

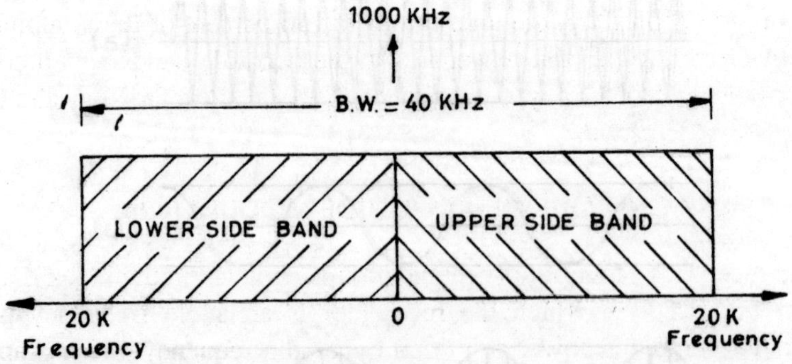

Fig. 15.2 Sidebands in *AM* wave for Carrier frequency = 1000 kHz and modulating frequency = 20 to 20 kHz.

15.2.2 Power distribution in Amplitude Modulated Wave

When an alternating voltage is applied across a resistance *R*, the power dissipated is

$$\text{Power dissipated} = \frac{(\text{R.M.S. value of voltage})^2}{R}$$

$$= \frac{(1/\sqrt{2}\ \text{peak value of voltage})^2}{R}$$

$$= \frac{(\text{peak value of voltage})^2}{2R} \tag{15.7}$$

The amplitude modulated wave has three frequency components with peak voltages V_C, $\frac{V_C m_a}{2}$ and $\frac{V_C m_a}{2}$. Hence, the total power dissipated is

$$P_{\text{total}} = \frac{V_C^2}{2R} + \frac{(V_C m_a/2)^2}{2R} + \frac{(V_C m_a/2)^2}{2R}$$

$$= \frac{V_C^2}{2R} + \frac{V_C^2}{2R}\left(\frac{m_a^2}{4}\right) + \frac{V_C^2}{2R}\left(\frac{m_a^2}{4}\right)$$

= (Power in carrier) + (power in upper side band) + (Power in lower side band)

$$= \frac{V_C^2}{2R}\left(1 + \frac{m_a^2}{2}\right) \tag{15.8}$$

If the modulation is 100 percent, i.e. $m_a = 1$, the power is

$$P = \frac{V_c^2}{2R}\left(1 + \frac{1}{2}\right) \tag{15.9}$$

This expression shows that the total power is 150 percent of the unmodulated carrier power and 50 percent power lies in the two side bands which contain the information. Hence, only one-third of the total power of the amplitude modulated wave is useful.

15.3 BASIC TECHNIQUE OF AMPLITUDE MODULATION

The basic technique of amplitude modulation is depicted in the block diagram shown in Fig. 15.3, in which the modulator is an audio frequency power amplifier. The carrier is power amplified by a radiofrequency power amplifier *(RFPA)*. The modulating voltage signal may be applied at the base, emitter, or collector electrode of the *RFPA* and accordingly the modulation is called base modulation, emitter modulation, or collector modulation respectively.

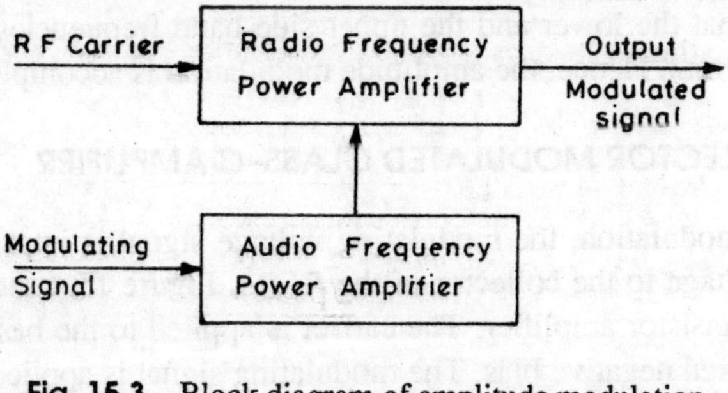

Fig. 15.3 Block diagram of amplitude modulation

Any nonlinear device may be used to achieve modulation but owing to high power output requirement, the transistors are usually used, which also show a nonlinear relationship between the output current and the input voltage as

$$I_c = A_0 + A_1 v + A_2 v^2 + A_3 v^3 + \ldots \tag{15.10}$$

where v is the resultant voltage obtained by the simultaneous application of carrier and modulating waves, i.e.

$$v = V_c \cos \omega_c t + V_m \cos \omega_m t \tag{15.11}$$

Using Eq. 15.11, Eq.15.10 becomes

$$\begin{aligned} I_c = A_0 &+ A_1(V_c \cos \omega_c t + V_m \cos \omega_m t) \\ &+ A_2(V_c \cos \omega_c t + V_m \cos \omega_m t)^2 \\ &+ A_3(V_c \cos \omega_c t + V_m \cos \omega_m t)^3 \\ &+ \ldots\ldots\ldots \end{aligned} \tag{15.12}$$

Solving Eq.15.12 trigonometrically,

$$I_c = A_0 I_0 + \frac{A_2(V_c^2 + V_m^2)}{2} + A_1 V_m \cos \omega_m t$$

$$+ \frac{A_2 V_m^2}{2} \cos 2\omega_m t + A_1 V_c \cos \omega_c t$$

$$+ A_2 V_c V_m \cos(\omega_c - \omega_m)t$$

$$+ A_2 V_c V_m \cos(\omega_c + \omega_m)t + \frac{A_2 V_c^2}{2} \cos 2\omega_c t$$

$$+ \ldots.. \tag{15.13}$$

This shows that the lower and the upper side-band frequencies are present in the output current. Hence, the amplitude modulation is accomplished.

15.4 COLLECTOR MODULATED CLASS-C AMPLIFIER

In collector modulation, the modulating voltage signal is impressed upon the dc supply voltage to the collector of the *RFPA*. Figure 15.4 shows a collector modulated transistor amplifier. The carrier is applied to the base of Q_1, which is kept at a fixed negative bias. The modulating signal is applied in series with the collector circuit. The output developed in the primary of the transformer T_3 is the desired modulated wave. In order to achieve a high level modulation the transistor Q_1 is operated in class-*C* operation. The overall efficiency is extremely great but high power of the modulating signal is needed, which requires several stages of amplifiers.

15.5 BALANCED MODULATOR

In order to obtain only the two side bands in the amplitude modulation, a circuit known as balanced modulator is used. The carrier is suppressed because it contains no information and consumes two-thirds of the power. The circuit for a transistor balanced modulator is shown in Fig. 15.5. The function of the circuit for carrier suppression is as follows:

Suppose only the radio frequency carrier is present, which is connected through C_2. The value of R_4 is so adjusted that the voltages at the bases of Q_1 and Q_2 are equal. This setting causes equal currents to flow in the outputs of Q_1 and Q_2, which in turn, flow in the primary of the transformer T_2 in opposite directions. Hence, the fields produced by the two equal and opposite currents cancel each other and no voltage is developed across the secondary of the

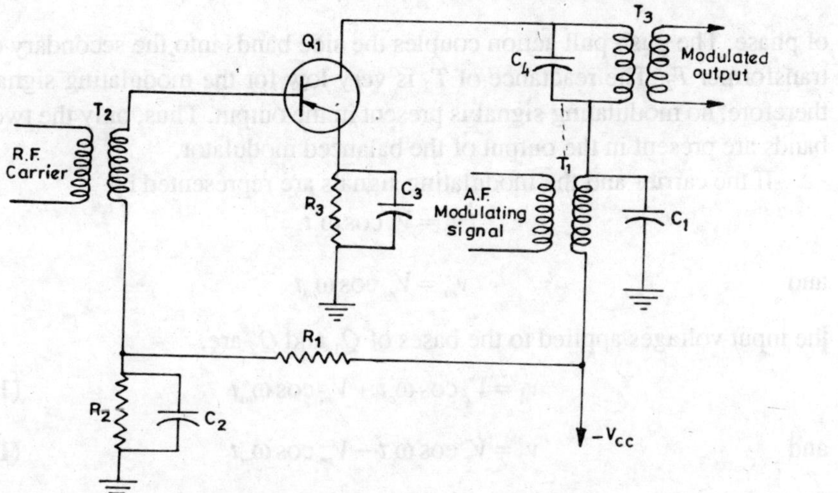

Fig. 15.4 Circuit configuration for collector modulated Class-C Amplifier

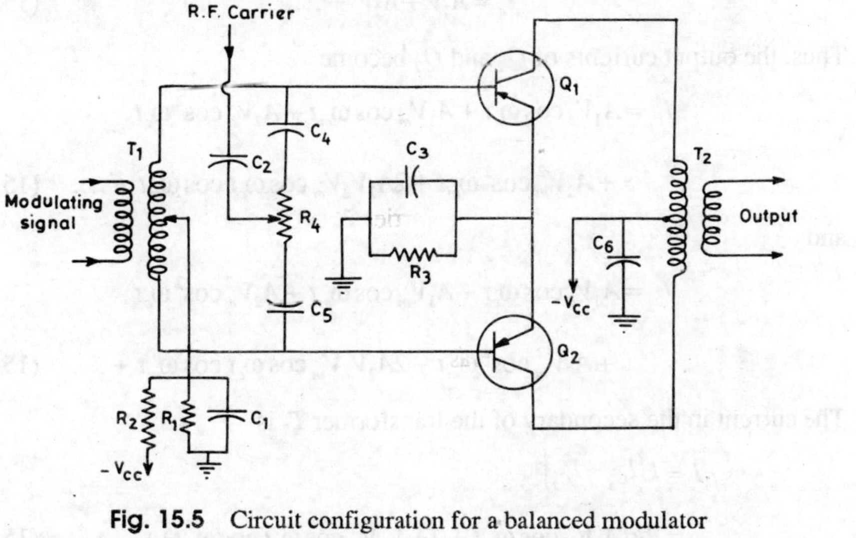

Fig. 15.5 Circuit configuration for a balanced modulator

transformer T_2. In this way, the carrier may be suppressed completely by properly adjusting R_4.

When both the RF and the modulating signal are applied simultaneously, the modulating voltage will appear at the bases of Q_1 and Q_2 in the opposite polarities because of the centre-tapped transformer T_1. The capacitors C_4 and C_5 are so chosen that they present high impedances to the modulating signal and therefore, the two bases of Q_1 and Q_2 become isolated for modulating signals. In the output of Q_1 and Q_2, the modulated waves each having carrier, lower, and upper side band frequencies are obtained.

The $R.F$ carrier is suppressed, as explained above. The sideband signals of Q_1 and Q_2 appear out of phase because modulating signals at Q_1 and Q_2 are out of phase. The push pull action couples the side bands into the secondary of the transformer T_2. The reactance of T_2 is very low for the modulating signal and therefore, no modulating signal is present in the output. Thus, only the two side bands are present in the output of the balanced modulator.

If the carrier and the modulating signals are represented by

$$v_c = V_c \cos \omega_c t$$

and

$$v_m = V_m \cos \omega_m t$$

the input voltages applied to the bases of Q_1 and Q_2 are,

$$v_1 = V_c \cos \omega_c t + V_m \cos \omega_m t \quad (15.14)$$

and

$$v_2 = V_c \cos \omega_c t - V_m \cos \omega_m t \quad (15.15)$$

The output current of a transistor may be represented by Eq.15.10 as

$$I_c = A_1 v + A_2 v^2 + \ldots. \quad (15.16)$$

Thus, the output currents of Q_1 and Q_2 become

$$I_{c_1} = A_1 V_c \cos \omega_c t + A_1 V_m \cos \omega_m t + A_2 V_c^2 \cos^2 \omega_c t$$

$$+ A_2 V_m^2 \cos^2 \omega_m t + 2A_2 V_2 V_m \cos \omega_c t \cos \omega_m t + \ldots. \quad (15.17)$$

and

$$I_{c_2} = A_1 V_c \cos \omega_c t - A_1 V_m \cos \omega_m t + A_2 V_c^2 \cos^2 \omega_c t$$

$$+ A_2 V_m^2 \cos^2 \omega_m t - 2A_2 V_c V_m \cos \omega_c t \cos \omega_m t + \ldots. \quad (15.18)$$

The current in the secondary of the transformer T_2 is

$$I = k\left(I_{C_1} - I_{C_2}\right)$$

$$= 2k(A_1 V_m \cos \omega_m t + 2A_2 V_c V_m \cos \omega_c t \cos \omega_m t) + \ldots \quad (15.19)$$

As the reactance of the T_2 is very low for ω_m, the term containing ω_m is negligible. Hence

$$I = 2K(2A_2 V_c V_m \cos \omega_c t \cos \omega_m t)$$

$$= 2KA_2 V_c V_m [\cos(\omega_c + \omega_m)t + \cos(\omega_c - \omega_m)t] \quad (15.20)$$

which shows the output current having only the lower and the upper sidebands.

15.6 SSB-SUPPRESSED MODULATION

The output of the balanced modulator contains the two sidebands only. As the information in the two sidebands is the same, only one sideband is necessary and sufficient for communicating the information. So, one of the side bands may also be filtered out by a suitable filter circuit. This modulation is known as a single sideband suppressed carrier modulation. The *SSB* system is helpful in saving power because only one sideband is transmitted. The *SSB* signal uses a very small range of the frequency spectrum and therefore, many *SSB* systems may be used in a small frequency spectrum without interference. The main disadvantage of the *SSB* system is that it depends upon the reinsertion of the carrier of the exact frequency and amplitude at the receiver in order to extract the information from the modulated signal.

A block diagram for *SSB*-suppressed modulation is shown in Fig. 15.6 where two balanced modulators are used. The carrier and the modulating signals fed to the two balanced modulators are out of phase by 90°. The output of the balanced modulator as in Eq.15.20 may be written as

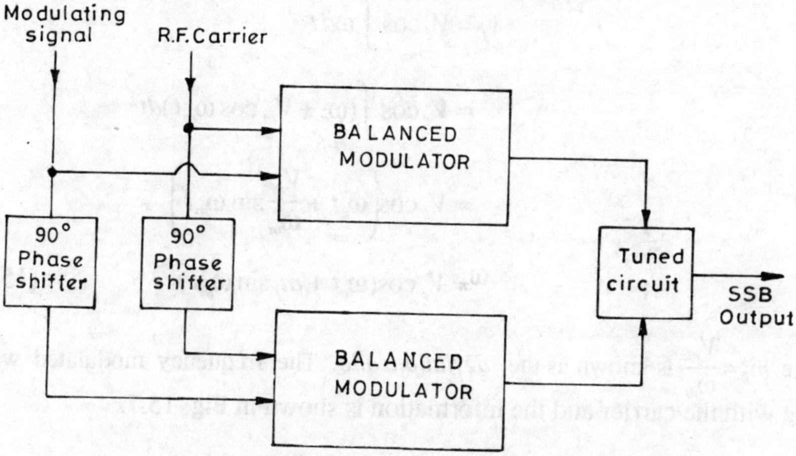

Fig. 15.6 Block diagram for *SSB*-suppressed modulation

$$I = 2K(2A_2V_cV_m \cos\omega_c t \cos\omega_m t)$$

This equation for the two modulators may be written as

$$I_1 = 4KA_2V_cV_m \cos\omega_c t \cos\omega_m t \tag{15.21}$$

and

$$I_2 = 4kA_2V_cV_m \cos(\omega_c t + 90)\cos(\omega_m t + 90)$$

$$= 4kA_2V_cV_m \sin\omega_c t \sin\omega_m t \tag{15.22}$$

The two currents are added to become the input of the tuned circuit as

$$I_{\text{total}} = I_1 + I_2$$

$$= 4KA_2V_cV_m[\cos\omega_c t\cos\omega_m t + \sin\omega_c t\sin\omega_m t]$$

$$= 4KA_2V_cV_m\cos(\omega_c - \omega_m)t \qquad (15.23)$$

This expression represents only the lower side band present in the output, which is obtained through a tuned circuit.

15.7 FREQUENCY MODULATION

In frequency modulation, the frequency of the modulated wave varies with the information. Thus

$$\omega = \omega_c + V_m\cos\omega_m t \qquad (15.24)$$

where $v_m = V_m\cos\omega_m t$ represents the information. The frequency modulated wave may be represented as

$$v_f = V_c\cos\int\omega dt$$

$$= V_c\cos\int(\omega_c + V_m\cos\omega_m t)dt$$

$$= V_c\cos\left(\omega_c t + \frac{V_m}{\omega_m}\sin\omega_m t\right)$$

$$= V_c\cos(\omega_c t + m_f\sin\omega_m t) \qquad (15.25)$$

where $m_f = \frac{V_m}{\omega_m}$ is known as the *deviationratio*. The frequency modulated wave along with the carrier and the information is shown in Fig. 15.7.

15.8 RADIOFREQUENCIES IN *FM* WAVE

In the frequency modulation also, new frequencies called sideband frequencies are produced but the number of sidebands for each modulating frequency is theoretically infinite. For example, the sidebands for f_c and f_m as the carrier and the modulating frequencies respectively are $f_c \pm f_m$, $f_c \pm 2f_m$, $f_c \pm 3f_m$, ... and so on. Mathematically, the generation of these frequencies may be explained as below:

The frequency modulated wave may be written as

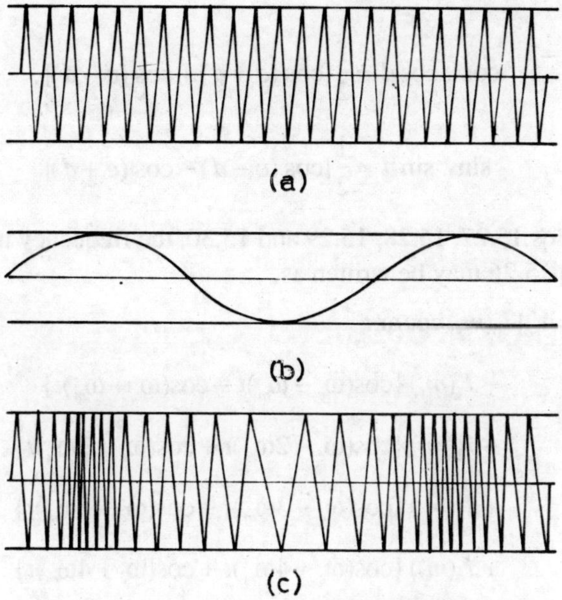

Fig. 15.7 (a) Carrier wave (b) Modulating wave (c) Frequency modulated wave

$$v_f = V_c \cos(\omega_c t + m_f \sin \omega_m t)$$

Using $\cos(c + d) = \cos c \cos d - \sin c \sin d$,

$$v_f = V_c[\cos \omega_c t \cos(m_f \sin \omega_m t) - \sin \omega_c t \sin(m_f \sin \omega_m t)] \quad (15.26)$$

The Bessel functions are defined by,

$$\cos(m_f \sin \omega_m t) = J_0(m_f) + 2J_2(m_f) \cos 4\omega_m t + 2J_4(m_f) \cos h\, \omega_m t + \ldots \quad (15.27)$$

and
$$\sin(m_f \sin \omega_m t) = 2J_1(m_f) \sin \omega_m t + 2J_3(m_f) \sin 3\omega_m t + 2J_5(m_f) \sin 5\omega_m t + \ldots \quad (15.28)$$

where $J_0, J_1, J_2, J_3, J_4, J_5, \ldots$ are the Bessel functions of the first kind. We also know that

$$\cos c \cos d = \frac{1}{2}[\cos(c+d) + \cos(c-d)] \tag{15.29}$$

and

$$\sin c \sin d = \frac{1}{2}[\cos(c-d) - \cos(c+d)] \tag{15.30}$$

Substituting Eqs.15.27, 15.28, 15.29 and 15.30, the frequency modulated wave defined by Eq.15.26 may be written as,

$$\begin{aligned} v_f = V_c[&J_0(m_f)\cos\omega_c t \\ &- J_1(m_f)\{\cos(\omega_c - \omega_m)t - \cos(\omega_c + \omega_m)t\} \\ &+ J_2(m_f)\{\cos(\omega_c - 2\omega_m)t + \cos(\omega_c + 2\omega_m)t\} \\ &- J_3(m_f)\{\cos(\omega_c - 3\omega_m)t - \cos(\omega_c + 3\omega_m)t\} \\ &+ J_4(m_f)\{\cos(\omega_c - 4\omega_m)t + \cos(\omega_c + 4\omega_m)t\} \\ &- J_5(m_f)\{\cos(\omega_c - 5\omega_m)t - \cos(\omega_c + 5\omega_m)t\} \\ &+ \ldots\ldots\ldots\ldots] \end{aligned} \tag{15.31}$$

This expression shows the presence of all the sidebands with frequencies $\omega_c \pm \omega_m$, $\omega_c \pm 2\omega_m$, $\omega_c \pm 3\omega_m$ ..... and so on.

15.9 OSCILLATOR CIRCUIT FOR FREQUENCY MODULATION

The frequency modulation may be accomplished by simply making the capacitance of the tank circuit of a *R-F* oscillator dependent on the modulating signal. Figure 15.8 shows such an oscillator circuit in which the output of the modulating amplifier is connected to the tank circuit of an *R.F* oscillator. As the collector potential of Q_1 varies with the modulating signal, the inherent capacitance present between the collector and emitter also varies and causes the reactance of the tank circuit to change. Thus, the resonant frequency of the *RF* oscillator also changes accordingly because $f_{\text{resonant}} = 1/2\pi\sqrt{LC}$. In this way the output of the *RF* oscillator becomes frequency modulated.

15.10 DEMODULATION OF *AM* WAVE BY DIODE DETECTOR

The demodulation of amplitude modulated waves may be accomplished by using a simple p-n junction diode. A typical circuit configuration of a diode detector is shown in Fig. 15.9. The values of C and R are so selected that the time constant of the circuit CR should be very small in comparison to the period of the modulating wave T_m and should be extremely large in comparison to the period of the carrier wave T_c, i.e.

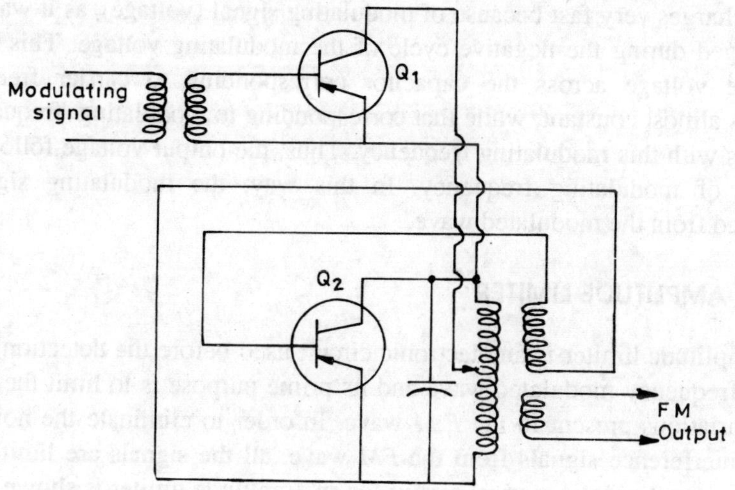

Fig. 15.8 Circuit configuration for frequency modulation

$$T_c \ll CR \ll T_m$$

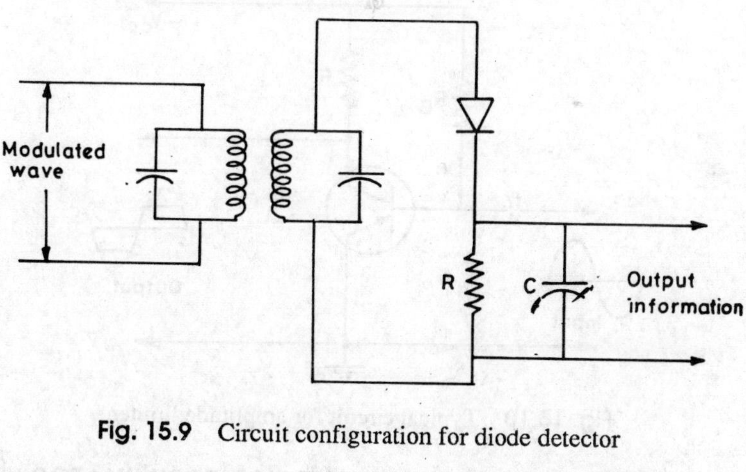

Fig. 15.9 Circuit configuration for diode detector

When *AM* wave is applied at the input, the diode is forward-biased for the positive half-cycle and remains cut-off during the negative half-cycle. During the positive half cycle, there is a flow of current in the circuit and the capacitor *C* charges towards the maximum applied voltage. In the negative half-cycle, the capacitor *C* discharges but this discharge is very little for the carrier frequency because $CR >> T_C$ and is very fast for the modulating frequency because $CR << T_m$.

During the next positive half cycle, the capacitor does not charge owing to carrier voltage, as it was not discharged during the negative cycle of the carrier.

But C charges very fast because of modulating signal (voltage), as it was fully discharged during the negative cycle of the modulating voltage. This means that the voltage across the capacitor corresponding to carrier frequency remains almost constant, while that corresponding to modulating frequency it changes with this modulating frequency. Thus, the output voltage follows the pattern of modulating frequency. In this way, the modulating signal is extracted from the modulated wave.

15.11 AMPLITUDE LIMITER

The amplitude limiter is an electronic circuit used before the detection circuit of the frequency modulated wave and its prime purpose is to limit the amplitude variations present in the $F.M$ wave. In order to eliminate the noise and other interference signals from the FM wave, all the signals are limited to a common amplitude. A typical circuit for an amplitude limiter is shown in Fig. 15.10 which is also known as saturation cut-off limiter. The transistor Q is so biased that it is driven to saturation by the first half of the signal and to cut-off by the second half of the signal. Hence, the output of the limiter has fixed positive and negative excursions.

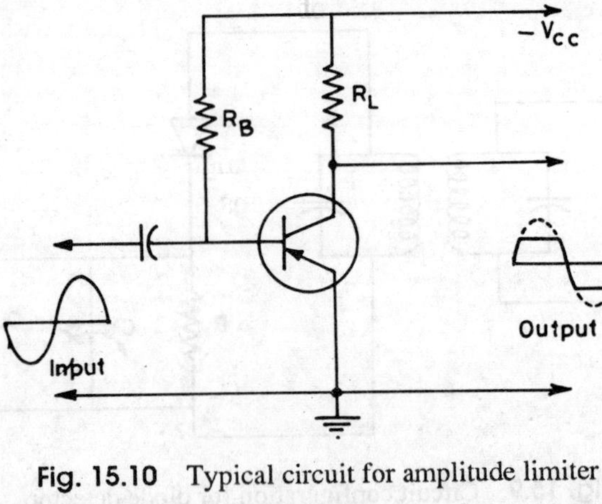

Fig. 15.10 Typical circuit for amplitude limiter

15.12 DEMODULATION OF *FM* WAVE BY DISCRIMINATOR

The demodulation or the detection of the frequency modulated waves is accomplished in two steps. Firstly, the FM wave is converted into an amplitude modulated wave and then the AM wave thus obtained is demodulated by any conventional method used for AM detection. Foster-Seeley or phase-shift discriminator is one of the commonly used FM detector. The circuit configuration of the discriminator is shown in Fig. 15.11, in which two inductively coupled tuned circuits are used and are tuned at the carrier frequency. The coil L has a high reactance to radio frequencies. Two voltages always appear across the diodes. One is owing to mutual inductance between the tuned circuits (V_3 across D_1 and V_4 across D_2) and the second is owing to the circuit of C and L, i.e. V_1. The function of the circuit is described on the basis of the phase diagram shown in Fig. 15.12.

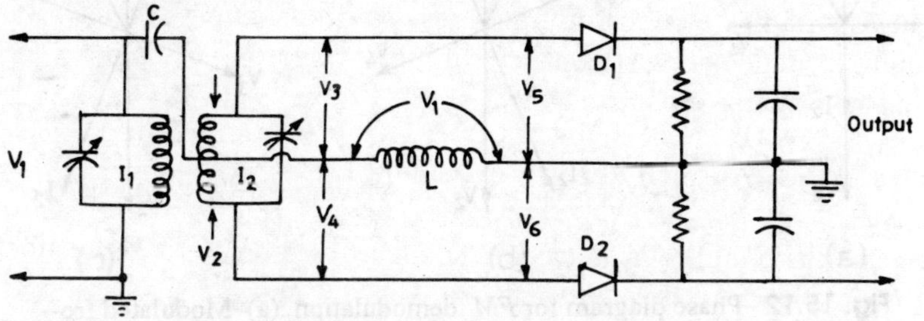

Fig. 15.11 Circuit configuration for *FM* demodulation Foster-Seeley discriminator

The voltage-current phase relationship dictates that the current leads the voltage for the capacitive load, while current lags the voltage for the inductive load.

(a) The input voltage V_1 leads I_1 by 90° and induced voltage V_2 lags I_1 by 90°. So, the voltage V_2 is 180° out of phase with V_1.

(b) For the carrier frequency which is also the resonant frequency of the tuned circuits, the induced current I_2 is in phase with the induced voltage V_2. The voltages V_3 and V_4 are equal and opposite in phase and each is 90° out of phase with I_2 and V_2 to centre tapping. The voltages applied to the two diodes V_5 and V_6 are the vector sums of V_1 and V_3 and V_1 and V_4 respectively. As these voltages V_5 and V_6 are equal and opposite in phase, the resultant output is zero.

(c) Above the carrier frequency, the frequency increases, making the capacitive reactance less than the inductive reactance and the current I_2 lags V_2. As V_3 and V_4 are developed owing to I_2, V_3 and V_4 also rotate with I_2 so that the phase between V_3 and V_4 remains 180°. This increases the phase angle between V_2 and V_3 from 90°, while the phase angle between V_2 and V_4 becomes less than 90°, as depicted in Fig. 15.12. This, in turn, makes voltage V_5 more than V_6 and positive output voltage signal is produced.

(d) When the modulated frequency decreases, making the capacitive reactance more than the inductive reactance the current I_2 leads V_2. The voltage vectors V_3 and V_4 also rotate with I_2 keeping their phase 180° apart. This decreases the phase angle between V_2 and V_3 from 90°, while the phase angle between V_2 and V_4 becomes more than 90° as shown in Fig. 15.12. This, further, makes voltage V_5 less than V_6 and the negative half of the voltage signal is produced.

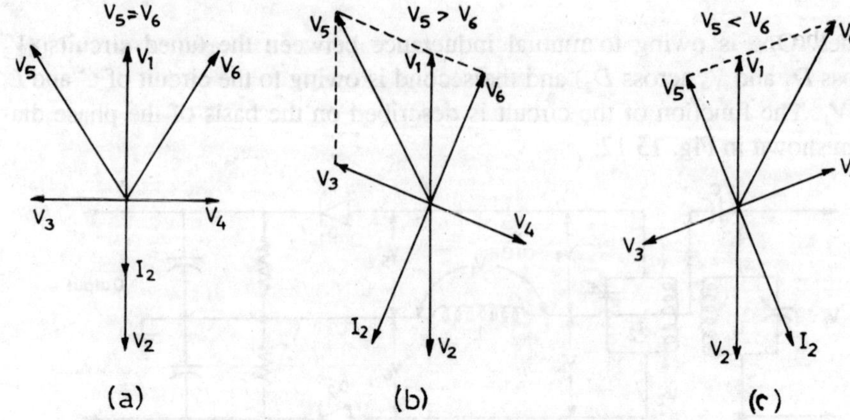

Fig. 15.12 Phase diagram for *FM* demodulation (a) Modulated frequency is equal to carrier frequency (b) Modulated frequency is more than carrier frequency (c) Modulated frequency is less than carrier frequency

Thus, the carrier frequency variation is converted into a voltage variation by using the discriminator circuit. Figure 15.13 shows the output voltage variation with the frequency. The output is zero for the carrier frequency, and positive and negative for the frequencies above and below the resonant carrier frequency respectively. As this discriminator circuit is sensitive to amplitude variations, an amplitude limiter is always connected before this circuit and amplitude variation in modulated wave is removed before feeding it to the discriminator circuit.

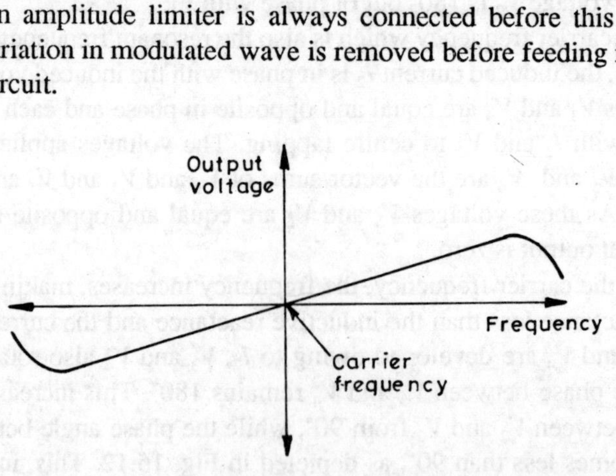

Fig. 15.13 Output of *FM* phase shift discriminator

15.13 DEMODULATION OF *FM* WAVE BY RATIO DETECTOR

The ratio detector is a modified form of the phase-shift discriminator. One of the two diodes is reversed in order to make them in series with the tuned circuit. A capacitor C_6 is also connected in the output circuit. The output is taken from the junction of C_4 and C_5 as shown in Fig. 15.14. At the carrier frequency, the charges on the capacitors C_4 and C_5 are the same, the output voltage is zero.

When the modulated frequency varies from the carrier frequency, one of the diodes conducts more while the other one conducts less, producing imbalance in the charges on C_4 and C_5. As the output voltage depends upon the ratio of the charges on C_4 and C_5 while the total voltage remains the same owing to the presence of C_6, the detector is known as the ratio detector. Since the capacitor C_6 keeps the total voltage steady by charging when the voltage increases, and discharging when the voltage decreases, the circuit becomes insensitive to the amplitude variations and there is no need of an amplitude limiter to be connected before the discriminator circuit.

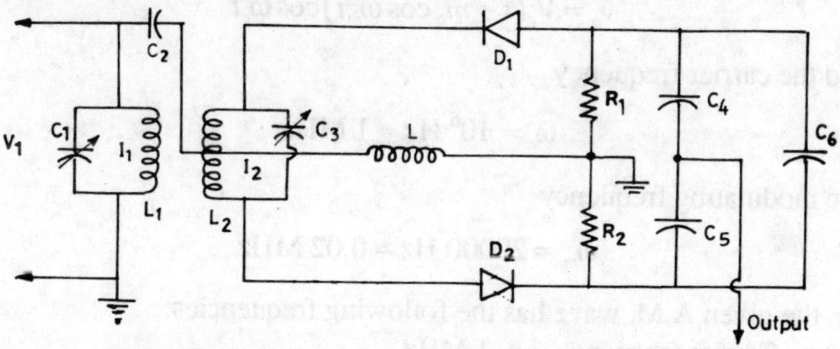

Fig. 15.14 Circuit configuration for ratio detector

15.14 SOLVED PROBLEMS

Problem 1

If the carrier and modulating waves are represented by $v_c = 100 \cos \omega_c t$ and $v_m = 50 \cos \omega_m t$, establish the amplitude modulated wave and find the modulation index.

Solution

The amplitude modulated wave may be written as

$$v_a = (V_c + V_m \cos \omega_m t) \cdot \cos \omega_c t$$

where $V_c = 100$ V and $V_m = 50$ V in this problem

$$\therefore \quad v_a = 100(1 + 0.5 \cos \omega_m t) \cos \omega_c t$$

The modulation index

$$m = \frac{V_m}{V_c} = 0.5 \text{ or } 50\%$$

Problem 2

The amplitude modulated wave is represented by

$$v_a = 1000(1 + 0.7 \cos 20{,}000t) \cos 10^6 t$$

Find the various frequencies present in the wave. Also, find the power contained in these frequencies if the wave is passed through a load resistor of 50 Ω.

Solution

The given amplitude modulated wave is

$$v_a = 1000(1 + 0.7\cos 20{,}000t)\cos 10^6 t$$

If we compare this wave with the standard *A.M.* wave,

$$v_a = V_c(1 + m_a \cos \omega_m t)\cos \omega_c t$$

we find the carrier frequency

$$\omega_c = 10^6 \text{ Hz} = 1 \text{ MHz}$$

and the modulating frequency

$$\omega_m = 20000 \text{ Hz} = 0.02 \text{ MHz}$$

Hence, the given A.M. wave has the following frequencies:

(i) Carrier frequency, i.e. 1 MHz.

(ii) Upper sideband frequency, i.e.

$$1 + 0.02 = 1.02 \text{ MHz.}$$

(iii) Lower sideband frequency i.e.

$$1 - 0.02 = 0.98 \text{ MHz}$$

Therefore, the bandwidth is $= 1.02 - 0.98$

$$= 0.,04 \text{ MHz}$$

$$= 40{,}000 \text{ Hz}$$

The power contained in carrier frequency,

$$P_c = \frac{V_c^2}{2R} = \frac{(1000)^2}{2 \times 50} = 10 \text{ kW}$$

and the power contained in each side band,

$$P_m = \frac{V_c^2}{2R} \cdot \frac{m_a^2}{4} = 10 \times \frac{(0.7)^2}{4} = 1.225 \text{ kW}$$

Problem 3

If the power contained in an *AM* wave is 15 kW and the carrier power is 10 kW, find the modulation index.

Solution

The powers contained in the *AM* wave and the carrier wave are

$$P_{\text{total}} = \frac{V_c^2}{2R}\left[1 + \frac{m_a^2}{4} + \frac{m_a^2}{4}\right]$$

and $$P_{\text{carrier}} = \frac{V_c^2}{2R}$$

∴ $$\frac{P_{\text{total}}}{P_{\text{carrier}}} = 1 + \frac{m_a^2}{2}$$

or $$\frac{15}{10} = 1 + \frac{m_a^2}{2}$$

or $$\frac{m_a^2}{2} = 0.5$$

or $$m_a = 1 \text{ or } 100\%$$

Problem 4

A transmitter is radiating waves with 12 kW power. Find the power contained in one sideband if $m_a = 0.5$.

Solution

The powers in *AM* wave and one sideband are given by

$$P_{\text{total}} = \frac{V_c^2}{2R}\left(1 + \frac{m_a^2}{2}\right)$$

and $$P_{\text{sideband}} = \frac{V_c^2}{2R}\left(\frac{m_a^2}{4}\right)$$

∴ $$P_{\text{sideband}} = P_{\text{total}} \times \frac{m_a^2}{4} / \left(1 + \frac{m_a^2}{2}\right)$$

$$= 12 \times \frac{1}{4 \times 4} / \left(1 + \frac{1}{8}\right) = 0.67 \text{ kW}$$

Questions and Problems

1. What do you mean by modulation and demodulation?
2. Find the expression representing the frequency spectrum of *AM* wave.
3. How is the power distributed in an *AM* wave?
4. How is the amplitude modulation achieved?
5. Describe a collector modulated class-*C* amplifier.

6. What is the use of a balanced modulator? Describe it.
7. What is an *SSB* system? Where is it used?
8. Describe frequency modulation. Find the various radio frequencies present in a *FM* wave.
9. How is frequency modulation achieved?
10. Describe a diode detector. What is the condition for the *C* and *R* values used in this detector?
11. What is the utility of an amplitude limiter?
12. Describe various techniques used for demodulating *FM* wave.
13. If the amplitudes of carrier and information are 150 V and 50 V with corresponding angular velocities as 1 MHz and 10 kHz, find the percentage of modulation and write the expression representing the amplitude modulated wave.
14. If the *AM* wave is represented by

$$v_a = 100(1 + 0.5\cos 10,000t - 0.7\cos 5,000t)\cos 10^7 t$$

find the various frequency bands present in the wave.
15. In the above Problem 14, calculate the various powers contained in each frequency band if the wave is passed through a resistor of 40 Ω.
16. Find the power required for the carrier wave to be modulated by a 20 kW A.M. wave at 70 percent modulation.

References

1. Mullar Richard. S and Karmins, Theodore I: *Device Electronics for Integrated Circuits*; John Wiley and Sons (1977).
2. Green, D.C.: *Electronics*; Pitman Publishing Ltd, (1978).
3. Millman, Jacob and Halkias, Christos C.: *Integrated Electronics;* Mc-Graw Hill Inc. (1972).
4. Polyakov, A.M.: *Semiconductor Made Simple*; Mir Publishers Moscow (1985).
5. Ryder, John D: *Electronic Fundamentals and Applications*, Prentice Hall (1979).
6. Coughlin, Robert F. and Driscoll Jr, Frederik F.: *Semiconductor Fundamentals*; Prentice Hall (1976).
7. Pascoe, Robert D.: *Fundamentals of Solid State electronics;* John-Wiley and Sons Inc. (1976).
8. Boylested, Robert and Nashelsky, Louis: *Electronic Devices and Circuit Theory*; Prentice Hall (1978).
9. Mannic, Horowtz: *Practical Design with Solid State Devices*: Reston Publishing Co. Inc. (1979).
10. Millman, Jacob and Halkias, Chistos C.: *Electronic Fundamentals and Applications* Mc. Graw-Hill Inc. (1976).
11. Terman F.E. : *Electronic and Radio Engineering*; McGraw-Hill Book Company Inc.(1955).
12. Fisher, J.E. amd Gatland, H.B.: *Electronics from Theory into Practice*; Pergamon Press Ltd. (1976).
13. Mottershed, Allen: *Electronic Devices and Circuits*; Prentice Hall of India (1985).
14. Millman, Jacob and Taub, Herbert: *Pulse, Digital, and Switching Waveforms.* Mc Graw-Hill Inc. (1965).
15. Manasse Fred K. : *Semiconductor Electronics Design;* Prentice Hall Inc. (1977).
16. Duarte, S.R. and Duarte, R.L.: *Electronics Assembly and Fabrication Methods;* McGraw-Hill Inc. (1973).
17. Bureau of Naval Personnel *Basic Electronics;* Dover Publications Inc. (1973).
18. Gottling,James, G: *Electronics-Models, Analysis, and Systems.* Marshal Dekker Inc. (1982).
19. Zambuto Mauro : *Semiconductor Devices.* McGraw-Hill Book Company (1989).
20. Franklin, C. Fitchen: *Electronic Integrated Circuits and Systems*; Van Nostrand Reinhold Company, (1970).
21. Adir, Bar-Lev: *Semiconductor and Electronic Devices.* Prentice Hall (1979).

Subject Index